METHODS IN MOLECULAR BIOLOGY

Series Editor
John M. Walker
School of Life and Medical Sciences
University of Hertfordshire
Hatfield, Hertfordshire, AL10 9AB, UK

For further volumes:
http://www.springer.com/series/7651

Tissue Proteomics

Methods and Protocols

Second Edition

Edited by

Minnie M. Sarwal, MD, FRCP, DCH, PhD

Professor, Department of Surgery, Medicine and Pediatrics,
Director, Kidney-Pancreas Transplant Program, University of California San Francisco, San Francisco, California, USA

Tara K. Sigdel, PhD

Asst. Professor, Department of Surgery, University of California San Francisco, San Francisco, California, USA

Humana Press

Editors
Minnie M. Sarwal, MD, FRCP, DCH, PhD
Professor, Department of Surgery
Medicine and Pediatrics
Director, Kidney-Pancreas
Transplant Program
University of California San Francisco
San Francisco, California, USA

Tara K. Sigdel, PhD
Asst. Professor, Department of Surgery
University of California San Francisco
San Francisco, California, USA

ISSN 1064-3745 ISSN 1940-6029 (electronic)
Methods in Molecular Biology
ISBN 978-1-4939-7852-6 ISBN 978-1-4939-7854-0 (eBook)
https://doi.org/10.1007/978-1-4939-7854-0

Library of Congress Control Number: 2018941896

Printed on acid-free paper

This Humana Press imprint is published by the registered company Springer Science+Business Media, LLC part of Springer Nature.
The registered company address is: 233 Spring Street, New York, NY 10013, U.S.A.

Preface

Publication of human genome data and subsequent advancements in molecular profiling methods, including high-throughput protein profiling technologies have helped us to understand complex biological systems that were out of routine analytical capabilities until a few decades ago. Given the enhanced ability to interrogate the proteome, we have complied this book with a series of related methods that clarify the different proteomic technologies and relate them to examples of specific clincial indications for advancing human health.

In this book, we have compiled methods to include a unique collection of detailed methods that represent advancement in protein sample preparation, protein profiling, peptide sample preparation and identification, and data analysis. The first chapter of this book provides a method for the depletion of myofibril-associated proteins using selective protein extraction as a tool in cardiac proteomics. Profiling of IgGs by using a high-density protein array in a high-throughput fashion is presented in Chapter 2. Chapter 3 provides a detailed method for the study of HLA class I- and class II-induced intracellular signaling and molecular associations in primary human endothelial cells. An advanced proteomics method for characterizing protein complexes using a novel chemical cross-linking coupled electrospray mass spectrometry is presented in Chapter 4. Chapters 5 and 6, offer detailed methods for peptide sample preparation in urinary proteomics, purification of targeted proteins from native tissues such as bovine testes and porcine brains. A critical aspect of every high-throughout proteomics in integrative manner is presented in Chapter 7. Chapter 8 details methods for untargeted screening of urinary peptides using offline nano-liquid chromatography—MALDI-TOF/TOF mass spectrometry which is very valuable in urine proteomic studies.

Considering the importance of sophistication and customization in proteomics methods for specific fields, in this book we have collected a number of relevant and technologically well developed methods that are of biological importance for basic and clinical sciences. Chapter 8 details methods for untargeted screening of urinary peptides using offline nano-liquid chromatography—MALDI-TOF/TOF mass spectrometry which is a very valuable tool for urine proteomic studies. Another method related with urine protein biomarkers is presented in Chapter 11 that offers a method for LC-SRM-based targeted quantification of urinary protein biomarkers.

Given the importance of identifying panels of potential biomarkers for clinical diagnostics, Chapter 9 is very relevant in identifying clinically relevant proteins in the context of multiplexed proteomic biomarker assays. Chapter 10 offers an advanced method for combined analysis of gene expression and protein expression in situ by performing in situ hybridization and double immunohistochemistry for the detection of proteins of interest in the context of kidney tissues.

Another related method related with urine protein biomarkers is presented in Chapter 11 that offers and discusses a method for LC-SRM-based targeted quantification of urinary protein biomarkers. In Chapters 12 and 13, we have presented protein samples fractionation and preparation methods. Chapter 14 provides detailed method for targeted verification for biomarkers for breast cancer aggressiveness. Chapter 15 offers a novel method for quantitative proteomics for colorectal cancer in mouse model. Chapter 16 of this book describes multiple reaction monitoring (MRM) using novel double isotopologue peptide standards for protein quantification.Chapter 17 includes a detailed method for isobaric labeling-based LC-MS/MS based proteomics for human pancreatic tissue. Similarly, a novel method for MALDI imaging mass spectrometry of N-glycans and tryptic peptides from the same FFPE tissue section is presented in Chapter 18. Chapter 19 details a method to analyze proteins in adipose tissue. Chapter 20 provides detailed method for quantification of breast cancer protein biomarkers at different expression levels in human tumors. Chapter 21 offers a very unique method for quantitative proteomic analysis of mass-limited tissue samples for spatially resolved tissue profiling. We have included highthroughput proteomics analysis of fresh-frozen biopsy tissue samples using pressure cycling technology coupled with SWATH mass spectrometry in Chapter 22. In addition, Chapter 23 offers a specific gel-free protocol for adipose tissue proteomics. In Chapter 24, MALDI imaging combined with laser microdissection-based microproteomics for protein identification is applied to intratumor heterogeneity studies which presents state-of-the-art technology in the field.

We are thankful to all the authors who have contributed to this book by providing methods that are believed to be instrumental in progressing cutting-edge research that is necessary to answer questions we face in basic and translational science. We are indebted to the help we received from Kaelyn Caspilo and Parhom Towfighi in the editing process. Parhom Towfighi's help with contacting the authors and publishers with regard to this book deserves special acknowledgement.

San Francisco, CA, USA ***Minnie M. Sarwal***
Tara K. Sigdel

Contents

Contributors

DEBORAH ALBERTS • *Department of Chemistry—Laboratory of Mass Spectrometry, University of Liége, MolSys, Liége, Belgium*

PEGGI M. ANGEL • *Department of Cell and Molecular Pharmacology, Medical University of South Carolina, Charleston, SC, USA*

CHARLES ANSONG • *Biological Sciences Division, Pacific Northwest National Laboratory, Richland, WA, USA*

DOMINIQUE BAIWIR • *Department of Chemistry—Laboratory of Mass Spectrometry, University of Liége, MolSys, Liége, Belgium*

OLIVER F. BATHE • *Cumming School of Medicine, University of Calgary, Calgary, AB, Canada; Department of Surgery, University of Calgary, Calgary, AB, Canada; Division of Surgical Oncology, Department of Oncology, Tom Baker Cancer Centre, University of Calgary, Calgary, AB, Canada*

CHRISTOPH H. BORCHERS • *University of Victoria—Genome BC Proteomics Centre, Victoria, BC, Canada; Department of Biochemistry and Microbiology, University of Victoria, Victoria, BC, Canada; Proteomics Centre, Segal Cancer Centre, Lady Davis Institute, McGill University, Montreal, QC, Canada; Gerald Bronfman Department of Oncology, Jewish General Hospital, Montreal, QC, Canada; Jewish General Hospital Proteomics Laboratory, Lady Davis Institute, McGill University, Montréal, QC, Canada*

PAVEL BOUCHAL • *Department of Biochemistry, Faculty of Science, Masaryk University, Brno, Czech Republic*

STEPHEN BRANTLEY • *M2Gen, Tampa, FL, USA*

DAVID BRITTON • *Proteome Sciences, plc, Cobham, UK*

DAVID G. CAMP II • *Biological Sciences Division, Pacific Northwest National Laboratory, Richland, WA, USA*

YI CHEN • *Molecular Oncology/Proteomics SRB3, Moffitt Cancer Center, Tampa, FL, USA*

GEREMY CLAIR • *Biological Sciences Division, Pacific Northwest National Laboratory, Richland, WA, USA*

TIMOTHY D. CUMMINS • *Medical Research Council Protein Phosphorylation and Ubiquitylation Unit, University of Dundee, Dundee, UK; Division of Nephrology and Hypertension, Clinical Proteomics Center, University of Louisville School of Medicine, Louisville, KY, USA*

DEJAN DOBI • *San Francisco, CA, USA*

RICHARD R. DRAKE • *Department of Cell and Molecular Pharmacology, Medical University of South Carolina, Charleston, SC, USA*

AZAD ESHGHI • *University of Victoria—Genome BC Proteomics Centre, Victoria, BC, Canada*

MICHELLE FOURNIER • *Molecular Oncology/Proteomics SRB3, Moffitt Cancer Center, Tampa, FL, USA*

YUQIAN GAO • *Biological Sciences Division, Pacific Northwest National Laboratory, Richland, WA, USA*

FENG GUO • *McArdle Laboratory for Cancer Research, WIMR II 6451, School of Medicine and Public Health, University of Wisconsin-Madison, Madison, WI, USA; Genencor International (DuPont), Palo Alto, CA, USA*

TIANNAN GUO • *Institute of Basic Medical Sciences, Westlake Institute for Advanced Study, Hangzhou, Zhejiang, People's Republic of China; Westlake University, Hangzhou, Zhejiang, People's Republic of China*
TERKEL HANSEN • *Natural Products and Medicinal Chemistry Research Group, Department of Pharmacy, UiT—The Arctic University of Norway, Tromsø, Norway*
GUIXUE HOU • *BGI-SHEHZHEN, Shenzhen, Guangdong, China*
YI-PING JIN • *UCLA Pathology and Laboratory Medicine, Immunogenetics Center, Los Angeles, CA, USA*
JOSEPH JOHNSON • *Molecular Oncology/Proteomics SRB3, Moffitt Cancer Center, Tampa, FL, USA*
JOHN M. KOOMEN • *Molecular Oncology/Proteomics SRB3, Moffitt Cancer Center, Tampa, FL, USA*
ZOLTAN G. LASZIK • *Department of Pathology, University of California, San Francisco, San Francisco, CA, USA*
JURAJ LENČO • *Department of Molecular Pathology and Biology, Faculty of Military Health Sciences, University of Defence, Hradec Kralove, Czech Republic*
FANG LI • *UCLA Pathology and Laboratory Medicine, Immunogenetics Center, Los Angeles, CA, USA*
CHIH-WEI LIU • *Center for Translational Biomedical Research, University of North Carolina at Greensboro, Kannapolis, NC, USA*
SIQI LIU • *BGI-Shenzhen, Shenzhen, Guangdong, China*
RÉMI LONGUESPÉE • *University of Heidelberg, Institute of Pathology, Heidelberg, Germany*
XIAOMIN LOU • *Beijing Institute of Genomics, Chinese Academy of Sciences, Beijing, China*
EMILY M. MACKAY • *Cumming School of Medicine, University of Calgary, Calgary, AB, Canada*
ANTHONY MAGLIOCCO • *Molecular Oncology/Proteomics SRB3, Moffitt Cancer Center, Tampa, FL, USA*
MANOUSOS MAKRIDAKIS • *Biotechnology Division, Biomedical Research Foundation, Academy of Athens (BRFAA), Athens, Greece*
PIERRE MARQUET • *Department of Pharmacology-Toxicology-Pharmacovigilance, INSERM UMR 850, CHU Limoges, University of Limoges, Limoges Cedex, France; U850 INSERM, University of Limoges, CHU Limoges, Limoges Cedex, France*
GABRIEL MAZZUCCHELLI • *Department of Chemistry—Laboratory of Mass Spectrometry, University of Liége, MolSys, Liége, Belgium*
ANAND MEHTA • *Department of Cell and Molecular Pharmacology, Medical University of South Carolina, Charleston, SC, USA*
PATRICK MENDEN • *McArdle Laboratory for Cancer Research, WIMR II 6451, School of Medicine and Public Health, University of Wisconsin-Madison, Madison, WI, USA*
RONALD J. MOORE • *Biological Sciences Division, Pacific Northwest National Laboratory, Richland, WA, USA*
NICK MORRICE • *ABSciex, Warrington, UK*
G. NICHOLAS EUROPE-FINNER • *Cardiovascular Research Centre, Institute of Genetic Medicine, Newcastle University, International Centre for Life, Newcastle upon Tyne, UK*
CARRIE D. NICORA • *Biological Sciences Division, Pacific Northwest National Laboratory, Richland, WA, USA*
KIM NORRIS-CANEDA • *Department of Cell and Molecular Pharmacology, Medical University of South Carolina, Charleston, SC, USA*

PAWEL PALMOWSKI • *Cardiovascular Research Centre, Institute of Genetic Medicine, Newcastle University, International Centre for Life, Newcastle upon Tyne, UK*
YVONNE PASING • *Tromsø Endocrine Research Group, Department of Clinical Medicine, UiT—The Arctic University of Norway, Tromsø, Norway; Division of Internal Medicine, University Hospital of North Norway, Tromsø, Norway*
SÉBASTIEN PASSERON • *Department of Pharmacology-Toxicology-Pharmacovigilance, INSERM UMR 850, CHU Limoges, University of Limoges, Limoges Cedex, France; U850 INSERM, University of Limoges, CHU Limoges, Limoges Cedex, France*
EDWIN DE PAUW • *Department of Chemistry—Laboratory of Mass Spectrometry, University of Liége, MolSys, Liége, Belgium*
PAUL D. PIEHOWSKI • *Biological Sciences Division, Pacific Northwest National Laboratory, Richland, WA, USA*
IAN PIKE • *Proteome Sciences, plc, Cobham, UK*
IVA PROCHÁZKOVÁ • *Department of Biochemistry, Faculty of Science, Masaryk University, Brno, Czech Republic; Masaryk Memorial Cancer Institute, Regional Centre for Applied Molecular Oncology, Brno, Czech Republic*
WEI-JUN QIAN • *Biological Sciences Division, Pacific Northwest National Laboratory, Richland, WA, USA*
ELAINE F. REED • *UCLA Pathology and Laboratory Medicine, Los Angeles, CA, USA*
NATHAN SALOMONIS • *Department of Pediatrics, Division of Biomedical Informatics, Cincinnati Children's Hospital Medical Center, University of Cincinnati, College of Medicine, Cincinnati, OH, USA*
GOPAL P. SAPKOTA • *Medical Research Council Protein Phosphorylation and Ubiquitylation Unit, University of Dundee, Dundee, UK*
MINNIE M. SARWAL • *Department of Surgery, University of California San Francisco, San Francisco, CA, USA*
FRANÇOIS-LUDOVIC SAUVAGE • *Department of Pharmacology-Toxicology-Pharmacovigilance, INSERM UMR 850, CHU Limoges, University of Limoges, Limoges Cedex, France; U850 INSERM, University of Limoges, CHU Limoges, Limoges Cedex, France*
ARMIN SCHNIERS • *Natural Products and Medicinal Chemistry Research Group, Department of Pharmacy, UiT—The Arctic University of Norway, Tromsø, Norway*
JACOB SHAW • *Stanford University, Stanford, CA, USA*
KELLY J. SHIELDS • *Department of Medicine, Lupus Center of Excellence—Autoimmunity Institute, Allegheny Health Network, Pittsburgh, PA, USA; Allegheny General Hospital, Pittsburgh, PA, USA*
TUJIN SHI • *Biological Sciences Division, Pacific Northwest National Laboratory, Richland, WA, USA*
TARA K. SIGDEL • *Department of Surgery, University of California San Francisco, San Francisco, CA, USA*
NICOLAS SMARGIASSO • *Department of Chemistry—Laboratory of Mass Spectrometry, University of Liége, MolSys, Liége, Belgium*
RICHARD D. SMITH • *Biological Sciences Division, Pacific Northwest National Laboratory, Richland, WA, USA*
NWE NWE SOE • *UCLA Pathology and Laboratory Medicine, Immunogenetics Center, Los Angeles, CA, USA*
JULIE TAGGART • *Cardiovascular Research Centre, Institute of Genetic Medicine, Newcastle University, International Centre for Life, Newcastle upon Tyne, UK*

Michael J. Taggart • *Cardiovascular Research Centre, Institute of Genetic Medicine, Newcastle University, International Centre for Life, Newcastle upon Tyne, UK*
Wing Chiu Tong • *Cardiovascular Research Centre, Institute of Genetic Medicine, Newcastle University, International Centre for Life, Newcastle upon Tyne, UK*
Parhom Towfighi • *University of California San Francisco, San Francisco, CA, USA*
Achim Treumann • *Newcastle University Protein and Proteome Analysis, Newcastle University, Cookson Building, Medical School, University of Newcastle, Newcastle upon Tyne, UK; Cardiovascular Research Centre, Institute of Genetic Medicine, Newcastle University, International Centre for Life, Newcastle upon Tyne, UK*
Nicole Valenzuela • *UCLA Pathology and Laboratory Medicine, Immunogenetics Center, Los Angeles, CA, USA*
Antonia Vlahou • *Biotechnology Division, Biomedical Research Foundation, Academy of Athens (BRFAA), Athens, Greece*
Hui Wang • *Biological Sciences Division, Pacific Northwest National Laboratory, Richland, WA, USA*
Yang Wang • *Beijing Institute of Genomics, Chinese Academy of Sciences, Beijing, China*
Vonetta L. Williams • *Molecular Oncology/Proteomics SRB3, Moffitt Cancer Center, Tampa, FL, USA*
Marek Wloch • *Molecular Oncology/Proteomics SRB3, Moffitt Cancer Center, Tampa, FL, USA*
Nathan Wlodarchak • *McArdle Laboratory for Cancer Research, WIMR II 6451, School of Medicine and Public Health, University of Wisconsin-Madison, Madison, WI, USA*
Elizabeth R. Wood • *Molecular Oncology/Proteomics SRB3, Moffitt Cancer Center, Tampa, FL, USA*
Changgong Wu • *Center for Advanced Proteomics Research, New Jersey Medical School Rutgers, The State University of New Jersey, Newark, NJ, USA*
Yongna Xing • *McArdle Laboratory for Cancer Research, WIMR II 6451, School of Medicine and Public Health, University of Wisconsin-Madison, Madison, WI, USA*
Qibin Zhang • *Center for Translational Biomedical Research, University of North Carolina at Greensboro, Kannapolis, NC, USA; Department of Chemistry and Biochemistry, University of North Carolina at Greensboro, Greensboro, NC, USA*
Xiaohai Zhang • *Cedars Sinai, Los Angeles, CA, USA*
Rui Zhao • *Environmental Molecular Sciences Laboratory, Pacific Northwest National Laboratory, Richland, WA, USA*
Yi Zhu • *Institute of Basic Medical Sciences, Westlake Institute for Advanced Study, Hangzhou, Zhejiang, People's Republic of China; Westlake University, Hangzhou, Zhejiang, People's Republic of China*

Methods in Molecular Biology (2018) 1788: 1–9
DOI 10.1007/7651_2017_73

Published online: 04 October 2017

Depletion of Myofibril-Associated Proteins Using Selective Protein Extraction as a Tool in Cardiac Proteomics

Achim Treumann, Pawel Palmowski, Wing Chiu Tong, Julie Taggart, Nick Morrice, G. Nicholas Europe-Finner, and Michael J. Taggart

Abstract

Muscle tissue poses a particular challenge to proteomic analysis due to a very wide range of protein abundances arising from the dominant expression of myofilament-related proteins. We address this issue by describing proteomic analysis with liquid chromatography–mass spectrometry (LC-MS) and sequential window acquisition of all theoretical mass spectra (SWATH), of guinea pig cardiac tissue prepared in two homogenization buffers: (1) An SDS-based buffer designed to extract "all" tissue proteins and (2) a long-established EDTA-containing buffer thought to preferentially extract non-myofibril-related proteins. We use gene ontology (GO) annotation-based assessment of subcellular localization to indicate if these enriched proteins congregate in the cytoplasm or in organellar lumens. This technique results in the preferential quantitation of less abundant non-myofibrillar proteins and, for future studies, offers the opportunity for more complete analyses of changes in heart tissue protein expression with biological circumstance.

Keywords Cardiac muscle, Myofibrils, Protein enrichment, Proteomics

1 Introduction

The regulation of protein expression, structure, and activity is a major determinant of biological function. In the last decade, there has been a dramatic increase in the use of liquid chromatography–-mass spectrometry (LC-MS)-based approaches with the intention of improving our understanding of the proteome that underlies the operations of many biological systems. So much so, it is now common to be able to identify and quantify tens of hundreds of proteins within, and between, biological samples. However, a particular issue in tissues such as muscle is the large dynamic range in protein expression with the molecular components of thick (myosin-containing) and thin (actin-containing) filaments. Estimates of total actin concentration of 750 μM place it as the most abundant muscle protein, occupying at least 10% of the proteome. The next most abundant protein is myosin II. Therefore, these proteins, and other myofilament-bound proteins, will dominate the proteome. As a result, most mass spectrometry (MS) scan time is taken up with detecting muscle tissue tryptic peptides from these proteins at the expense of detecting those derived from lesser abundant molecules. In turn, this may result in the experimenter losing a lot of important biological information.

For tissue proteomics, it would be beneficial to have procedures that reliably increase the breadth of protein constituents detected. A common method of preparing tissue for LC-MS analysis is to homogenize material in an SDS-based solution with the "catch-all" intention of extracting as much protein as possible [1, 2]. However, this does not overcome the aforementioned problem of a large dynamic range. We noted that in many early muscle biochemistry studies designed to isolate individual cytoplasmic proteins, or proteins bound loosely to myofilaments, tissue was homogenized in a non-detergent EDTA-based solution [3–5]. This allows for the possibility of utilizing two homogenization procedures—one all-encompassing and one with selectivity for non-myofilamentous proteins—to increase accessibility to the wide range of proteins within the muscle proteome. Therefore, in this study, we apply LC-MS/MS with sequential window acquisition of all theoretical mass spectra (SWATH) [6] to guinea pig cardiac tissue to report the enrichment of separate protein constituents from homogenizations in SDS- or EDTA-based solutions (*see* **Notes 7** and **8**). In particular, these methods demonstrate that the use of EDTA-based extraction results in the exposure of non-myofibrillar protein components that may otherwise have remained undetected.

2 Materials

2.1 Cardiac Tissue Preparation

1. Cardioplegic solution: glucose 227.5 mM, mannitol 34.5 mM, KCl 30 mM, $NaHCO_3$ 25 mM, pH 7.4).
2. 10 ml plastic syringe and 18G gauge bulbous tip cannula.

2.2 Cardiac Tissue Homogenization and Tryptic Digestion

1. SDS-containing buffer: 100 mM Tris base pH 7.6, 4% SDS, 0.1 M DTT, 20 μl per ml protease inhibitor (Sigma, P8340).
2. EDTA-containing buffer: 4 mM EDTA, 2 mM EGTA, 5 mM DTT, 150 mM sucrose, 20 μl per ml protease inhibitor (Sigma, P8340).
3. 2 ml volume homogenization tubes containing a mixture of 1.4 mm and 2.8 mm ceramic beads (VWR International, BERTKT03961-1-0092).
4. Minilys homogenizer (Bertin Technologies).
5. Trypsin, sequencing grade modified (Sigma, v5111).
6. UA: 8 M Urea (Sigma, U5128) in 0.1 M Tris–HCl pH 8.5.
7. UB: 8 M Urea (Sigma, U5128) in 0.1 M Tris–HCl pH 8.0.

2.3 Liquid Chromatography

1. Buffer A: 2% acetonitrile/0.1% formic acid in LCMS grade water.
2. Buffer B: 0.1% formic acid in acetonitrile.
3. HPLC column: Waters nanoAcquity BSS T3 1.8 mm 250 × 0.075 mm.
4. Trap column: 20 × 0.18 mm 2 g–v/v 5 mm C_{18}.

2.4 Mass Spectrometry

ABSciex TripleTOF 6600 mass spectrometer with a nanospray 3 source fitted with a NewObjective FS360-20-10 emitter.

3 Methods

3.1 Overview

A schematic overview of the methodological procedure is given in Fig. 1.

3.2 Heart Isolation and Tissue Homogenization

Whole hearts were obtained from adult non-pregnant guinea pigs (Dunkin Hartley from Harlan, UK) sacrificed by anaesthetic overdose in accordance with the Animals (Scientific Procedures) Act 1986 under UK Home Office project license approval.

1. Immediately flush excised hearts with plenty of cardioplegic solution by retrograde perfusion via the aorta to clear all blood (*see* **Note 1**).
2. Dissect out the left ventricular wall and freeze with liquid nitrogen. Frozen tissues can be stored at −80 °C.
3. Homogenize samples (10 μl buffer per mg of tissue) for 25 s at for 5 cycles, centrifuge at 16,000 × *g* for 5 min.
4. Collect and heat the supernatant to 95 °C for 4 min, which can then be stored at −80 °C until use.

3.3 Tryptic Digestion

1. Take an equal amount (~250 μg of total protein) of tissue from each sample.
2. Reduce and denature lysates in a solution containing DTT and SDS to final concentrations of, respectively, 0.1 M and 4%.
3. Incubate for 5 min at 95 °C with continuous shaking in a Thermomixer (Eppendorf).
4. Mix protein extracts with 200 μl of UA buffer, transfer to centrifugal spin filters, and centrifuge for 40 min at 14,000 × *g*. The same conditions will be used for all of the following centrifugations.
5. Wash samples two times with 200 μl of UA buffer.
6. Add 100 μl of alkylating agent (50 mM iodoacetamide in UA buffer) to each sample and shake for 1 min. Incubate for 5 min in the dark at room temperature.

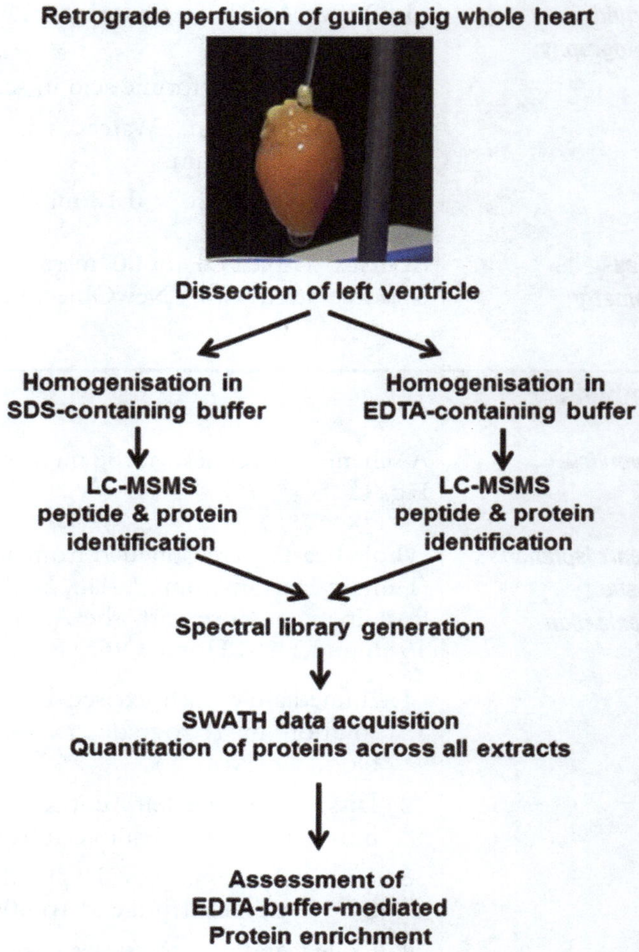

Fig. 1 Schematic diagram of the experimental workflow

7. Centrifuge the protein mixture and wash twice with UB buffer to remove the excess of alkylating agent.
8. Perform tryptic digestion overnight at 37 °C. Use 1 μg of trypsin in 0.05 M NH_4HCO_3 per 100 μg of protein and a final volume of 150 μl) (*see* **Note 2**).
9. Centrifuge and wash the digests. The eluents can be stored frozen at −20 °C.

3.4 LC-MS/MS

1. Automatically inject 1 μg of protein digest into mass spectrometer through the Eksigent ekspert nanoLC425 HPLC system.
2. Separate peptides at 300 nl/min using a 120 min gradient from 5% acetonitrile/0.1% formic acid in water to 40% acetonitrile/0.1% formic acid in water on the Waters nanoAcquity BSS T3 1.8 μm 250 × 0.075 mm column.

3. Load samples are onto the 20 × 0.18 mm 2 g-v/v 5 μm C_{18} trap column at 5 μl/min in 0.1% TFA/water. Wash the trap column for 10 min at 5 μl/min. Direct the column eluent to the Ab Sciex 6600 QqTOF mass spectrometer through the ABSciex nanospray 3 source, fitted with the NewObjective FS36020-10 emitter.
4. For information-dependent data acquisition (IDA) MS, use data between mz 400–2000 (250 ms), followed by Top30 precursors with charge states 2–5 (total cycle time 1.8 s) (*see* **Note 3**).
5. Acquire product ion spectra (50 ms) from m/z 100–2000 using a rolling collision energy for precursors that exceeds 150 cps. Exclude precursor ions for 15 s after one occurrence.
6. Based on a manual analysis of the IDA data, a SWATH acquisition range needs to be chosen. In the case of the example data shown in Fig. 2, the acquisition range is 400–1000 m/z, with 27 variable SWATH bins for a total duty cycle of 1.33 s (40 ms scan time per bin). Each sample should be run at least twice (technical replicates). The size of the variable SWATH bins can be calculated using the Microsoft Excel-based AB Sciex variable window assay calculator (*see* **Note 4**).

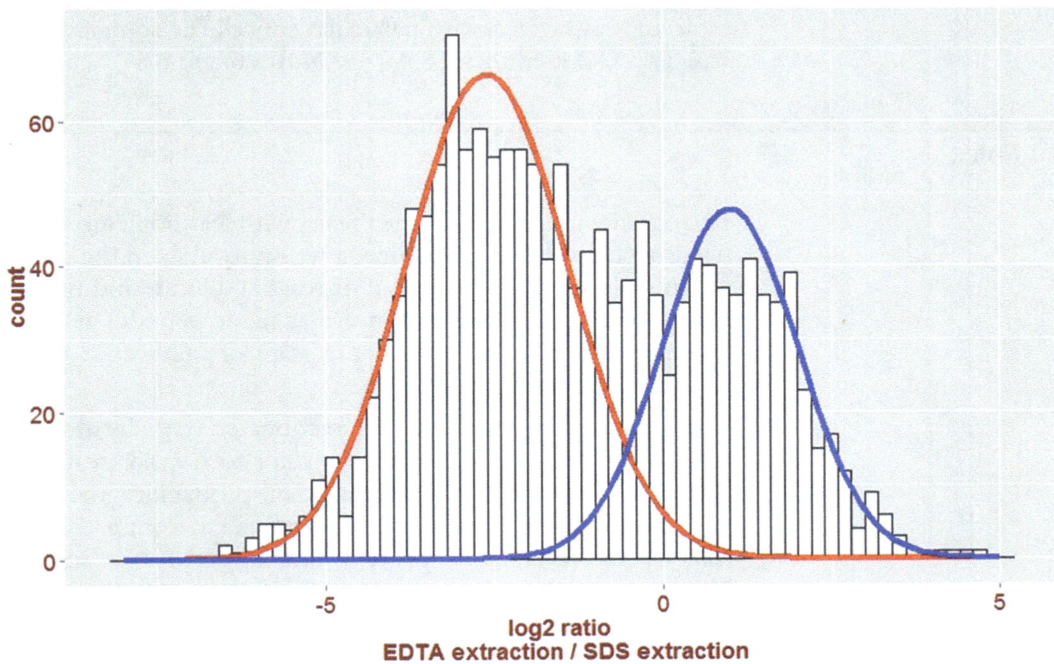

Fig. 2 Example histogram (bin width 0.2) depicting the $\log_2$ ratio of the total peptide intensity of proteins extracted using the EDTA-based extraction buffer to the total peptide intensity of proteins extracted using an SDS-based extraction buffer. The *blue* curve is a Gaussian curve depicting the distribution of ratios of proteins enriched in the EDTA-based extraction buffer and the *red* curve shows the distribution of ratios of proteins depleted from the EDTA based extraction buffer

3.5 Data Analysis

1. Search *.wiff files from the IDA experiments using Protein Pilot 5 against the uniprot guinea pig proteome (version 2014_04, 20511 sequences) concatenated to the common Repository for Adventitious Proteins (cRAP, ftp://ftp.thegpm.org/fasta/cRAP).
2. For SWATH experimentation and analysis, an in-house spectral library for guinea pig heart can be created from a combination of IDA MS/MS data obtained across several SDS- and EDTA-based guinea pig heart extracts. Using this library, process SWATH data using Peakview 2.1 (AB Sciex) with the SWATH MSMSall microapp. Export data from Peakview 2.1 as markerview files (*.mrkvw) and then process through Markerview 1.2.1 (AB Sciex) or export as a tab-separated ASCII file and further process using in-house written R scripts or the Perseus software framework (http://www.perseus-framework.org/).
3. Derive protein intensities from technical replicates (SWATH runs). Average and apply a two sample T-test (permutation-based FDR <5%, s0 = 1 [3]) to identify proteins of which the total peptide intensity significantly differs between the two extraction methods. Submit resulting protein lists (EDTA-enriched, ETDA-depleted and Common for proteins not differing significantly) to STRING v10 [7] to assess for enrichment of molecules associated with particular subcellular compartments based on GO identifiers [8, 9] (*see* **Notes 5** and **6**).

4 Notes

1. Retrograde perfusion of the heart with cardioplegic buffer needs to take place immediately after removal from the guinea pig in order to flush the organ of residual blood (and thereby minimize the reporting of noncardiac tissue peptides and proteins) and to place the heart in a state of quiescence before dissection and homogenization.
2. The tissue homogenates are not fractionated (e.g., by ultracentrifugation, or gel electrophoresis) prior to trypsin treatment. Therefore, this procedure details, in all probability, the minimum information that can be discriminated from a complex mixture of heart tissue proteins utilizing the two separate homogenization procedures.
3. After generating heart tissue extracts using parallel SDS- or EDTA-based buffers, analyze all fractions from three biological replicates using information-dependent acquisition (IDA). This results in the identification of almost twice as many proteins from the SDS-based extraction as from the EDTA-based extraction.

4. Following IDA data acquisition, and the establishment of a spectral library, subject all samples to quantitative SWATH analysis. This yields quantitative information on proteins from both extracts in all three biological replicates. The results of this quantitative analysis (Fig. 2) should illustrate that the EDTA-based buffer does preferentially extract a different set of proteins from the SDS-based extraction.
5. GO analysis of myofibrillar-associated proteins (MFAP) in the EDTA-based fraction, in comparison with the SDS-based extract, should confirm that myofilament- and myofibril-related proteins are significantly depleted from the former fraction (Fig. 3). As a likely consequence of depletion of these very abundant proteins, several other groups of proteins appear to be significantly enriched in the EDTA-based extract (green bars in Fig. 3).

   Due to the depletion of myofibrillar proteins, there is a statistically significant enrichment of mitochondrial matrix proteins and of extracellular proteins in the EDTA-based heart

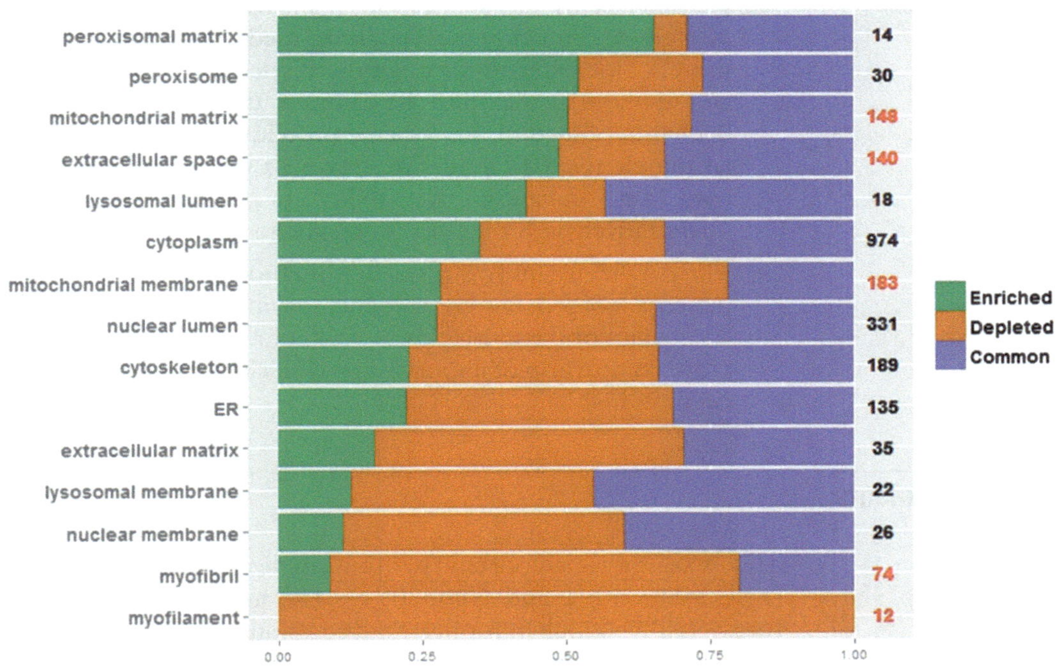

Fig. 3 Diagram showing the distribution of various gene ontology cell components (on the *left* of the diagram) in the EDTA-based extraction buffer. The *green bar* indicates proteins that are enriched in the EDTA-based extraction buffer, the *orange bar* are proteins that are depleted in this buffer, and the *purple bar* are proteins that do not differ in relative abundance between this extract and that of SDS-based buffer. To the *right* of each bar is an indication of the number of proteins in each cell component. If the suggested distribution of a particular cell component differs significantly (FDR <5%) from that in the SDS-based extraction buffer, this number is depicted in *red*

muscle extracts. Mitochondrial membrane proteins and myofibrillar proteins are significantly depleted from this fraction. There is also a trend toward the enrichment of proteins localized within the matrix/lumen of other subcellular organelles, such as peroxisomes and lysosomes, but not from nuclei or the ER. It is noteworthy, that membrane proteins, including proteins derived from the membranes of subcellular organelles, are also depleted from the EDTA-based extracts. The latter might be caused either by the proximity of these organelles to myofilaments, and the interactions of their membranes with the myofilaments, or the lack of solubility of membrane proteins in the detergent-free extraction buffer.

6. It is advantageous to keep the selected number of GO annotations for cellular components that are most relevant to the muscle cells investigated to a minimum. The temptation to assess proteins "expression" appearing in lists across all GO annotations for subcellular localizations should be countered with the realization that many proteins are listed in several GO categories.
7. We describe the (re)application of an EDTA-based non-detergent tissue extraction method, long-used in the field of muscle protein biochemistry for isolation of selected soluble proteins, and apply it to the study of cardiac muscle proteomics. The depletion of myofibril-associated proteins from heart ventricular muscle tissue homogenates reveals a specific subset of proteins and enhances the ability of characterizing them quantitatively in proteomics experiments.
8. The parallel use of two extraction methods that show separate patterns of protein enrichments promises to broaden the assessment of the muscle tissue proteome and facilitate quantitative comparison of important protein changes with biological circumstance.

Acknowledgments

This work was supported by the Medical Research Council (MR/L009560/1 and G0902091). The authors would like to thank Sir Philip Cohen for helpful discussions.

References

1. Manza LL, Stamer SL, Ham AJ, Codreanu SG, Liebler DC (2005) Sample preparation and digestion for proteomic analyses using spin filters. Proteomics 5(7):1742–1745. https://doi.org/10.1002/pmic.200401063
2. Wisniewski JR, Zougman A, Nagaraj N, Mann M, Wiśniewski JR (2009) Universal sample preparation method for proteome analysis. Nat Methods 6:359–362. https://doi.org/10.1038/nmeth.1322

3. Tusher VG, Tibshirani R, Chu G (2001) Significance analysis of microarrays applied to the ionizing radiation response. Proc Natl Acad Sci U S A 98(9):5116–5121. https://doi.org/10.1073/pnas.091062498
4. Cohen P, Duewer T, Fischer EH (1971) Phosphorylase from dogfish skeletal muscle. Purification and a comparison of its physical properties to those of rabbit muscle phosphorylase. Biochemistry 10(14):2683–2694
5. Krebs EG, Fischer EH (1955) Phosphorylase activity of skeletal muscle extracts. J Biol Chem 216(1):113–120
6. Gillet LC, Navarro P, Tate S, Röst H, Selevsek N, Reiter L, Bonner R, Aebersold R (2012) Targeted data extraction of the MS/MS spectra generated by data-independent acquisition: a new concept for consistent and accurate proteome analysis. Mol Cell Proteomics 11: O111.016717. doi: https://doi.org/10.1074/mcp.O111.016717
7. Szklarczyk D, Franceschini A, Wyder S, Forslund K, Heller D, Huerta-Cepas J, Simonovic M, Roth A, Santos A, Tsafou KP, Kuhn M, Bork P, Jensen LJ, von Mering C (2015) STRING v10: protein-protein interaction networks, integrated over the tree of life. Nucleic Acids Res 43(Database issue):D447–D452. https://doi.org/10.1093/nar/gku1003
8. Gene Ontology Consortium (2015) Gene Ontology Consortium: going forward. Nucleic Acids Res 43(Database issue):D1049–D1056. https://doi.org/10.1093/nar/gku1179
9. Ashburner M, Ball CA, Blake JA, Botstein D, Butler H, Cherry JM, Davis AP, Dolinski K, Dwight SS, Eppig JT, Harris MA, Hill DP, Issel-Tarver L, Kasarskis A, Lewis S, Matese JC, Richardson JE, Ringwald M, Rubin GM, Sherlock G (2000) Gene ontology: tool for the unification of biology. The Gene Ontology Consortium. Nat Genet 25(1):25–29. https://doi.org/10.1038/75556

Methods in Molecular Biology (2018) 1788: 11–21
DOI 10.1007/7651_2017_92

Published online: 08 November 2017

Discovery of Immune Reactive Human Proteins by High-Density Protein Arrays and Customized Validation of Potential Biomarkers by ELISA

Tara K. Sigdel and Minnie. M. Sarwal

Abstract

Because of our access to human genome data and ever-improving genome sequencing and proteome analysis methods, we are much better in terms of our understanding of biological processes. In addition to genomics, proteomics, and other "omics" methods, availability of more sophisticated molecular assaying methods has augmented our knowledge about immune processes toward auto- and allogeneic targets. High-density protein arrays are developed to analyze protein-small molecule interactions, enzyme-substrate profiling, protein-protein interaction, and immune monitoring by assessing antibodies in the serum.

Keywords Antibodies, Biomarker, ELISA, Immune monitoring, Protein arrays

1 Introduction

Organ transplantation is the optimal treatment for most end-stage organ failure. However, due to lack of a specific and sensitive means to monitor graft injury, long-term outcomes are not optimal [1]. Given the current status of organ shortage, advanced methods of organ monitoring and improving graft life expectancy are an unmet need in organ transplantation. Antibodies are known to be involved in acute and chronic rejection of transplanted organs, but their ability to predict graft injury events is poorly understood. Antibody's role in causing graft injury in different injury types has not been studied. Recently, microarray technology has evolved beyond nucleic acid hybridization to serve as a platform for detecting protein-specific antibodies [2]. Protein array technology that can be used to analyze protein-small molecule interactions for identification of small molecules that have specific affinity to human antigens, enzyme-substrate profiling to identify novel substrates for enzymes, and protein-protein interaction has also been used to assess antibody level in the blood collected from transplant patients to identify reactive antigens activated in different graft injury phenotypes. Schematics of several steps involved in this process is presented in Fig. 1.

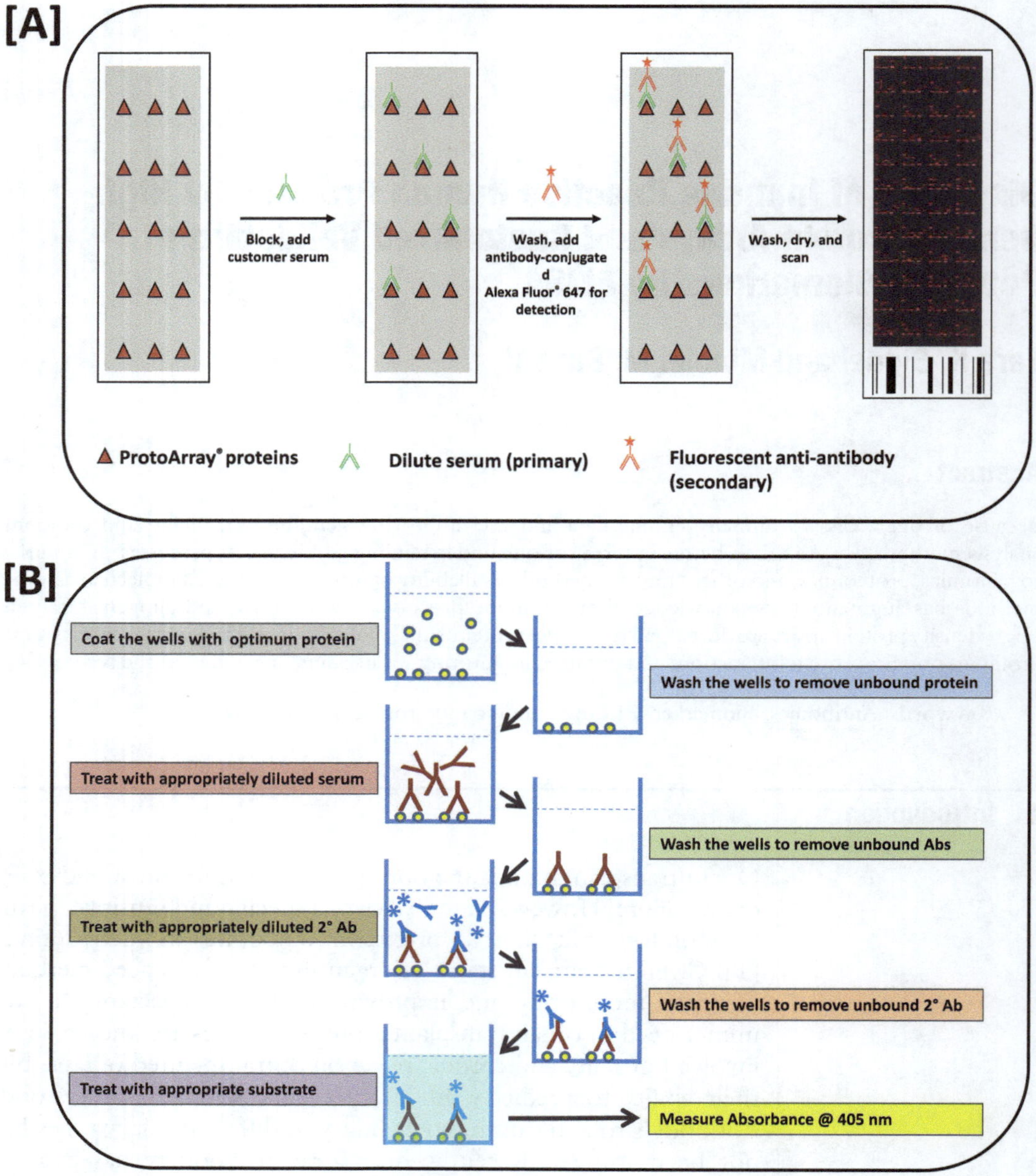

Fig. 1 (**a**) Basic steps of ProtoArray assay; (**b**) summary of indirect ELISA

We have used the protein array platform to develop an immunogenic and anatomic roadmap of the most likely non-HLA antigens that result in serological responses in kidney transplantation [3]. We have also identified a number of antigens that are associated with kidney transplantation, end-stage renal disease, and IgA nephropathy [4–8]. A successful study requires well-designed experiments and optimized methods, sophisticated data analysis, and subsequent validation of findings made by high-throughput "omic" platform. This chapter provides two independent platforms for identification and validation of antibody levels in human sera. First, we have provided a detailed method of identification of increased antibody response against auto/alloantigens by assaying antibodies against ~9500 human antigens using high-density protein arrays (ProtoArray®, Life Technologies, Carlsbad, CA). Then we have presented protocols for antibody ELISA methods to validate the discovery made by protein arrays. Protein array and MSD ELISA protocol (Meso Scale Discovery, Gaithersburg, MD) include methods that are based on our protocol adopted from corresponding manufacturer's protocols.

2 Materials

2.1 Protein Arrays

1. Human serum or plasma sample (dilute the sample 1:150 in washing buffer, store on ice until use) (*see* **Note 4.1.3**).
2. ProtoArray® Human Protein Microarray v5.0 (Life Technologies, Carlsbad, CA).
3. Alexa Fluor® 647 Goat anti-Human IgG (Life Technologies, Cat. no. A21445).
4. Blocking buffer*: either ProtoArray® Blocking Buffer Kit (Life Technologies, Cat no. PA055) or 10× Synthetic Block (Life Technologies, Cat no. PA017).

   *Blocking buffer composition (50 mM HEPES, pH 7.5, 200 mM NaCl, 0.08% Triton® X-100, 25% glycerol, 20 mM reduced glutathione, 1× synthetic block, 1 mM DTT).
   Recipe for 50 mL blocking buffer fresh is 1 M HEPES, pH 7.5, 2.5 mL; 5 M NaCl, 2 mL; 10% Triton® X-100, 0.4 mL; 50% glycerol, 25 mL; reduced glutathione, 305 mg; 10× synthetic block 5 mL; deionized water to 50 mL; adjust pH to 7.5 with NaOH. (For volume requirement, *see* **Note 4.1.9**).
   *Washing buffer composition (1× PBS, 0.1% Tween 20, 1× synthetic block). Recipe for 600 mL washing buffer is 10× PBS 60 mL, 10% Tween 20 6 mL, 10× synthetic block 60 mL, deionized water to 600 mL. (For volume requirement, *see* **Note 4.1.9**).
5. Clean, 4-chamber incubation tray with cover (QuadriPERM* Culture Dish. 4-chamber polystyrene culture dish, VWR, San Francisco, CA), chilled on ice forceps and deionized water shaker (capable of circular shaking at 50 rpm; set the shaker at 4 °C).

6. Microarray slide holder and centrifuge equipped with a plate holder.
7. Fluorescence microarray scanner (e.g., GenePix 4000B Microarray Scanner or equivalent).
8. Microarray data acquisition software (e.g., GenePix® Pro from Molecular Devices).
9. Data analysis software (ProtoArray® Prospector recommended, available from www.invitrogen.com/protoarray, Significance Analysis of Microarray (SAM) software [9], GraphPad Prism (GraphPad Software, La Jolla, CA) or equivalent software for basic statistics and visualization and presentation).

2.2 ELISA Assays

2.2.1 Conventional ELISA

1. Nunc-Immuno™ Plates (Thermo Scientific, Cat no. 449824).
2. Coating buffer: 15 mM Na_2CO_3, 30 mM $NaHCO_3$, 0.02% NaN_3.
3. Nonfat dry milk powder (BioRad, Blotting Grade Blocker Non Fat Dry Milk Cat no. 170-6404XTU).
4. PBST buffer: 1× PBS, 0.05% Tween 20.
5. Primary antibody (positive control) and secondary antibody (Goat anti-Human IgG Secondary Antibody, AP conjugate, Life Technologies, Cat # 62-8422).
6. AP-PNPP liquid substrate (Sigma, cat # 7998).
7. The standard for UV-visible microplate reader absorbance (absorbance detection in the UV-visible wavelength range 190–1000 nm).

2.2.2 MSD Platform

1. The SECTOR Imager 2400 reader or 6000 reader (Meso Scale Discovery, Gaithersburg, MD).
2. Standard and/or High Bind Plates (Meso Scale Discovery, Gaithersburg, MD).
3. Blocker A (Meso Scale Discovery, Gaithersburg, MD).
4. Read Buffer T (Meso Scale Discovery, Gaithersburg, MD).
5. SULFO-TAG labeled antibody (Meso Scale Discovery, Gaithersburg, MD).

3 Methods

3.1 Serum and Plasma Processing and Sample Preparation

3.1.1 Serum Processing

1. Collect a blood sample in a red-top tube with blood using standard venipuncture technique. Let the tube stand in upright position for 20–30 min until clot forms.
2. Centrifuge blood at 800 × *g* for 10 min and aliquot serum and store at −80 °C until ready to use.

3.1.2 Plasma Processing

1. Collect blood sample into commercially available anticoagulant-treated tubes, e.g., EDTA-treated (lavender tops) or citrate-treated (light-blue tops). Heparinized tubes (green tops) are indicated for some applications.
2. Centrifuge tubes for 10 min at 1000–2000 × *g* using a refrigerated centrifuge, and aliquot the upper plasma portion, and store at −80 °C until ready to use.

3.1.3 Sample Preparation (Serum or Plasma)

1. Prior to use, process the sample to remove any aggregates by centrifugation (12,000 × *g* for 30 s in a microcentrifuge).
2. Recommended dilution is 1:500 by the manufacturer, but in our hands 1:150 dilution in washing buffer works the best. Users may have to optimize dilution based on their initial results.

3.2 ProtoArray

A summary of the probing method is presented in Fig. 1a.

3.2.1 Blocking and Detecting

1. Thaw the protein array slides by placing them at 4 °C for at least 15 min.
2. Place the protein array slides with barcoded side facing up into each well of a 4-chamber tray.
3. Pipet 5 mL blocking buffer (cooled to 4 °C) into each chamber, avoiding any direct pipetting onto the slides.
4. Incubate the slides for 1 h at 4 °C on a shaker set at 50 rpm (circular shaking preferred).
5. After the incubation step, aspirate blocking buffer using vacuum or a pipette.
6. Wash the slides with 5 mL washing buffer by incubating the tray for 5 min at 4 °C on a shaker set at 50 rpm (circular shaking).
7. Aspirate the buffer using vacuum or pipette.
8. Add 5 mL serum or plasma sample diluted (1:150 or 1:500 or project-specific optimized dilution) in washing buffer without touching the slide surface.
9. Incubate the tray for 90 min at 4 °C on a shaker set at 50 rpm (circular shaking).
10. Aspirate the sample using vacuum or pipette.

11. Wash each array with 5 mL washing buffer with gentle shaking on a shaker set at 50 rpm for 5 min at 4 °C. Aspirate the washing buffer.
12. Repeat wash step four more times using fresh washing buffer each time to obtain a total of five washes.
13. Prepare detection antibody by mixing 2.5 μl Alexa Fluor® 647 Goat anti-Human IgG Antibody with 5 mL washing buffer per array to obtain a final antibody concentration of 1 μg/mL. Store on ice until use.
14. Add 5 mL Alexa Fluor® 647 antibody solution to the incubation tray.
15. Incubate the tray for 90 min at 4 °C on a shaker set at 50 rpm (circular shaking).
16. Aspirate the antibody solution.
17. Wash each array with 5 mL washing buffer with gentle shaking on a shaker set at 50 rpm for 5 min at 4 °C. Aspirate the washing buffer and repeat wash four more times.

3.2.2 Drying of Slides and Scanning

1. Remove slides from the 4-chamber incubation tray by inserting the tip of the forceps into the indentation at the numbered end of the slides, and gently pry the array upward, and pick up the array by holding the array by its edges only.
2. Insert the array into a slide holder, and quickly rinse by dipping the slides into a large beaker filled with deionized water five times. It is important to properly place the slides in the slide holder to prevent damage to the array during centrifugation.
3. Immediately centrifuge the array in the slide holder or 50 mL conical tube at 200 × *g* for 1 min in a centrifuge (equipped with a plate rotor, if you are using the slide holder) at room temperature. Ensure the array is completely dry.
4. After drying, store the arrays vertically or horizontally in a slide box protected from light and avoid prolonged exposure to light. To obtain the best results, scan the array within 24 h of probing.
5. To scan the array, start the appropriate array acquisition and analysis software on the computer connected to the fluorescence microarray scanner (GenePix 4000B Microarray Scanner or equivalent).
6. Follow the instrument manufacturer's protocol to scan the slides with the following setting: wavelength, 635 nm; pixel size, 10 μm; PMT gain, 700; lines to average; 1.0 laser power, 100%; and focus position, 0 μm.
7. Save the image to a suitable location as "multi-image TIFF" file.

3.2.3 Data Generation

1. Use GenePix® Pro microarray data acquisition software or equivalent on the computer. Open the saved image (.tiff), and open the .GAL files downloaded from ProtoArray® Central.
2. Adjust the subarray grid to ensure the grid is in proper location for each subarray. After the grid is properly adjusted and all features are aligned, save .GPS file.
3. Once the gridding is completed, acquire the pixel intensity data for each feature by clicking the Analyze button in GenePix® Pro, and save/export the results as a .GPR (GenePix® Results) file.

3.2.4 Data Analysis (Using Prospector Analyzer®, Life Technologies)

1. Install ProtoArray® Prospector (www.invitrogen.com/protoarray).
2. Start ProtoArray® Prospector from the desktop icon. Set the application to Immune Response Profiling (for serum samples) or Immune Response Profiling with Plasma (for plasma samples).
3. Select the Load and Analyze button from the Tool Bar to normalize single protein arrays by selecting .GPR files with immune response profiling option selected from Application pulldown menu.
4. Single array analysis generates a list of human proteins having significant IgG antibodies against them.
5. After single array analysis, group normalize group of array data by grouping the arrays from same sample type (i.e., disease group or control).
6. Once multiple arrays of the same phenotype/class are group normalized, perform group comparison analysis. This allows identifying significantly increased antibodies in case of disease or injury.
7. Group normalization provides normalized data for further analysis.
8. Once the result is generated, users can see what antibodies are increased in how many samples per phenotype (% increase) and Chebyshev's inequality (CI) P value.

3.2.5 Data Analysis Using Significant Analysis of Microarrays (SAM) [9]

1. The SAM software is available at http://www-stat.stanford.edu/~tibs/SAM/.
2. Normalized protein array data output from Prospector Analyzer (Life Technologies, Carlsbad, CA) can be used.
3. Data should be cube root transferred as the software does not handle a wide range of data such as that from ProtoArray very well.
4. SAM allows performance of a number of analyses such as two class unpaired (control and treatment groups with samples from

different patients), two class paired (samples before and after treatment from the same patients), multiclass (more than two groups with each containing different experimental units), time course (each experimental units is measured at more than one time point; experimental units fall into a one or two class design), etc.

3.3 Conventional ELISA Method

3.3.1 Schematics of ELISA Steps Is Presented in Fig. 1b

1. Coat the plates by pipetting 50 μL protein/coating buffer mixture onto each well of Nunc-Immuno™ Plates. Seal the plate with clear sealing film.
2. Store the plates at 4 °C at least overnight.
3. Dump the protein and the coating buffer out of the plates by dabbing it on a pad of absorbant paper stack. Briefly blot on paper to remove excess protein on the surface.
4. Wash the plates five times with PBST.
5. To block nonspecific protein binding, coat wells with 100 μL per well 2% milk in PBST. Allow the plate to incubate at room temperature for at least 1 h.
6. Add primary antibody as a positive control and standard or diluted serum in 2% dry milk in PBST and incubate at 4 °C overnight or 1 h at room temperature.
7. Wash the plates five times with PBST with 200 μL buffer per well.
8. Without letting the plate go dry, add the secondary antibody, anti-IgG-alkaline phosphatase (AP) diluted 1:1000 in 2% milk in PBST.
9. Incubate for 1 h at room temperature and wash the plates five times with PBST.
10. Prepare developing reagent with PNPP tablets using manufacturer's protocol.
11. Add 75 μL of developing reagent per well and incubate the plates for 30 min.
12. Read the plate using a standard UV-visible microplate reader at 405 nm.

3.4 MSD ELISA

3.4.1 Protocol Overview of MSD Antibody ELISA

1. Coat Hi Bind microwell plate with target protein diluted in PBS. If it is spot coat, pipet 5 μL of target protein and dry on plate. If it is solution, coat the wells by pipetting 30 μL of target protein per well (*see* **Note 4.2.2.3**).
2. Block with 150 μL per well of blocking solution (either 5%(w/v) Blocker A or 3% nonfat dry milk in PBS) 1 h at RT with shaking (300–600 rpm).
3. Wash the plate once with PBS or PBST.

4. Perform a titration for serum dilution optimization by comparing different fold dilutions (1:10-, 1:100-, 1:500-fold dilution in sample dilution buffer, e.g., 1% Blocker A in PBST, LowCross-Buffer (Candor Bioscience).
5. Seal the plate and shake for 1–2 h at RT.
6. Wash the plate three times with PBST.
7. Add 25 μL per well detection antibody (0.25 or 1 μg/mL) diluted in 1% Blocker A, seal, and shake 1 h at RT.
8. Wash the plates three times with PBST.
6. Add 150 μL Read Buffer T per well and read plate immediately using the SECTOR Imager 2400 reader or 6000 reader (Meso Scale Discovery, Gaithersburg, MD) following manufacturer's manual.

4 Notes

4.1 Protein Arrays

1. If multiple experiments are to be run for a project, it is very critical that all experiment conditions are maintained as consistent as possible. Especially, the incubation time, number and wash techniques and times, shaker speed, and sample and antibody dilutions must be maintained consistent.
2. Proper laboratory precautions are critical such as wearing a laboratory coat, disposable gloves, and eye protection.
3. Temperature of the arrays and reagents should be kept at 2–8 °C throughout the experiment.
4. Never touch the surface of the array as it will scratch and damage surface of the array which will contribute to smears and scratches.
5. It is critical to keep the arrays wet all the time during the experiment. Make sure the slides are submerged in the appropriate reagent.
6. A clean dust-free experiment location helps to reduce high background signals.
7. Arrays must be dried prior scanning, and scanning must be performed immediately after the experiment is completed.
8. Protect the arrays from direct light after probing with the fluorescent detection reagent and before scanning.
9. Each ProtoArray requires 5 mL of blocking buffer and 60 mL of wash buffer.
10. *Probing other immunoglobulins such as IgA, IgM*: The ProtoArray® design favors measurement of IgGs as the platform contains internal control IgGs that can be used for linear normalization. However, measurement of other

immunoglobulins using this platform is possible. The measurement of other immunoglobulins requires labeling of detection antibodies such as anti-human IgA, anti-human IgM, etc., by an Alexa Fluor tag. These antibodies and Alexa Fluor labeling kits are commercially available. Once the detection antibodies are labeled, the experiment described in this protocol are equally applicable to measure other antibodies other than IgG.

11. *Strengths*: Proto Array enables us to assay antibodies against almost 10,000 human antigens in one single experiment. The method is robust and reproducible and has a potential to provide us with information that otherwise would have taken years by conventional low-throughput methods. This sophistication has provided researchers a quick screening of reactive antigens and move on to validation steps which only uses fewer number of antigen/antibody pairs.
12. *Issues*: The arrays and reagents involved are expensive. There is no way to determine specificity of antibody-antigen. Even though it includes almost 10,000 antigens, it still doesn't provide information on reactivity of other remaining thousands of antigens.

4.2 ELISA

4.2.1 Conventional ELISA Method

13. It is important to determine the appropriate dilution of protein in coating buffer to achieve desired protein concentration.
14. Coated plates can be kept for a prolonged period up to 12 months at 4 °C. However, it is recommended to assess quality of the plate using a positive control.
15. It is important that once plates are coated, the plates are never allowed to sit dry for more than a minute or two as this could disrupt binding of protein with the surface at the bottom of the well. If there is a possibility of either of the blocking buffer or the primary antibody remaining on the plates for more than a week, use 0.1% sodium azide.
16. *Strength*: It is simple to set up and does not require specific equipment. Reagents and scanner are readily available.
17. *Issues*: Since protein arrays use fluorescence detection system, discovery made by protein array is difficult to reproduce using conventional chromophores AP or HRP. Since the detection is not sensitive, it requires relatively concentrated sample which contributes to high background.

4.2.2 MSD ELISA Method

18. MSD method is preferred for validation because of its sensitivity to quantify low-abundance analytes and its electrochemiluminescence detection system.
19. While coating standard plates, the coating buffer is PBS + 0.03% Triton X-100. No Triton should be added to

the PBS coating buffer for spot coating standard plates or for solution coating high bind or standard plates.

20. The amount of protein to be tested for optimal amount is 10, 40, and 160 ng protein per well.
21. *Strength*: Because of its electrochemiluminescence detection system, it is sensitive and is compatible with ProtoArray detection system. MSD provides multiplexed ELISA option that enables users to perform ELISA on as many as ten antibodies on a single plate thereby saving time and reagents to be used in the assays.
22. *Issues*: It requires specific reagents compatible with electrochemiluminescence detection system. The scanner for this detection system is expensive and may not be available to many researchers readily.

References

1. Sigdel TK, Sarwal MM (2011) Recent advances in biomarker discovery in solid organ transplant by proteomics. Expert Rev Proteomics 8 (6):705–715
2. Mattoon D et al (2005) Biomarker discovery using protein microarray technology platforms: antibody-antigen complex profiling. Expert Rev Proteomics 2(6):879–889
3. Li L et al (2009) Identifying compartment-specific non-HLA targets after renal transplantation by integrating transcriptome and "antibodyome" measures. Proc Natl Acad Sci U S A 106 (11):4148–4153
4. Li L et al (2010) Compartmental localization and clinical relevance of MICA antibodies after renal transplantation. Transplantation 89 (3):312–319
5. Li L et al (2010) Differential immunogenicity and clinical relevance of kidney compartment specific antigens after renal transplantation. J Proteome Res 9(12):6715–6721
6. Sigdel TK et al (2012) Non-HLA antibodies to immunogenic epitopes predict the evolution of chronic renal allograft injury. J Am Soc Nephrol 23(4):750–763
7. Sigdel TK et al (2011) Profiling of autoantibodies in IgA nephropathy, an integrative antibiomics approach. Clin J Am Soc Nephrol 6 (12):2775–2784
8. Sutherland SM et al (2009) Protein microarrays identify antibodies to protein kinase Czeta that are associated with a greater risk of allograft loss in pediatric renal transplant recipients. Kidney Int 76(12):1277–1283
9. Tusher VG, Tibshirani R, Chu G (2001) Significance analysis of microarrays applied to the ionizing radiation response. Proc Natl Acad Sci U S A 98(9):5116–5121

Methods in Molecular Biology (2018) 1788: 23–41
DOI 10.1007/7651_2017_79

Published online: 10 October 2017

HLA Class I and Class II-Induced Intracellular Signaling and Molecular Associations in Primary Human Endothelial Cells

Nicole Valenzuela, Nwe Nwe Soe, Fang Li, Xiaohai Zhang, Yi-Ping Jin, and Elaine F. Reed

Abstract

The signaling capacity of HLA molecules in vascular cells has been well established. Intracellular signaling and association with the coreceptor integrin β4 has been well-studied for HLA class I. However, little is known regarding HLA class II intracellular signaling in human endothelial cells. Investigation of HLA class II has been challenging due to the loss of HLA class II expression in cultured primary cells. Herein, we describe methods for inducing expression of endogenous alleles and loci of HLA class II molecules, as well as for studying intracellular signaling. This includes siRNA knockdown of proteins and coimmunoprecipitation of putative coreceptors for HLA in primary human aortic endothelial cells.

Keywords Adenovirus, CIITA, Coimmunoprecipitation, Endothelial cells, HLA, siRNA

1 Introduction

Alloimmune responses in transplant recipients are primarily targeted against the nonself alleles of human leukocyte antigen (HLA) molecules. Production of antibodies against donor HLA class I and class II is significantly associated with the risk of rejection and graft loss [1]. Antibody-mediated rejection can be caused by the canonical effector functions of antibodies. This includes activation of the classical complement cascade and conscription of Fc gamma receptor (FcγR)-bearing NK cells, which are neutrophils and monocytes/macrophages that can mediate cytopathic and proinflammatory effects. In addition to these Fc-mediated functions, many investigations have revealed that antibodies to HLA have novel agonistic effects on vascular endothelial and smooth muscle cells. For example, crosslinking of HLA class I by the $F(ab')_2$ region of antibodies causes intracellular signaling, in which mTOR is a central regulator, leading to changes in cell function including proliferation, motility and cytoskeletal reorganization, and leukocyte recruitment [2–6].

To date, most research has focused on the effects of antibodies to HLA class I molecules, because HLA class I loci are constitutively and highly expressed on cultured vascular cells. In contrast, although evidence suggests that HLA class II is present on human microvascular endothelial cells (EC) in situ [7–9], expression of HLA-DR, DP, and DQ is rapidly lost on such cells cultured *ex vivo*. Consequently, little is known about the signaling pathways downstream of HLA class II. Studying the effects of HLA class II antibodies is particularly critical, as clinical reports have demonstrated that antibodies to HLA class II, especially HLA-DQ, predominate in chronic rejection and predict worse outcome [10]. Fundamental molecular biology assays, combined with newer genetic engineering and molecular association technologies, are essential for dissection of the intricate intracellular signaling pathways in vascular cells in vitro.

EC upregulate both HLA class I and HLA class II molecules in response to stimulation with TNFα and/or IFNγ [11], an approach which can be used to study HLA class II antibodies in vitro. Limited reports using this approach have suggested that HLA class II molecules can indeed induce signaling after ligation with antibodies [12]. However, extensive signaling and transcription is also induced by inflammatory cytokines such as TNFα and IFNγ, and we have found that these parallel pathways often confound interpretation of HLA class II-specific responses. Recently, Taflin et al. described a method of lentiviral transduction of HLA-DR genes into EC, which enabled the study of endothelial immunogenicity to allogeneic CD4 T cells [13]. An advantage of this approach is the relatively pure induction of an HLA class II without increased expression of cytokines, chemokines, and adhesion molecules. However, several important drawbacks include expression of only one locus of HLA class II, without DP and DQ (so locus-specific effects cannot be studied), and expression of only one allele of HLA-DR (so allele differences cannot be determined). Herein, we describe an approach utilizing adenoviral delivery of the class II transactivator (CIITA), the transcription factor which controls expression of HLA class I and HLA class II. This method triggers transcriptional upregulation of the cell's endogenous alleles of HLA-DR, DQ, and DP genes.

HLA class I molecules have short cytoplasmic tails and no known signaling motifs; thus the initial pathway used by these molecules to transduce intracellular signaling was not well understood. Recently, our group used detailed molecular analyses to identify a coreceptor, integrin β4, which is required for HLA class I to cause phosphorylation of downstream signaling mediators leading to functional changes [5]. Whether HLA class II also requires coreceptors or association with intracellular adaptors for transduction of intracellular signaling is not known. In this chapter, we describe our protocols for in vitro induction of endogenous HLA class II expression, and for determining intracellular signaling pathways and the molecular associations between HLA and other molecules in primary human EC, using Western blotting, small interfering RNA (siRNA), and coimmunoprecipitation.

2 Materials

2.1 CIITA Adenoviral Vector (See Note 1)

1. Plasmid pcDNA3 myc CIITA (Addgene, Cambridge, MA).
2. Plasmid pENTR™ 4 (Invitrogen, Carlsbad CA).
3. Plasmid pAd/PL-DEST Gateway® (Invitrogen, Carlsbad CA).
4. DH5™ competent cells (Invitrogen, Carlsbad CA).
5. HEK293A cell line (Invitrogen, Carlsbad CA).
6. Adeno-X™ Rapid Titer Kits (Clontech, Mountain View CA).
7. Gateway LR Clonase II enzyme mix (Invitrogen, Carlsbad CA).
8. Lipofectamine 2000 (Invitrogen, Carlsbad CA).
9. Opti-MEM medium (Invitrogen, Carlsbad CA).

2.2 Western Blotting

1. Human aortic endothelial cells (HAEC, ~80% confluence) (*see* **Note 2** *for culture methods*).
2. Endothelial cell culture medium: 80% Medium 199, 20% heat-inactivated fetal bovine serum (FBS), 84 μg/ml heparin (~18 U/ml), 1 mM sodium pyruvate, 33 μg/ml endothelial cell growth supplement, 100 U/ml penicillin/streptomycin (*see* **Note 3**).
3. Lysis buffer: 50 mM Tris–HCl, pH 7.4, 1% 3-[(3-Cholamido-propyl) dimethylammonio]-1-propanesulfonate hydrate (CHAPS), 150 mM NaCl, 50 mM NaF (*see* **Note 4**).
4. Protease inhibitors: 1.5 mM Sodium orthovanadate (Na_3VO_4), 0.5 mM phenylmethylsulfonyl fluoride (PMSF), 0.2 μg/ml Aprotinin, 0.5 μg/ml Leupeptin (*see* **Note 5**).
5. Primary antibodies against phosphorylated and total target proteins (*see* **Note 6**).
6. HRP-conjugated secondary antibodies.
7. 1× PBS.
8. 10× TBS: 12.1 g Tris base, pH 7.6, 80 g NaCl, Milli-Q H_2O to make 1 l, Each 100 ml 10× TBS add 900 ml ddH_2O and 1.0 ml Tween 20 to become 1× TBST.
9. 1× running buffer (10×): 30.2 g Tris base, 144 g Glycine, 10 g SDS, Milli-Q H_2O to make 1 l.
10. Milli-Q water.
11. Precast gradient gels or premade polyacrylamide gels.

12. Power supply.
13. Cell scrapers.
14. Gel loading tips.
15. Protein marker (such as PageRuler Plus prestained protein Ladder, Cat #26619, Thermo Scientific; Prestained broad range protein marker premixed format, Cat # 7720, Cell Signaling Technology).
16. PVDF membrane.
17. Methanol.
18. Filter paper.
19. Foam pads.
20. Gel Electrophoresis unit.
21. Transfer apparatus.
22. 1× transfer buffer (10×): 30.5 g Tris base, pH 8.3, 144 g Glycine, Milli-Q H_2O to make 1 l.
23. Tween 20.
24. 10% Sodium Azide.
25. Ponceau S working solution: 0.1% w/v Ponceau S, 5% v/v acetic acid solution, 1× ready-to-use solution (*see* **Note** 7).
26. Dry nonfat milk.
27. Bovine serum albumin (BSA).
28. Chemiluminescent kit (ECL).
29. Film developer and film or Chemiluminescent imager.

2.3 siRNA

1. OPTI-MEM® Medium.
2. Mirus TransIT-TKO® Transfection Reagent (*see* **Note 8**).

2.4 Coimmuno precipitation

1. Ice-cold lysis buffer (*see* Sect. 2.2, **item 3**).
2. Ice-cold wash buffer: 40 mM HEPES, pH 7.5, 120 mM NaCl, 1 mM EDTA, 10 mM β-glycerophosphate, 50 mM NaF (*see* **Note 9**).
3. Protease inhibitors (*see* Sect. 2.2, **item 4**).
4. 6× SDS buffer: 300 mM Tris–HCl, pH 6.8, 600 mM dithiothreitol, 10% SDS, 0.0012% bromophenol blue, 60% Glycerol, ddH_2O (*see* **Note 10**).
5. Ice-cold 1× PBS without Ca^{2+} or Mg^{2+}.
6. Protein A or G or A/G agarose beads.
7. Antibody against protein of interest for pull-down.

3 Methods

3.1 Expression of Endogenous HLA Class II in EC Using Adenoviral Vector Delivery of CIITA

3.1.1 Generate CIITA-Adenovirus Vector (See Note 11)

1. Subclone CIITA gene from plasmid pcDNA3-myc-CIITA into pENTR 4 vector. It is highly recommended that a negative control, such as LacZ, be generated in parallel (*see* **Note 12**).
2. Transfer DNA fragment encoding CIITA from pENTR 4 vector into adenovirus-based vector pAd/PL-DEST by using the Gateway LR Clonase II enzyme mix according to the manufacturer's instructions.
   (a) Midi-prep pENTR™ 4 vector and measure DNA concentration.
   (b) Add.
   Entry vector: 50–150 ng.
   pAD/PL-DEST: 150 ng.
   Add TE buffer (pH 8.0) to 8 μl.
   Vortex briefly.
   (c) Thaw Gateway LR Clonase II on ice for about 2 min, vortex for 2 s twice. Add 2 μl of LR Clonase II into the tube from **Step b**, vortex briefly and spin down. Return Gateway LR Clonase II immediately to a −20 °C freezer.
   (d) Incubate at room temperature (RT) for 1.5–2 h.
   (e) Add 1 μl Proteinase K (provided with Gateway LR Clonase II), and incubate at 37 °C for 10 min. Cool down on ice for 2 min.
3. Transform 2 μl of the product from **Step 2** into 50 μl DH5α competent cells according to datasheet.
4. Spread cells on LB plate with Ampicillin for selection.
5. Pick colonies and inoculate in 4 ml LB medium for 16 h.
6. Purify the plasmids using a miniprep kit (Qiagen) from 3 ml of culture. Elute plasmid with 20 μl TE. Sequence the plasmids to confirm they carry the CIITA gene.

Caution: Adenovirus based vector pAd/PL-DEST *is >30 kb long. Be gentle when manipulating it.*

3.1.2 Transfection of CIITA-Adenovirus Vector into HEK293A Cells

7. The day before transfection, plate HEK293A cells at the concentration of 5×10^5/well in a 6-well tissue culture treated plate. Save one well for control without transfection.

8. Linearize 15 μl CIITA-adenovirus vector with restriction enzyme Pac I.
9. Purify the digestion product using a Qiagen miniprep column. Elute with 50 μl TE (prewarmed to 70 °C). Alternatively Pac I can be inactivated by heating to 65 °C for 20 min.
10. Store at −20 °C.

Caution: Do not use phenol to purify linearized CIITA-adenovirus vector, as phenol will interfere with the downstream transfection step.

11. On the same day of restriction digestion, transfect the linearized vector into HEK293A cells using Lipofectamine 2000.
    (a) Warm Opti-MEM medium for 20 min to RT. Dilute 1 μg of the linearized vector in 250 μl Opti-MEM and mix by pipetting up and down 3 times. RT for 5 min.
    (b) Vortex Lipofectamine 2000 briefly and dilute it in 250 μl Opti-MEM. Mix by pipetting up and down. Incubate at RT for 5 min.
    (c) Combine the diluted vector and lipofectamine 2000. Mix by pipetting up and down, then incubate at RT for 20 min.
    (d) Remove culture medium from HEK293A cell culture plate.
    (e) Add 1.5 ml cell culture medium (w/o antibiotics) to vector and lipofectamine from step c. Mix well by pipetting up and down 3 times and add mixture dropwise to HEK293A cells.
    (f) Incubate at 37 °C for 24 h. Then, remove medium containing Lipofectamine–vector mixture and replenish HEK293A cells with complete cell culture medium containing antibiotics.

Caution: You are working with adenoviruses now. The laboratory must meet Biosafety Level 2 (BSL-2) standards to continue the following procedures.

12. 48 h post-transfection, trypsinize cells and seed them on a 10-cm dish at 5×10^5/well.
13. Change medium every 2–3 days until visible regions of cytopathic effect (CPE) are observed, usually about 7 days. Change medium for the last time. Usually another 2–3 days later, 50–80% cells round up. Collect cells by forcefully pipetting and squirting medium on the bottom of the dish.
14. Spin down and add 8 ml PBS to the cell pellet. Freeze and thaw four times. Spin down.

15. Amplification of virus: Add 2 ml supernatant onto HEK293A cells per 150-mm dish, totaling 3 dishes.
16. About 2 days later, 50–80% HEK293A cells become rounded up. Squirt cells. Spin down and add 10–15 ml PBS. Freeze and thaw four times. Add 10% glycerol into supernatant, aliquot and store vials in a −80 °C freezer.
17. Titer adenovirus using Adeno-X™ Rapid Titer Kits according to the manufacturer's instructions.

3.1.3 Infection of Primary EC with Adenovirus Carrying CIITA

1. Wash actively growing EC in a T75 flask once with PBS and add 5.3 ml medium M199.
2. Add CIITA-adenovirus at a multiplicity of infection (MOI) of 2:1, or suitable amount of virus which can induce HLA class II expression in 80–90% of cells.
3. Shake every 5 min for 30 min at 37 °C.
4. Add 9.4 ml complete medium without antibiotics.
5. Change medium next day.
6. 48–72 h post- infection, analyze HLA-class II expression by flow cytometry using relevant antibodies (Fig. 1).

3.2 Western Blotting

3.2.1 Protein Lysate Preparation

1. Arrest EC at 60–80% confluence with M199 containing 1% FBS for 6–16 h, treat with stimuli (e.g., HLA class I and II

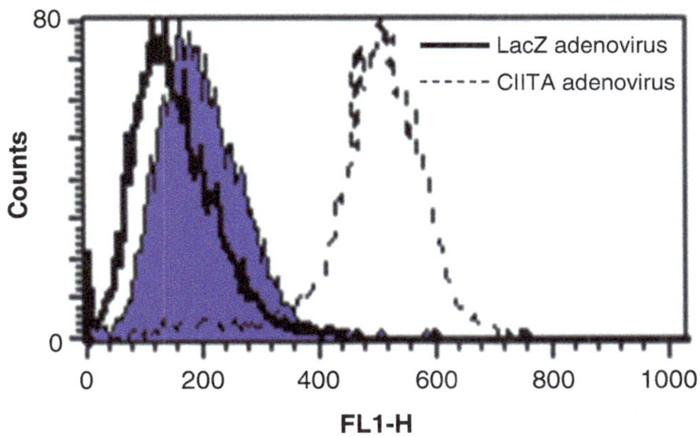

Fig. 1 Representative flow cytometry histograms showing expression of HLA class II on primary endothelial cells transfected with CIITA adenovirus. Primary endothelial cells were transfected with adenovirus encoding LacZ (negative control, *black line*) or adenovirus encoding CIITA (*dashed line*). Cells were stained with HLA class II antibodies and expression was analyzed by flow cytometry. *Solid peak* indicates unstained cells

murine Ab) for desired time and wash cells with ice-cold PBS (*note:* PBS containing 1 mM sodium orthovanadate is optimal) three times.

2. Remove the PBS and lyse the cells with an appropriate amount of cold lysis buffer (60 μl for 35 mm^2 dish, 100 μl for 60 mm^2 dish, and 350 μl for 100 mm^2 dish). Incubate the dishes with lysis buffer on ice for 30 min.
3. Scrape cells from dish and transfer to a 1.5 ml Eppendorf tube, keep the tube on ice for 10 min and sonicate the samples for 5 min on ice.
4. Centrifuge at (20,817 × *g*) for 10 min at 4 °C. Retain the supernatant.
5. (*Optional*) If the desired target protein will be detected with murine primary Ab and its weight is around 50 or 25 kDa, preclear the cell lysates with protein A/G PLUS-agarose beads for 30 min at 4 °C. Spin the samples at 1000 × *g* for 2 min at 4 °C, collect the supernatants, and transfer to new Eppendorf tubes. (*See* **Note 13**).
6. Measure protein concentration using the Bradford method, or BCA method with the bicinchoninic acid protein assay kit, according to manufacturer instructions.
7. Add 6× loading buffer for 20–30 μg protein to make samples with a final concentration of 1× loading buffer.
8. Vortex samples and heat at 95 °C for 5 min. Make sure the cap is tightly closed to prevent evaporation or opening of tubes.
9. Put samples back on ice and centrifuge at (20,817 × *g*) for 1 min.

3.2.2 Electrophoresis

10. Load the protein marker and samples into the gel using gel loading pipette tips.
11. Add 1× running buffer in the bottom and top chamber of electrophoresis unit. Solution should cover the bottom wire of electrode.
12. Cover the top and connect the anodes; Run at 25 mA/gel around 2 h for min-gel and 6 h for large gels.
13. Presoak PVDF membrane in methanol for 15 s. Also soak Western foam pads and filter paper in 1× transfer buffer for over 45 min.
14. Place plastic plate (black plate downward) and foam pad in the buffer tank containing 1× transfer buffer.
15. Disassemble gel by prying back gel plate from gel. Cut bottom strip and top wells off gel (*see* **Note 14**).
16. Place presoaked filter paper on gel; Flip plastic plate and put on the top of the foam pad soaked in 1× transfer buffer; pry gel

away letting the gel pull away from the plastic plate and onto the filter paper.

17. Place presoaked membrane on gel.
18. Place second presoaked filter paper over the membrane.
19. Roll out air bubbles with pipette, assemble the apparatus (black plate, pad, filter paper, gel, membrane, filter paper, pad, clear/red plate) and place the apparatus into the transfer tank to ensure the electric current runs through the gel (−) to the membrane (+).
20. Place a stir bar in the tank, then place an ice-pack in the tank; put apparatus on the stir plate; cover the tank and connect anodes (*see* **Note 15**).
21. Run at 100 V for 1 h 15 min, or at 30 V or 100 mA overnight (*see* **Note 16**).
22. When transfer is complete, remove membrane.

3.2.3 Protein Lysate Immunoblotting and Quantitation

1. Pour a little Ponceau S working solution onto membrane to observe protein bands and transfer efficacy; Rinse membrane with water until red color is mostly gone, then rinse with TBST for 2 min before adding blocking solution.
2. Block membrane at RT for 15–60 min with 5% nonfat milk in TBST solution (*see* **Note 17**).
3. (*Optional*) Wash 1× with 1× TBST for 1 min.
4. Incubate membrane with primary Ab solution (normally 1:500–1:2000 dilution in 1% BSA or milk in TBST) 2 h at RT or overnight at 4 °C with agitation (e.g., on a rocker). (*See* **Note 18**).
5. Wash 5× for 5 min with 1× TBST.
6. Incubate membrane with corresponding secondary Ab (normally 1:3000–1:10,000 dilution in 1% milk or BSA in TBST) solution 1–2 h at RT with agitation (e.g., on a rocker).
7. Wash 5× for at least 5 min each with 1× TBST.
8. Use chemiluminescent kit for detection according to manufacturer's instructions.
9. Develop by film developing machine or chemiluminescent image system.
10. Quantitate band intensity by using Image J (www.imagej.nih.gov) to calculate the intensity of each band. Normalize the intensity of phosphorylation of target proteins to their individual total proteins or to housekeeping loading controls (such as actin or tubulin). (*See* **Notes 19** and **20**).
11. Representative Western blotting data are given in Fig. 2.

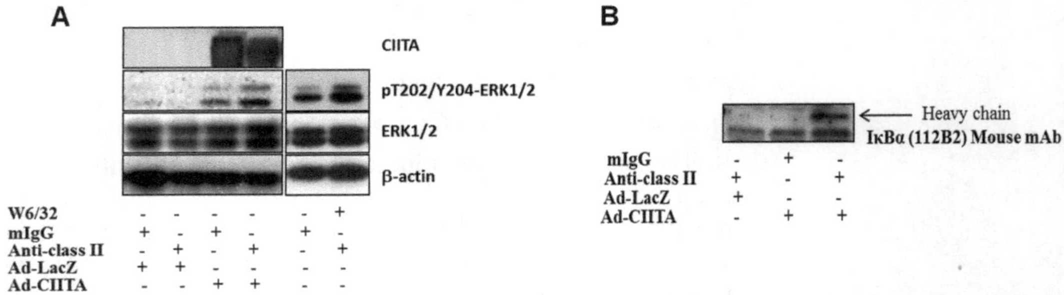

Fig. 2 Illustrative Western blot showing induction of CIITA in transfected cells, and detection of HLA class II antibody-induced ERK1/2 phosphorylation. Primary aortic endothelial cells were transfected with negative control LacZ adenovirus or CIITA adenovirus. Cells were stimulated with negative control mIgG or with antibody to HLA class II (anti-class II). (**a**) Expression of CIITA and phosphorylation of ERK1/2 was measured by Western blotting. (**b**) Expression of IκBα was measured by Western blot using mouse monoclonal IκBα primary antibody and HRP-conjugated goat anti-mouse secondary. Note that the anti-mouse-HRP secondary also detects the heavy chain of anti-class II antibody which is bound to endothelial cells (50 kDa), and which may interfere with interpretation of proteins near 50 kDa

3.3 siRNA

Before reconstituting the siRNA products, pulse-centrifuge the tube with siRNA to collect the pellet at the bottom of the tube. Resuspend siRNA to a stock concentration, such as 20 μM, using siRNA buffer.

1. Seed human aortic endothelial cells (HAEC) in 35 mm dishes in 2 ml complete medium until 70% confluent (Fig. 3).
2. Bring OPTI-MEMI medium, transfection reagent, and siRNA to RT; warm up M199 (without antibiotics, FBS, and ECGS) in 37 °C water bath.
3. Prepare siRNA complex: Add 200 μl OPTI-MEMI medium to 5 ml tube, add 6 μl transfection reagent, mix, and incubate for 5 min at RT. Add 10 μl control siRNA or 10 μl targeted siRNA for a final concentration of 100 nM siRNA, mix by pipetting, and incubate for 5 min at RT.
4. Remove culture medium, wash cells with 2 ml M199 (without antibiotics and FBS) (*see* **Note 21**).
5. siRNA transfection: add 800 μl M199 (without antibiotics and FBS) into cell culture, and add 200 μl transfection complex onto HAEC. Incubate for 6 h in 37 °C, 5% CO_2. Then add 1 ml of HAEC complete medium (without antibiotics) into transfected HAEC and continue to incubate for up to 48 h (*see* **Note 22**).
6. Effeciency and specificity of siRNA: to determine the effeciency of siRNA on target protein expression, transfect HAEC with different concentrations of siRNA and detect protein expression with Western blot. As show in Fig. 4, total FAK expression

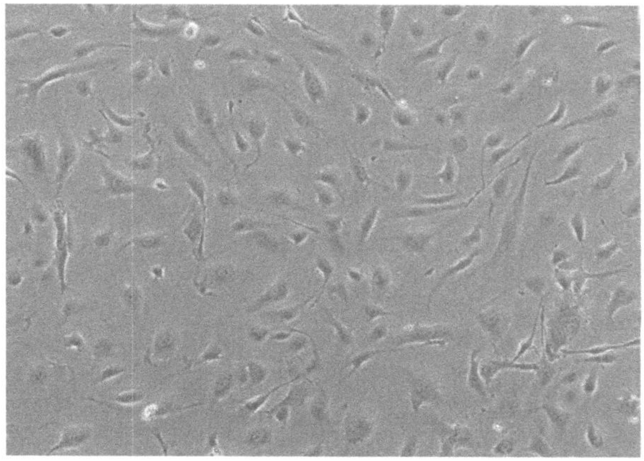

Fig. 3 Micrograph of primary human aortic endothelial cells at approximately 70% confluence

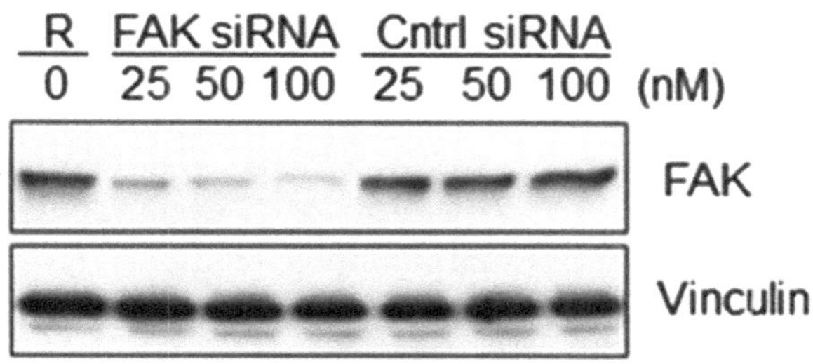

Fig. 4 Effect of control and FAK siRNA on protein expression of FAK and unrelated proteins. Endothelial cells were transfected with 25, 50, or 100 nM of control or FAK-specific siRNA, and total protein expression of FAK or vinculin as a loading control was determined by Western blotting

is reduced in a manner that is dependent upon siRNA dose. Specificity of the siRNA against FAK is determined by blotting for other proteins, which are not affected by FAK siRNA, and by addition of control nontargeting siRNA. Transfection of HAEC with FAK siRNA or control siRNA does not substantially change the expression of other proteins [14].

3.4 Coimmunoprecipitation

3.4.1 Preparation of Lysates

1. Seed EC at approximately ~90%, and serum starve in M199 containing 0.2% FBS for 4 h.
2. Treat endothelial monolayer with stimulus of interest (e.g., HLA class I antibody, W6/32 or control antibody of same species) for the desired experimental time point.

3. Place the cell culture dish on ice and wash the cells with ice cold 1× PBS without Ca^{2+} and Mg^{2+} three times.
4. Drain the PBS, then add ice-cold lysis buffer (0.5–0.75 ml per 10^7 cells/100 mm^2 dish/150 cm^2 flask) (*see* **Note 23**).
5. Scrape adherent cells off the dish using a cold plastic cell scraper, then gently transfer the cell suspension into a pre-cooled microcentrifuge tube.
6. Centrifuge in a micro centrifuge at 4 °C (*see* **Note 24**).
7. Gently remove the tubes from the centrifuge and place on ice. Gently aspirate the supernatant and place in a new tube kept on ice, and discard the pellet (*see* **Note 25**).
8. The protein concentration can be measured by Bradford or another assay as above.

3.4.2 Preclearing of the Lysates (Optional)

Preclearing is recommended for samples which contain abundant IgG, such as tissue lysate, or that are contaminated with a protein that nonspecifically binds with the immunoprecipitated complex. These conditions will interfere with visualization of the target protein.

9. Incubate the sample with a nonspecific control of the same species of antibody that will be used for the immunoprecipitation.
10. Incubate for 1–2 h with gentle agitation at 4 °C.
11. Add 25–50 μl of protein A/G-agarose beads into the lysate containing 0.5–1 ml (250–500 μg protein) (*see* **Note 26**).
12. Incubate for 1 h at 4 °C with gentle agitation.
13. Spin the sample at 950 × *g* for 5 min at 4 °C, collect supernatant and transfer to a new Eppendorf tube and proceed to the immunoprecipitation step.

3.4.3 Protein Agarose Beads Preparation

14. Always keep samples on ice.
15. Add the required amount of agarose beads into the Eppendorf tube kept on ice (*see* **Note 27** and Fig. 5a).
16. Centrifuge at 950 × *g* for 5 min at 4 °C to remove the preserved solution. Remove the supernatant carefully.
17. Resuspend the pellet agarose beads with half the volume of ice-cold 1× PBS (e.g., initial 200 μl agarose beads should be resuspended with 100 μl 1× PBS). Vortexing can damage the beads. Therefore, gently mix the beads by tapping with a fingertip (*see* **Note 28**).

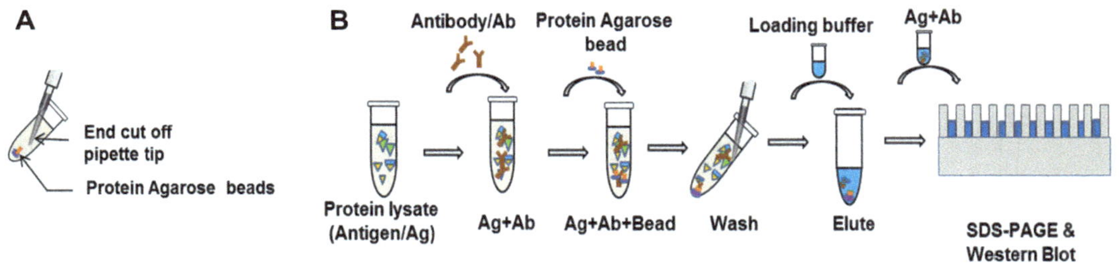

Fig. 5 Schematic diagram of coimmunoprecipitation method. (**a**) Preparation of protein A/G agarose beads using a wide-mouth or trimmed pipet tip. (**b**) Steps of coimmunoprecipitation. Protein lysate is incubated with primary antibody to the protein of interest, which is precipitated with its associated proteins using protein A/G agarose beads. Protein complexes are eluted from beads and analyzed by Western blotting

3.4.4 Immunoprecipitation

Always keep samples on ice.

18. Incubate the lysate with an antibody overnight at 4 °C with gentle agitation or rotation (*see* **Note 29**).
19. Add 25–50 μl of appropriate protein agarose beads into the lysate and antibody mixture (*see the protein agarose bead section for in-detail bead preparation*)
20. Incubate the lysate and antibody–bead mixture at 4 °C with gentle agitation for 2–4 h (the optimal incubation time can be determined by end user).

    It is recommended to wash three times to remove nonspecific proteins. All samples should be kept on ice.
21. When the lysate and antibody–bead mixture incubation time is over, centrifuge the tubes at 950 × *g* for 5 min at 4 °C.
22. Remove the supernatant and discard (*see* **Note 30**).
23. Add approximately the 0.75–1.0 ml wash buffer. Mix the bead complex by gently tapping with a fingertip.
24. Centrifuge at 950 × *g for* 5 min at 4 °C. Remove the supernatant and discard.
25. At final wash, carefully remove as much wash buffer as possible from the beads complex. Then, proceed to the elution step.

    The Ag-Ab complex can be eluted from the beads by heating the samples in loading buffer containing denaturant sodium dodecyl sulfate (SDS).
26. Add 25 μl of 3× SDS buffer for 25 μl of the bead complex.
27. Heat for 5 min at 100 °C in a heating block. Cool tubes to RT.
28. Centrifuge the eluted samples at 10,500 × *g* for 5 min at RT.
29. Transfer the supernatant (eluted protein complex) into a new tube and discard the beads.

3.4.5 Western Blot Analysis

Analyze samples for the presence of your proteins of interest by Western blot as described above.

30. Load the sample (eluted protein complex) on SDS-PAGE gel.
31. Detect coprecipitated and target proteins. (*See* **Notes 31** *and* **32**).

An overview of the coimmunoprecipitation process is given in Fig. 5b. Representative results are shown in Fig. 6.

4 Notes

1. Study of HLA antibody signaling commonly relies on commercially available murine monoclonal antibodies, such as W6/32. It is essential to include negative control antibody stimuli that are host and isotype matched for HLA antibody (isotype/host matched). Moreover, it is highly recommended to confirm findings with different HLA antibody sources, such as other monoclonals or human serum containing HLA antibodies, to demonstrate that the response is not an idiosyncratic phenomenon with one clone of HLA antibody.
2. Endothelial cell culture:
   - Coat cell culture dishes with 0.1% Gelatin for at least 30 min at 37 °C: Rinse them once with 1× PBS.
   - EC cells are usually split in a 1:3 ratio and cultured in complete medium.
   - Culture 2–3 days at 37 °C in incubator with 5% CO_2 and 100% moisture.
3. Heparin should be dissolved with LPS free sterile water and passed through 0.45 μm filter to make 4.0 mg/ml stock solution; ECGS dissolved in sterile PBS; 33 μg/ml endothelial cell growth supplement (ECGS, Corning Discovery Labware, Ca# 356006, Bedford, MA).

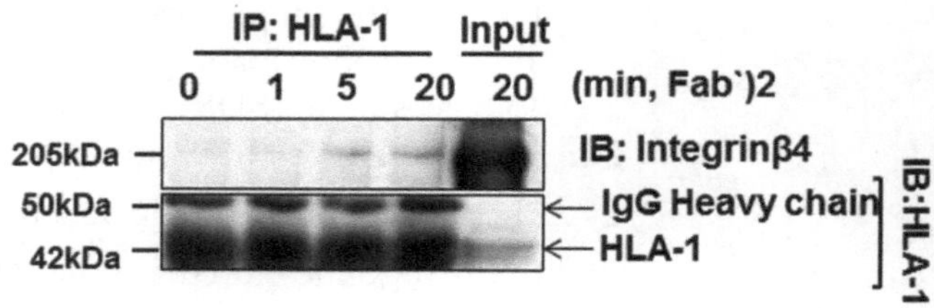

Fig. 6 Representative Western blot demonstrating coimmunoprecipitation of integrin β4 with HLA. HLA was immunoprecipitated with $F(ab')_2$ of W6/32, and pull-down lysates were analyzed for the presence of integrin β4 subunit by Western blotting. Note the band at 50 kDa representing the heavy chain of IgG

4. Store up to 6 months at 4 °C. Immediately before use, add protease inhibitors (Na_3VO_4, PMSF, Aprotinin, and Leupeptin). The choice of lysis is better determined by the protein of interest. CHAPS lysis buffer is recommended for integral membrane proteins such as integrins. For cytosolic proteins, RIPA lysis buffer can be used. A high concentration of detergent can interfere with protein complex formation. Therefore, it is recommended to choose the appropriate lysis buffer depending on the target protein.
5. Commercially available protease inhibitor cocktails can also be used.
6. For detection of a protein by Western blotting with a molecular weight near 50 kDa or 25 kDa, use a primary antibody from a different species of immunoprecipitated antibodies and the appropriate secondary antibody.
7. 1× ready-to-use solution (aMResco, Solon, OH, Cat# K793-500ML)
8. We have tested several transfection reagents for siRNA knockdown, including FuGENE 6, Lipofectamine 2000, and Dharmacon transfection reagents. While other transfection agents work for primary HAEC, we found that Mirus TransIT-TKO® Transfection Reagent yielded optimal results. Most siRNAs are commercially available, such as from Dharmacon, or can be designed against your target of interest.
9. Store up to 6 months at 4 °C. Immediately before use, add protease inhibitors (Na_3VO_4, PMSF, Aprotinin, and Leupeptin). (The lysis buffer can also be used for the washing step in Co-IP).
10. For 3× SDS buffer, dilute with ddH_2O.
11. Note that CIITA controls transcription of other genes that may be modulated by transfection. It is recommended that the effect of CIITA transfection alone on proteins of interest be determined in preliminary experiments.
12. Use of a negative control adenovirus, such as LacZ, is recommended to control for nonspecific signaling/increased background in adenoviral-transfected cells.
13. During gel analysis of antibody-induced EC activation by monoclonal HLA class I and II antibodies, IgG often masks the presence of other proteins with similar molecular weight when using murine monoclonal primary antibody. Therefore, removal of IgG from lysate by preclearing with protein A/G agarose beads is optimal when detecting target proteins with murine primary antibody prior to SDS-PAGE.
14. The bottom of the gel is thicker. If it is not removed, it causes bubbles during the transfer process. In addition, removing the

dye strip at the bottom of the gel will make the membrane clean after transfer.

15. Pay attention to the connection of anodes to power supply: usually the red electrode goes to positive.
16. For high molecular weight proteins and phosphorylated proteins, it is better to transfer overnight.
17. For some phosphorylation antibodies, blocking with 5% BSA-TBST is desirable. See antibody datasheet for specific requirements.
18. For phosphorylated proteins, overnight incubation is desirable.
19. Weak or no signals or bands in Western blotting may be due to a variety of factors.
    - Primary antibody may be too old, or may not recognize target protein in the cell system. Buy new antibody or try another antibody from a different source or clone.
    - May use BSA as blocking reagent instead of milk, or shorten blocking time.
    - Decrease Tween 20 amount in TBST and wash time.
    - Load more protein in each well if due to insufficient antigen.
    - Ensure transfer occurred by rolling bubbles out, using protein marker to monitor transfer, and predetect with Ponceau S staining.
    - Increase primary/secondary antibody concentration and/or eliminate Tween 20 in antibody solution.
    - After transfer, but before blocking, let the membrane completely dry to aid in retaining protein on the membrane.
20. High background or many nonspecific bands in Western blotting may be improved by:
    - BSA solution as blocking buffer may increase high background. Try using 5% milk-TBST instead or adding 0.01–0.02% SDS into the blocking solution.
    - Increase blocking time.
    - Optimize primary and secondary antibody dilutions; reduce antibody incubation time: increase Tween 20 in the antibody solution.
    - Increase wash time and make sure using TBST instead of TBS as wash buffer.
    - Milk contains IgG that may cross-react with goat secondary antibodies; in this case, using BSA or commercial blocking buffer without mammalian proteins, such as Odyssey™ Blocking Buffer.

- After blocking, do not allow the membrane to dry; for reprobing the membrane, wet the membrane with methanol before washing with TBST.
- Allow enough antibody volume to cover membrane and keep agitation during incubation.
- Always handle membrane carefully and with clean forceps; always use clean lab wares for membrane.

21. For primary cells transfection, use medium containing no antibiotics, and perform a wash of cells to remove serum before siRNA transfection.
22. If this is the first time using a particular siRNA, it is recommended to optimize the transfection conditions, including siRNA delivery time, such as 24, 48, and 72 h; and the dose of siRNA, for instance 25, 50, or 100 nM.
23. The culture dish with adherent cells and lysis buffer can be stored at −80 °C if the experiment cannot proceed to the next step on the same day. Later, it can be thawed and the experiment can proceed to **Step 5**.
24. The centrifugation force and time depends on the cell type. A guideline is 10 min at 10,000 × *g* for EC, but this must be determined by the end user.
25. The harvested cell lysate can be stored at −80 °C if the experiment cannot proceed to the next step on the same day. Later, it can be thawed and the experiment can proceed to the immunoprecipitation steps.
26. Detailed preparation of protein A/G agarose beads can be found in the protein agarose bead preparation section.
27. Protein agarose beads are very fragile. Therefore, it is recommended to use a wide tip or to cut the end off of pipette tip to prevent damage to the beads (Fig. 5a).
28. Choosing the correct beads is important for successful Co-IP. Whether protein A or G or A/G PLUS Agarose beads should be used can be determined on the species of immunoglobulin and isotype of immunoprecipitated antibody. Generally, protein G has the strongest binding affinity for mouse IgG, whereas protein A has the strongest binding affinity for all rabbit isotypes [15]. Protein A/G PLUS may have the strongest and broadest binding affinity for isotypes of different species. In addition, one should be aware of the molecular weight of the target protein complex. Protein A has a 17 kDa molecular weight and Protein G is 47 kDa. Therefore, to avoid an unnecessary overlapping effect of the Protein A and G band in Western blotting, the correct protein agarose beads should be determined by the end user.

29. The amount of protein and antibody are determined by the abundance of the protein and the affinity of the antibody for the protein. Approximately 100–1000 μg of protein is required for co-IP. The antibody datasheet can be used as a reference for recommended antibody concentration. It is important to adjust the volume of the lysate. Typically, each sample should have 500–1000 μl to get enough volume for rotation.
30. It is important to aspirate the supernatant carefully in order to avoid aspirating the bead complex.
31. Failure to detect specific antigen by coimmunoprecipitation may be due to low expression of either the target protein or coprecipitated protein. It is recommended to use input (total cell lysate) in Western blotting as a positive control to estimate the expression level.
32. The presence of a suspected nonspecific band in coimmunoprecipitation might be due to too much antibody concentration. This can be improved by optimizing the antigen–antibody ratio in a preliminary experiment. Alternatively, the washing step may not be sufficient. For each wash, thoroughly mix the bead complex with the wash buffer by gently tapping with a fingertip. Finally, a positive control (input) and a negative control (IP with same species of antibody) should be included in each experiment to detect the estimated molecular weight of target proteins or nonspecific proteins binding (Fig. 6).

References

1. Everly MJ et al (2013) Incidence and impact of de novo donor-specific alloantibody in primary renal allografts. Transplantation 95:410–417
2. Valenzuela NM et al (2013) Blockade of p-selectin is sufficient to reduce MHC I antibody-elicited monocyte recruitment in vitro and in vivo. Am J Transplant 13:299–311
3. Jin YP et al (2014) Everolimus inhibits anti-HLA I antibody-mediated endothelial cell signaling, migration and proliferation more potently than sirolimus. Am J Transplant 14:806–819
4. Li F et al (2011) Antibody ligation of human leukocyte antigen class I molecules stimulates migration and proliferation of smooth muscle cells in a focal adhesion kinase-dependent manner. Hum Immunol 72:1150–1159
5. Zhang X et al (2010) HLA class I molecules partner with integrin β4 to stimulate endothelial cell proliferation and migration. Sci Signal 3:ra85
6. Ziegler ME et al (2012) HLA class I-mediated stress fiber formation requires ERK1/2 activation in the absence of an increase in intracellular Ca2+ in human aortic endothelial cells. Am J Physiol Cell Physiol 303:C872–C882
7. Muczynski KA et al (2001) Unusual expression of human lymphocyte antigen class II in normal renal microvascular endothelium. Kidney Int 59:488–497
8. Muczynski KA et al (2003) Normal human kidney HLA-DR-expressing renal microvascular endothelial cells: characterization, isolation, and regulation of MHC class II expression. J Am Soc Nephrol 14:1336–1348

9. Barbatis C et al (1987) Immunocytochemical analysis of HLA class II (DR) antigens in liver disease in man. J Clin Pathol 40:879–884
10. Wiebe C et al (2012) Evolution and clinical pathologic correlations of de novo donor-specific HLA antibody post kidney transplant. Am J Transplant 12:1157–1167
11. Wedgwood JF et al (1988) Effect of interferon-gamma and tumor necrosis factor on the expression of class I and class II major histocompatibility molecules by cultured human umbilical vein endothelial cells. Cell Immunol 111:1–9
12. Jane-wit D et al (2013) Alloantibody and complement promote T cell–mediated cardiac allograft vasculopathy through noncanonical nuclear factor-κB signaling in endothelial cells. Circulation 128:2504–2516
13. Taflin C et al (2013) Study of the allogeneic response induced by endothelial cells expressing HLA class II after lentiviral transduction. Methods Mol Biol 960:461–472
14. Jin YP et al (2007) RNA interference elucidates the role of focal adhesion kinase in HLA class I-mediated focal adhesion complex formation and proliferation in human endothelial cells. J Immunol 178:7911–7922
15. Bonifacino JS et al (2001) Immunoprecipitation. In: Ausubel FM et al (eds) Current protocols in molecular biology. Chapter 10, Unit 10 16

Methods in Molecular Biology (2018) 1788: 43–61
DOI 10.1007/7651_2017_85

Published online: 25 October 2017

Characterization of Protein Complexes Using Chemical Cross-Linking Coupled Electrospray Mass Spectrometry

Timothy D. Cummins and Gopal P. Sapkota

Abstract

Identification and characterization of large protein complexes is a mainstay of biochemical toolboxes. Utilization of cross-linking chemicals can facilitate the capture and identification of transient or weak interactions of a transient nature (Huang and Kim, PloS One 8:e61430, 2013; Gao et al., J Vis Exp doi: 10.3791/51387, 2014). Here we describe a detailed methodology for a cell culture-based proteomic approach. We describe the generation of cells stably expressing green fluorescent protein (GFP)-tagged proteins under the tetracycline-inducible promoter and subsequent proteomic analysis of GFP-interacting proteins. We include a list of proteins that were identified as interactors of GFP.

Keywords GFP affinity purification, Inducible expression, Interactome, Mass spectrometry, Protein complex, Protein–protein interaction

1 Introduction

Ectopic introduction of an inducible GFP-tagged protein in cells is a useful technique for characterization of protein–protein interactions. Introduction of a pcDNA-5/FRT/TO™ expression vector (Invitrogen) GFP-tagged gene with a pOG44 Flip-recombinase expression plasmid into Flp-IN™ T-REx™ 293 cells (Invitrogen) will integrate the tagged gene into a single pre-designated FRT site. Following integration of the GFP-tagged gene, cells are treated with hygromycin to select for cells that have successfully integrated the pCDNA 5/FRT/TO vector containing the gene of interest. Cells are then maintained in growth medium supplemented with hygromycin. The stable integration of GFP-tagged genes in T-REx™ 293 cells allows for real-time detection of protein expression and troubleshooting for appropriate induction time and dose of doxycycline or tetracycline. Once stable cells are established, a time course of doxycycline or tetracycline can be performed to determine appropriate duration to achieve desired levels of protein expression (Fig. 1).

Purification of GFP or GFP-tagged proteins in complex is performed in concert with the addition of dithiobis succinimidyl

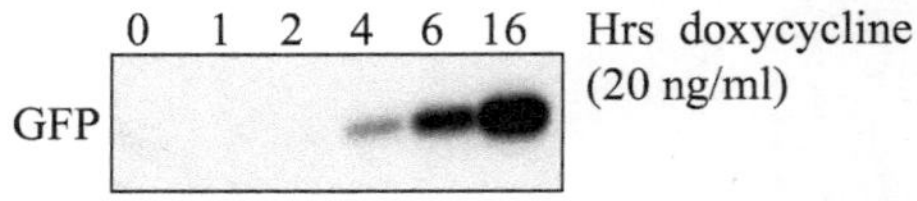

Fig. 1 A time course of doxycycline induction of GFP expression in T-REx 293 GFP cells. This establishes a time course of robust induction of the target protein to determine appropriate amount of protein production for interaction analysis. Here, 6–16 h of doxycycline induction is appropriate to perform protein isolation and downstream mass spectrometry analysis

propionate (DSP) during lysis. DSP has dual NHS esters (see DSP structure in Fig. 2), thus targeting primary amines in protein complexes and forming a conjugated network of proteins [3, 4]. Lysis buffer is gently supplemented with DSP prior to lysis in a dropwise fashion to minimize precipitation. Appropriate amount of lysis buffer is then added to cells and allows DSP to interact with primary amines in proteins and complexes of the proteins. Figure 3 illustrates a graphical representation of the pulldown workflow and protein identification by mass spectrometry.

2 Materials

1. T-REx 293 cell lines (Invitrogen, R780-07). Maintain cells in DMEM supplemented with 10% fetal bovine serum (FBS, Hyclone), 2 mM L-glutamine (Gibco), 100 U/ml penicillin, 200 μg/ml streptomycin, 15 μg/ml blasticidin, and 100 μg/ml zeocin.
2. Recombinase Flp-IN pOG44 (V6005-20) and pCDNA5 FRT/TO (V6520-20) (Invitrogen).
3. Clone individual genes of interest into pCDNA5 FRT/TO using convenient multiple cloning sites.
4. Lysis buffer: 40 mM HEPES, pH 7.4, 120 mM NaCl, 1 mM EDTA, 10 mM sodium pyrophosphate, 50 mM NaF, 1.5 mM sodium orthovanadate, 1% Triton, 1 tablet complete protease inhibitor cocktail per 25 ml.
5. Prepare dithiobis succinimidyl propionate (DSP, Pierce) 200 mg/ml in dimethyl sulfoxide (prepare fresh). DSP forms an NHS ester with primary amines and is cleaved upon reduction with dithiothreitol (DTT). This property allows for purification of conjugated DSP-protein complexes and release upon addition of the reducing agent in the final steps just prior to performing SDS-PAGE.

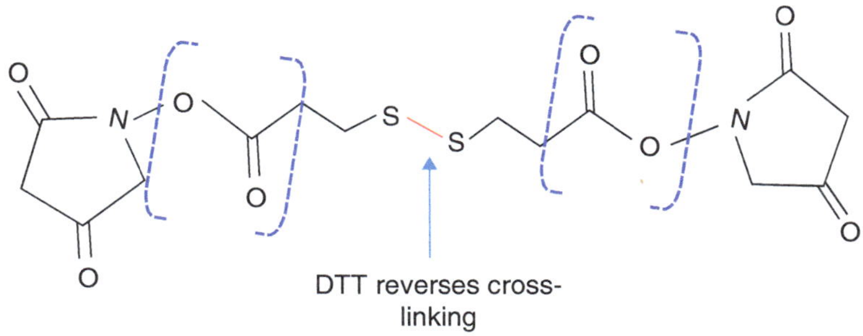

Fig. 2 Dithiobis (succinimidyl propionate): *Blue* brackets indicate carbodiimide that reacts with primary amines releasing the NHS ring while cross-linking polypeptides; *red* disulfide bond indicates bond that reacts with reducing reagent to cleave linkage and release cross-linked proteins or polypeptides

6. Equilibrate GFP-trap beads (ChromoTek) in lysis buffer (above) in a 50% slurry.
7. 1.5 mm NuPAGE Bis-Tris 10% precast gels (Invitrogen).
8. MOPS SDS-PAGE running buffer: 50 mM MOPS, 50 mM Tris Base, 0.1% SDS, 1 mM EDTA, pH 7.7.
9. Spin X columns (Corning® Costar® Spin-X®), centrifuge tube filters, cellulose acetate membrane, pore size 0.45 μm, sterile.
10. SimplyBlue™ SafeStain (Invitrogen) or Coomassie blue dye.
11. Triethylammonium bicarbonate (Sigma).
12. Sequencing grade modified trypsin (Promega).
13. Opti-MEM serum-free medium (Life Technologies).
14. 25 μl of polyethyleneimine (1 mg/ml) (Polysciences) or other transfection reagent.
15. 100 μg/ml hygromycin.
16. Doxycycline (20 ng/ml) or tetracycline (1 μg/ml).
17. 6.25 ml of 1 M Tris–HCl, pH 7.4 per 25 ml lysis buffer (200 mM Tris–HCl final).
18. 50 μl packed sepharose or agarose beads.
19. 10 ml of 50 mM Tris–HCl, pH 7.5, 0.1 mM EGTA.
20. 60 μl of 1× LDS sample buffer with 0.1 M DTT final.
21. 500 μl of (a) water, (b) 50% acetonitrile, (c) 50 mM NH_4HCO_3 (ammonium bicarbonate; prepared fresh), and (d) 50% acetonitrile/50 mM NH_4HCO_3.
22. 10 mM DTT/50 mM NH_4HCO_3.
23. 75 μl of 100 mM iodoacetamide/50 mM NH_4HCO_3.

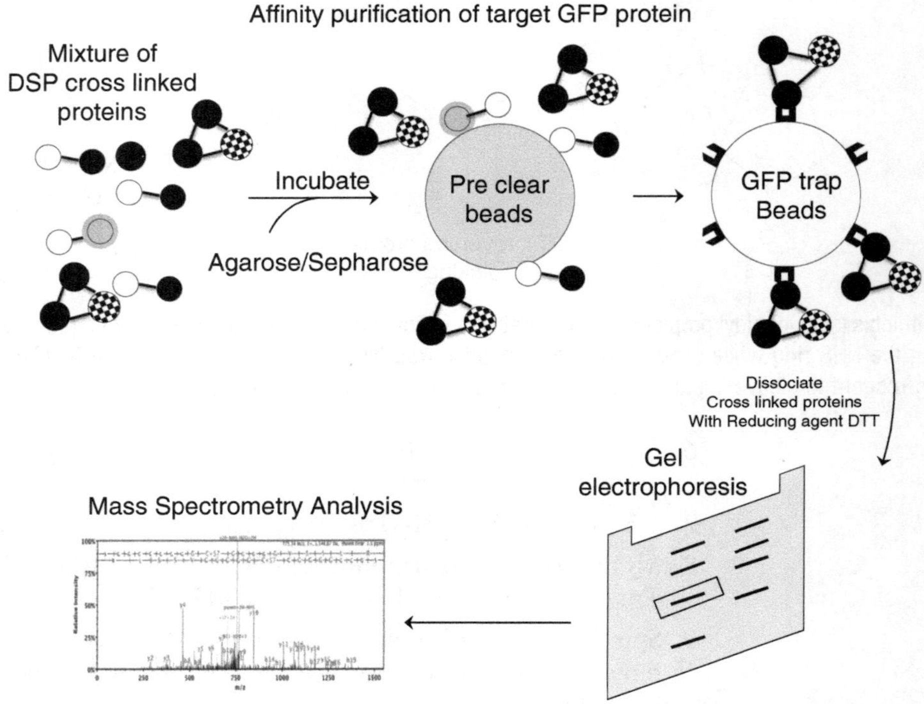

Fig. 3 Schematic workflow for GFP-trap purification and mass spectrometric analysis of co-purified proteins

24. 2.5% formic acid.
25. HPLC grade 5% acetonitrile/0.1% formic acid.
26. 3-μm C_{18} reversed-phase (RP) resin column (Acclaim PepMap100, 75 μm × 15 cm).
27. Thermo LTQ Orbitrap.
28. Poly-Prep filter columns (Biorad).

3 Methods

3.1 Transfection and Establishing a GFP Stable Cell Line

1. Plate ~1 × 10^6 (~60% confluent) in a 10-cm dish in complete medium lacking zeocin/blasticidin. Place cells in a 37 °C, 5% CO_2 incubator overnight.
2. Pre-warm 1 ml of Opti-MEM serum-free medium in a 15 ml conical tube.
3. Add 9 μg pOG44 plasmid and 1 μg of pCDNA5/FRT/TO/GFP vector (*Note:* maintain a 9:1 ratio of pOG44 with pCDNA5/FRT/TO/GFP when scaling target cell dish size).

4. Add 25 μl of polyethyleneimine (1 mg/ml) or other transfection reagent to transfection medium and mix by tapping or vortexing for 15 s.
5. Leave the transfection mix for 15 min at room temperature.
6. Add the entire transfection mixture to the target cell dish in a dropwise fashion and return cells to the incubator.
7. 24 h post-transfection split cells into three dishes, and maintain in DMEM, 10% FBS, L-glutamine, penicillin, and streptomycin for 24 h.
8. Exchange medium with fresh medium containing 15 μg/ml blasticidin and 100 μg/ml hygromycin. Hygromycin selection will cause the death of un-transfected cells lacking the integration of pCDNA5/FRT/TO/GFP vector into the FRT site. Replace medium every 3 days. As a negative control, use the selection medium in un-transfected control cells.
9. Check cells daily until majority of cells are dead and a number of colonies have formed. Then begin the expansion of cells for protein expression analysis.
10. Test doxycycline (20 ng/ml) or tetracycline (1 μg/ml) induction time course (Fig. 1) for GFP expression. Fluorescence microscopy can also be used to ensure that every cell expresses GFP upon induction with doxycycline or tetracycline.

3.2 Purification of GFP-Interacting Proteins from Stable Cell Lines Using DSP Cross-Linking Agent

1. Induce GFP protein expression with 20 ng/ml doxycycline for a defined time.
2. Wash cells once in ice-cold PBS. Gently add PBS to prevent dislodging cells during washing steps.
3. Lyse by scraping the desired number and size (10–15 cm) of semi-confluent dishes in complete lysis buffer with 2.5 mg/ml DSP. Transfer lysates onto a falcon tube.
4. Keep the lysates on ice for 30 min; then quench cross-linking reaction by adding 6.25 ml of 1 M Tris–HCl, pH 7.4 per 25 ml lysis buffer (200 mM Tris–HCl final); and incubate for further 30 min. (Lysates can be snap frozen in liquid nitrogen and stored at −80 °C this stage.)
5. Spin lysates at 10,000 × g for 30 min at 4 °C. Collect a small aliquot of cleared lysate for immunoblot analysis.
6. Filter the lysates through a filter column to clear the lysate of any debris.
7. To clarified lysate, add 50 μl packed sepharose or agarose beads for 1 h at 4 °C. This preclearing step minimizes non-specific interactions to solid phase.

8. To the cleared lysate (10–60 mg total protein), add 50 μl of equilibrated GFP-trap beads and incubate for 4 h on a roller at 4 °C.
9. Wash beads in 4 × 10 ml of complete lysis buffer containing 0.5 M NaCl. Addition of NaCl removes weak or low affinity spurious interactions.
10. Wash beads with 10 ml of 50 mM Tris–HCl, pH 7.5, and 0.1 mM EGTA.
11. Reconstitute beads in 60 μl of 1× LDS sample buffer with 0.1 M DTT final, incubate for 1 h at 37 °C, and boil for 5 min at 95 °C. DTT reduces disulfide bond from DSP to remove cross-linking of proteins.
12. Isolate denatured proteins from GFP-trap beads by spinning through spin X columns.

3.3 Gel Electrophoresis of GFP Purified Protein Complexes

1. Prepare a 10% NuPAGE Bis-Tris (1.5 mm thick) gel for electrophoretic separation of proteins. Rinse wells thoroughly with MOPS running buffer.
2. Load equal amounts of control GFP and GFP protein of interest into alternating wells (leave empty wells between samples to minimize potential cross contamination during loading).
3. Resolve proteins at ~160 V for ~45 min.
4. Stain using mass spectrometry grade staining reagent containing Coomassie blue dye or Simply Blue.
5. Image gel in clean petri dish, and avoid unnecessary handling to minimize keratin contamination.

3.4 In-Gel Trypsin Digestion

1. Excise bands of interest for each sample lane (see example Fig. 4a).
2. Cut bands into small 1–2 mm cubes with a clean scalpel. Place into clean 1.5 ml centrifuge tubes, and minimize keratin contamination by changing gloves frequently.
3. Wash gel bands sequentially in 500 μl of (a) water, (b) 50% acetonitrile, (c) 50 mM NH_4HCO_3 (ammonium bicarbonate; prepared fresh), and (d) 50% acetonitrile/50 mM NH_4HCO_3. Place samples on shaker for 10 min at room temperature for each wash step. Aspirate wash solution after each step.
4. In-gel alkylation and reduction steps:

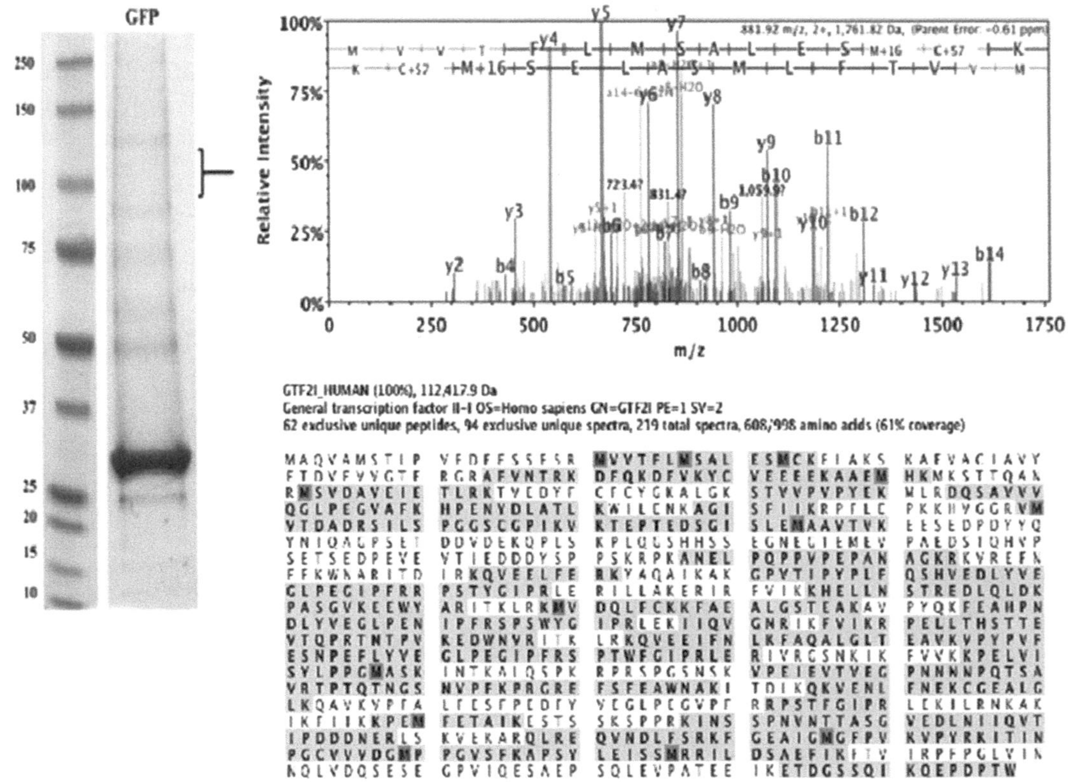

Fig. 4 (**a**) Affinity-purified GFP was resolved by SDS-PAGE with molecular weight markers (kDa) indicated in the first lane. Brackets indicate one excised gel piece, which was tryptically digested for mass spectrometric electrospray analysis. (**b**) Identification of a prominent protein from the 100 kDa excised band in GFP lane. GFP interactors are compared to target proteins of interest to exclude non-specific GFP, or bead-interacting proteins included in Table 1 are highly enriched promiscuous GFP contaminants that can be excluded from target protein pulldown analysis

(a) To each tube add 75 μl of freshly prepared 10 mM DTT/50 mM NH_4HCO_3 and incubate for 20 min at 37 °C. Remove solution and proceed to alkylation step.

(b) Add 75 μl of freshly prepared 100 mM iodoacetamide/ 50 mM NH_4HCO_3 and incubate in dark at room temperature for 20 min.

5. Remove liquid and wash bands in 50 mM NH_4HCO_3 for 10 min at room temperature.
6. Remove liquid and replace with 50% acetonitrile/50 mM NH_4HCO_3 to remove any dye stain. Repeat wash steps until gel pieces are colorless.
7. Add 300 μl of acetonitrile to each tube for 10–15 min. Speed vacuum away additional acetonitrile. Dry gel pieces will appear opaque.

8. Re-swell gel pieces in 25 mM triethylammonium bicarbonate containing 5 μg/ml of sequencing grade modified porcine trypsin. Place on shaker at 30 °C. After 20–30 min, check if gel pieces are adequately soaked in solution. If not, add more triethylammonium bicarbonate to assure coverage of bands. Digest gel bands overnight.
9. Following overnight digest, add an equivalent volume of acetonitrile to each gel piece and shake for 10–15 min.
10. Transfer supernatant onto a clean tube. Place directly into vacuum centrifuge and spin until dry.
11. To gel pieces, add additional 100 μl of 50% acetonitrile/2.5% formic acid and shake. Add supernatant to dried tubes from step 10. Transfer tubes to vacuum centrifuge until dry. Samples are now ready for one-dimensional HPLC electrospray ionization. Samples can be stored at −80 °C until ready to perform mass spectrometry analysis.

3.5 Mass Spectrometry

1. Solubilize digested peptide samples into a working volume of HPLC grade 5% acetonitrile/0.1% formic acid for injection into the HPLC autosampler, typically 20–30 μl volumes are reasonable.
2. Resulting peptides are injected onto a 3-μm C_{18} reversed-phase (RP) resin column. Peptides are eluted into a Thermo LTQ Orbitrap over a 40-min gradient of 2% acetonitrile/0.1% formic acid to 50% of a 90% acetonitrile/0.08% formic acid solvent.
3. Spectra are collected under data-dependent analysis selecting for the five most intense MS ions to be fragmented for MS/MS sequencing (see, e.g., Fig. 4b). Depending on instrumentation, additional ion fragmentation can be integrated into the data-dependent analysis. Exclusion lists for repeat ions can be developed to increase spectral analysis.
4. Raw data files/spectra can be submitted to in-house proprietary (Mascot, Sequest Sorcerer) or nonproprietary (Global proteome machine analysis www.gpm.org) database search softwares. Once spectra have been assigned and proteins identified, the lists of interacting proteins can be interrogated using ontological and cluster analysis softwares (see Sect. 3.6). Quantitative analysis can be performed using Scaffold, a statistical package for determination of data quality and spectral assignments [5].

3.6 Bioinformatics Analysis of Dataset

Analysis of shotgun proteomic lists can be conducted using proprietary software such as Ingenuity Pathways Analysis or public databases such as the Database for Annotation, Visualization and Integrated Discovery (DAVID, http://david.abcc.ncifcrf.gov)

[6, 7]. These software packages allow for delving into the dataset and assigning molecular function and disease correlation along with developing various other ontological connections within the dataset. Gene ontology (geneontology.org) can also be used to interrogate and determine term enrichment within a dataset [8]. Cell localization, function, and disease associations can be assigned to groups of proteins to better understand characteristics of the proteome cohort [9]. Solid-phase interactors (i.e., GFP binders) can be compared to the CRAPome database of non-specific interactors as well (crapome.org) [10]. This repository is useful toward ruling out non-specific interactions.

4 Notes and Conclusions

Affinity purification of target protein complexes provides a convenient way to identify novel binding partners involved in biological processes. Inclusion of GFP tags allows for determination of transfection efficiency by visualization. One caveat is the effect of the large tag on native interactions. Considering this caveat, a strategy to tag target proteins either at the N or C terminus, or alternatively at an internal site, could provide insight into the effects of GFP fusion on native protein interactions. Careful determination of the induction time for the GFP fusion protein should be considered for potential spurious interactions due to heightened overexpression and effects of high abundance peptides masking identification of lower abundance but biologically important interactions. Using another method of fractionation can improve proteome identification and minimize dampening of signal for low abundance peptides (i.e., strong cation or strong anion HPLC fractionation of peptides or size exclusion chromatography of proteins prior to tryptic digestion). The identification of protein complexes can yield a great deal of information in terms of protein function and involvement in biological processes. The above protocol provides a detailed approach toward conducting protein network characterization using mass spectrometry-based technologies. The inclusion of a list of human proteins that interact with GFP under the experimental conditions described above will allow researchers to exclude these as contaminants in similar proteomic approaches looking at potential interactors of GFP-tagged proteins (Table 1).

We have employed the above methodology routinely for many proteomic studies [11–14] and continue to do so. Affinity purification of protein complexes and analysis by mass spectrometry is a widely used technique, and so variations to the above methodology also exist.

Table 1
GFP-binding proteins in a DSP cross-linked pulldown experiment

Gene symbol	Spectral count (normalized)	MW (kDa)
ACTN4	233.77	105
EPPK1	204.55	556
PLEC	204.55	532
GTF2I	175.33	112
PRDX1	146.11	22
YWHAZ	146.11	28
DSG1	146.11	114
EEF1A1	116.89	50
YWHAE	116.89	29
GRP78	116.89	72
RPS27A	116.89	18
UBA52	116.89	15
ENO1	116.89	47
GNB2L1	116.89	35
ALB	116.89	69
KRT77	116.89	62
EEF1A2	87.664	50
ACTN1	87.664	103
TUFM	87.664	50
RPS3	87.664	27
NONO	87.664	54
SF3B3	87.664	136
PCMT1	87.664	25
ACLY	87.664	121
VCL	87.664	124
PDIA6	87.664	48
GRP94	87.664	92
MCM6	87.664	93
NUP98	87.664	198
PRDX4	87.664	31
PCCB	87.664	58

(continued)

Table 1
(continued)

Gene symbol	Spectral count (normalized)	MW (kDa)
TPI1	87.664	31
HNRNPUL1	87.664	96
PDAP1	87.664	21
CLTC	58.443	192
ACTG1	58.443	42
SPTAN1	58.443	285
EPRS	58.443	171
SPTBN1	58.443	275
NUP214	58.443	214
RANBP2	58.443	358
HNRNPM	58.443	78
GART	58.443	108
HSPD1	58.443	61
NUP155	58.443	155
PRDX2	58.443	22
MTHFD1	58.443	102
HNRNPU	58.443	91
CCT4	58.443	58
TCP1	58.443	60
HNRNPA1	58.443	39
YWHAH	58.443	28
KHSRP	58.443	73
SND1	58.443	102
HNRNPA2B1	58.443	37
MARS	58.443	101
NUP93	58.443	93
NUMA1	58.443	238
PHGDH	58.443	57
GDI1	58.443	51
YWHAG	58.443	28
PSMA7	58.443	28

(continued)

Table 1
(continued)

Gene symbol	Spectral count (normalized)	MW (kDa)
TPM3	58.443	33
USP5	58.443	96
NUP107	58.443	106
RAE1	58.443	41
PRPF6	58.443	107
TAGLN2	58.443	22
RPL11	58.443	20
SDHB	58.443	?
AARS	58.443	107
HIST1H4A	58.443	11
RAB5C	58.443	23
MIF	58.443	12
MOV10	58.443	114
UHRF1	58.443	90
FUBP1	58.443	68
ACTC1	29.221	42
ACTA1	29.221	42
ACTBL2	29.221	42
POTEE	29.221	121
USP9X	29.221	292
CSNK1D	29.221	47
CNOT1	29.221	267
CCT2	29.221	57
ATP5A1	29.221	60
CCT8	29.221	60
EIF4A1	29.221	46
E1F4A2	29.221	46
MCM3	29.221	91
MCM7	29.221	81
DDB1	29.221	127
PABPC4	29.221	71

(continued)

Table 1
(continued)

Gene symbol	Spectral count (normalized)	MW (kDa)
PABPC1	29.221	71
NUP205	29.221	228
SMC3	29.221	142
PARP1	29.221	113
NUP188	29.221	196
PSMD1	29.221	106
RAD50	29.221	154
GAPDH	29.221	36
PSMC5	29.221	46
MCM4	29.221	97
VCP	29.221	89
LARS	29.221	134
IKBKAP	29.221	150
DARS	29.221	57
SFPQ	29.221	76
SFPQ	29.221	76
AHCY	29.221	48
SMC4	29.221	147
KRT18	29.221	48
DDX5	29.221	69
MYO6	29.221	150
UBA1	29.221	118
YWHAQ	29.221	28
SDHA	29.221	?
RPS4X	29.221	30
DNAJC7	29.221	56
CSDE1	29.221	89
UBAP2L	29.221	115
LDHB	29.221	37
NUP88	29.221	84
RPSA	29.221	33

(continued)

Table 1
(continued)

Gene symbol	Spectral count (normalized)	MW (kDa)
CDC5L	29.221	92
PCNA	29.221	29
UPF1	29.221	124
GDI2	29.221	51
COPB2	29.221	102
PHB2	29.221	33
HDLBP	29.221	141
PDIA3	29.221	57
ADAR	29.221	136
RTCB	29.221	55
ADE2/PAICS	29.221	47
ALDH18A1	29.221	87
PSMB5	29.221	28
SHMT2	29.221	56
DDX1	29.221	82
FKBP4	29.221	52
RPS3A	29.221	30
MST4	29.221	47
FAM115A	29.221	102
OGT1	29.221	117
HNRNPH1	29.221	49
TPM4	29.221	29
NPEPPS	29.221	103
KIF11	29.221	119
RPL4	29.221	48
PUM1	29.221	126
TARS	29.221	83
ALDOA	29.221	39
RAN	29.221	24
RPS11	29.221	18
DIAPH1	29.221	141

(continued)

Table 1
(continued)

Gene symbol	Spectral count (normalized)	MW (kDa)
HNRNPA3	29.221	40
NAP1L1	29.221	45
BUB3	29.221	37
PDCD6IP	29.221	96
RPL10A	29.221	25
RPS8	29.221	24
LDHA	29.221	37
ABCE1	29.221	67
ILF3	29.221	95
DNAJC10	29.221	91
KRT31	29.221	47
KRT35	29.221	50
KRT33B	29.221	46
KRT33A	29.221	46
C14orf166	29.221	28
HSD17B10	29.221	27
RPL17	29.221	21
SERPINB4	29.221	45
PRDX6	29.221	25
SERPINB6	29.221	43
SARS2	29.221	58
EPS15	29.221	99
HNRNPH2	29.221	49
DPYSL2	29.221	62
PRDX3	29.221	28
DDX42	29.221	103
TBCE	29.221	59
CSRP2	29.221	21
CKAP4	29.221	66
RBM39	29.221	59
EIF6	29.221	27

(continued)

Table 1
(continued)

Gene symbol	Spectral count (normalized)	MW (kDa)
PMFCB	29.221	54
RPL13A	29.221	24
CSTF1	29.221	48
PPA1	29.221	33
MICAL1	29.221	118
TFAM	29.221	29
SEC22B	29.221	25
JUP	29.221	82
PPIF	29.221	22
TBCB	29.221	27
ST13	29.221	41
SYNCRIP	29.221	70
CBX3	29.221	21
RPL19	29.221	23
RANBP1	29.221	23
RPL28	29.221	16
PFN1	29.221	15
RBM10	29.221	104
LMNB2	29.221	68
RPL17	29.221	21
SERPINB4	29.221	45
PRDX6	29.221	25
SERPINB6	29.221	43
SARS2	29.221	58
EPS15	29.221	99
HNRNPH2	29.221	49
DPYSL2	29.221	62
PRDX3	29.221	28
DDX42	29.221	103
TBCE	29.221	59
CSRP2	29.221	21

(continued)

Table 1
(continued)

Gene symbol	Spectral count (normalized)	MW (kDa)
CKAP4	29.221	66
RBM39	29.221	59
EIF6	29.221	27
PMPCB	29.221	54
RPL13A	29.221	24
CSTF1	29.221	48
PPA1	29.221	33
MICAL1	29.221	118
TFAM	29.221	29
SEC22B	29.221	25
JUP	29.221	82
PPIF	29.221	22
TBCB	29.221	27
ST13	29.221	41
SYNCRIP	29.221	70
CBX3	29.221	21
RPL19	29.221	23
RANBP1	29.221	23
RPL28	29.221	16
PFN1	29.221	15
RBM10	29.221	104
LMNB2	29.221	68
HIST1H1E	29.221	22
ACAA2	29.221	42
HNRNPR	29.221	71
PDE12	29.221	67
SF3B4	29.221	44
UBE2N	29.221	17
LIG3	29.221	113
ERAL1	29.221	48
CARS	29.221	85

(continued)

Table 1 (continued)

Gene symbol	Spectral count (normalized)	MW (kDa)
UCHL3	29.221	26
SCYL2	29.221	104
CBX1	29.221	21
ERO1L	29.221	54

Proteins are rank ordered by normalized spectral count enrichment using Scaffold software (v.4.2.1)

Acknowledgments

We thank Kirsten McLeod and Janis Stark for helping with tissue culture, Thomas Macartney for cloning pCDNA5/FRT/TO/GFP, and the staff at the Sequencing Service (School of Life Sciences, University of Dundee, Scotland) for DNA sequencing. We thank Dave Campbell and the mass spectrometric team at the MRC Protein Phosphorylation and Ubiquitylation Unit for optimization of protocols relating to sample processing for mass spectrometry. TDC is supported by the UK MRC Career Development Fellowship. GS is supported by the UK Medical Research Council and the pharmaceutical companies supporting the DSTT (AstraZeneca, Boehringer-Ingelheim, GlaxoSmithKline, Merck-Serono, Pfizer, and Johnson & Johnson).

References

1. Huang BX, Kim HY (2013) Effective identification of Akt interacting proteins by two-step chemical crosslinking, co-immunoprecipitation and mass spectrometry. PLoS One 8:e61430
2. Gao M, McCluskey P, Loganathan SN, Arkov AL (2014) An in vivo crosslinking approach to isolate protein complexes from drosophila embryos. J Vis Exp 86
3. Kim KM, Yi EC, Kim Y (2012) Mapping protein receptor-ligand interactions via in vivo chemical crosslinking, affinity purification, and differential mass spectrometry. Methods 56:161–165
4. Corgiat BA, Nordman JC, Kabbani N (2014) Chemical crosslinkers enhance detection of receptor interactomes. Front Pharmacol 4:171
5. Searle BC (2010) Scaffold: a bioinformatic tool for validating MS/MS-based proteomic studies. Proteomics 10:1265–1269
6. Yang LJ, Ma DQ, Cui H (2014) Proteomic analysis of immature rat pups brain in response to hypoxia and ischemia challenge. Int J Clin Exp Pathol 7:4645–4660
7. Huang da W, Sherman BT, Lempicki RA (2009) Systematic and integrative analysis of large gene lists using DAVID bioinformatics resources. Nat Protoc 4:44–57
8. Carnielli CM, Winck FV, Paes Leme AF (2015) Functional annotation and biological interpretation of proteomics data. Biochim Biophys Acta 1854:46–54
9. Pawar H, Kashyap MK, Sahasrabuddhe NA, Renuse S, Harsha HC, Kumar P, Sharma J, Kandasamy K, Marimuthu A, Nair B, Rajagopalan S, Maharudraiah J, Premalatha CS, Kumar KV, Vijayakumar M, Chaerkady R, Prasad TS, Kumar RV, Kumar RV, Pandey A (2011) Quantitative tissue proteomics of

esophageal squamous cell carcinoma for novel biomarker discovery. Cancer Biol Ther 12:510–522

10. Mellacheruvu D, Wright Z, Couzens AL, Lambert JP, St-Denis NA, Li T, Miteva YV, Hauri S, Sardiu ME, Low TY, Halim VA, Bagshaw RD, Hubner NC, Al-Hakim A, Bouchard A, Faubert D, Fermin D, Dunham WH, Goudreault M, Lin ZY, Badillo BG, Pawson T, Durocher D, Coulombe B, Aebersold R, Superti-Furga G, Colinge J, Heck AJ, Choi H, Gstaiger M, Mohammed S, Cristea IM, Bennett KL, Washburn MP, Raught B, Ewing RM, Gingras AC, Nesvizhskii AI (2013) The CRAPome: a contaminant repository for affinity purification-mass spectrometry data. Nat Methods 10:730–736
11. Vogt J, Dingwell KS, Herhaus L, Gourlay R, Macartney T, Campbell D, Smith JC, Sapkota GP (2014) Protein associated with SMAD1 (PAWS1/FAM83G) is a substrate for type I bone morphogenetic protein receptors and modulates bone morphogenetic protein signalling. Open Biol 4:130210
12. Al-Salihi MA, Herhaus L, Macartney T, Sapkota GP (2012) USP11 augments TGFbeta signalling by deubiquitylating ALK5. Open Biol 2:120063
13. Herhaus L, Al-Salihi M, Macartney T, Weidlich S, Sapkota GP (2013) OTUB1 enhances TGFbeta signalling by inhibiting the ubiquitylation and degradation of active SMAD2/3. Nat Commun 4:2519
14. Herhaus L, Al-Salihi MA, Dingwell KS, Cummins TD, Wasmus L, Vogt J, Ewan R, Bruce D, Macartney T, Weidlich S, Smith JC, Sapkota GP (2014) USP15 targets ALK3/BMPR1A for deubiquitylation to enhance bone morphogenetic protein signalling. Open Biol 4:140065

Methods in Molecular Biology (2018) 1788: 63–72
DOI 10.1007/7651_2017_90

Published online: 06 April 2018

Optimization for Peptide Sample Preparation for Urine Peptidomics

Tara K. Sigdel, Carrie D. Nicora, Wei-Jun Qian, and Minnie M. Sarwal

Abstract

Analysis of native or endogenous peptides in biofluids can provide valuable insight into disease mechanisms. Furthermore, the detected peptides may also have utility as potential biomarkers for noninvasive monitoring of human diseases. The noninvasive nature of urine collection and the abundance of peptides in the urine make analysis by high-throughput "peptidomics" methods an attractive approach for investigating the pathogenesis of renal disease. However, urine peptidomics methodologies can be problematic with regard to difficulties associated with sample preparation. The urine matrix can provide significant background interference in making the analytical measurements, in that it hampers both the identification of peptides and the depth of the peptidomics read when utilizing LC-MS-based peptidome analysis. We report on a novel adaptation of the standard solid-phase extraction (SPE) method to a modified SPE (mSPE) approach for improved peptide yield and analysis sensitivity with LC-MS-based peptidomics, in terms of time, cost, clogging of the LC-MS column, peptide yield, peptide quality, and number of peptides identified by each method. The mSPE method provides significantly improved efficiencies for the preparation of samples from urine. The mSPE method is found to be superior to the conventional, standard SPE method for urine peptide sample preparation when applying LC-MS peptidomics analysis, due to optimized sample cleanup that provides improved experimental inference from confidently identified peptides.

Keywords Biomarker, Biomarker discovery, Peptidomics, Proteomics, Transplantation, Urine

1 Introduction

Disease-specific biomarkers remain as an unmet need in an overwhelming majority of cases for monitoring human health. Though there is an ever-increasing effort to identify specific and sensitive biomarkers for monitoring human health and for the early detection of disease onset, many hurdles still exist in identifying effective biomarkers [1]. In this context, urine could prove to be an important proximal fluid in providing biomarkers that could be tested noninvasively [2]. The use of urine is particularly useful in the context of kidney and urinary system-related diseases such as acute and chronic kidney diseases, diseases with either the urinary bladder or prostate, and kidney transplantation [2–9]. Circulating peptides in urine are to be considered potentially useful as biomarkers, as they are both abundant and easily accessible. Increased efforts to identify and analyze these valuable peptides have been undertaken, as they are thought to contribute insight into disease mechanisms as well as provide potential biomarkers for diagnosis, prognosis, therapeutic intervention, and treatment outcome [10]. The emergence of new and sophisticated methods of molecular profiling has aided our ability to analyze peptides in complex biological mixtures, in combination with the analysis of their degradation patterns. This information may provide important clues about underlying (patho)physiological processes [11, 12]. Different technology platforms have been applied for the study of urine proteins and peptides such as LC-ESI-MS [13], LC-MALDI-MS [14], CE-MS [15], and SELDI-TOF [16]. While SELDI-TOF and CE-MS are relatively easy platforms to operate, their weakness is that they do not allow for peptide identification resulting in incomplete biomarker discovery and identification. A previous study has included the application of an LC-MALDI-MS approach, where a standard solid-phase extraction (SPE) method for peptide enrichment and purification was utilized [14]. Unlike urine peptide analysis by MALDI [17–19], urine peptidomics analysis by LC-MS is highly sensitive to the presence of unidentified contaminants that exist in urine, which, if not effectively removed by further purification, impair the performance of the LC-MS analysis. When the standard solid-phase extraction (SPE) [20] method of urine peptide extraction and purification is utilized for sample preparation, persisting contaminants in the urine, inclusive of urobilin and urobilinogen, clog the LC column and interfere with the assay performance (Sigdel et al., unpublished data). Herein we provide a protocol for a modification of the SPE method, termed the modified SPE (mSPE) method, which provides optimal peptide extraction and purification for urinary peptide analysis by LC-MS. Overall schematics of the method used in this study are summarized in Fig. 1.

2 Materials

1. Urine sample/native, small MW (<10 kDa) peptide fraction.
2. Hydrophilic-lipophilic-balanced reversed-phase sorbent-based solid phase extraction: a mixture of two monomers, hydrophilic *N*-vinylpyrrolidone and lipophilic divinylbenzene.
3. Processed silicon carbide resin.
4. Low pH buffer: pH 3.5–4.
5. Na-phosphate buffer: pH 12.5.

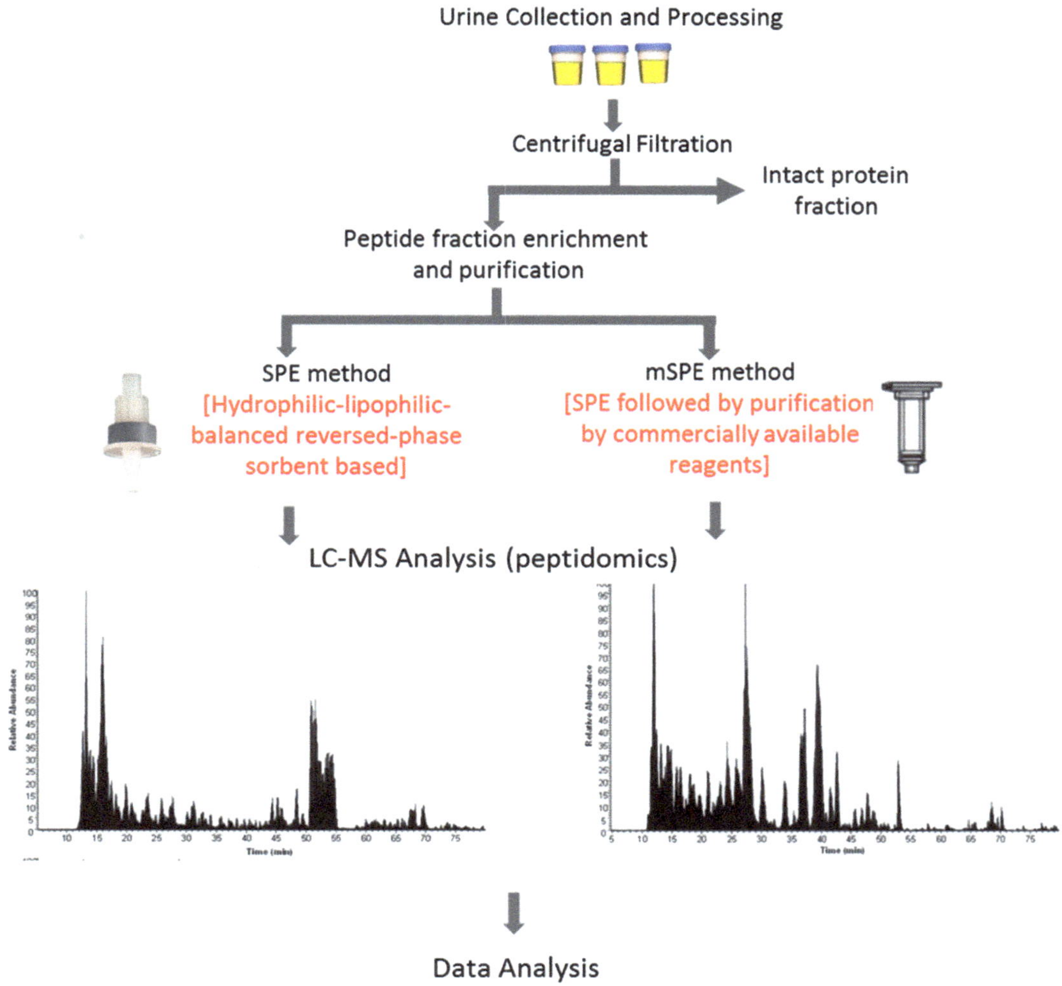

Fig. 1 Schematics of peptide extraction and purification strategy. Urine samples were passed through a centrifugal filtration column to separate intact (larger MW proteins) from native peptides (smaller MW peptides). The peptide fraction was subjected to solid-phase extraction (SPE) and modified SPE (mSPE) methods described in Sect. 3. The resulting peptides were analyzed by LC-MS and identified for comparison of robustness of the two isolation and purification methods

2.1 Urine Collection, Initial Processing, and Storage

1. 1 M Tris–HCl.
2. Amicon ultracentrifugal filtration tubes (10 K molecular weight cutoff, Millipore, Bedford, MA).
3. 15 mL Falcon tube.

2.2 Sample Preparation

1. ProteoSpin column (Norgen Biotek, Thorold, Canada).
2. Oasis HLB extraction cartridge (Waters, Milford, MA).
3. 100% acetonitrile.

4. 0.1% trifluoroacetic acid (TFA).
5. 50% TFA.
6. 70% acetonitrile.
7. SpeedVac.
8. Ethyl acetate.

2.3 Solid-Phase Extraction (SPE)

1. Oasis HLB extraction cartridge.
2. 100% acetonitrile.
3. 0.1% trifluoroacetic acid (TFA).
4. 50% TFA.
5. 70% acetonitrile.

2.4 Enrichment of Low-MW Endogenous Peptides (mSPE)

1. ProteoSpin™ Urine Protein Concentration Micro Kit (#17400, NOrgen Biotek, Thorold, Canada).
2. Oasis HLB extraction cartridge.
3. pH binding buffer (pH 3.5).
4. Collection tube.
5. Column activation buffer (pH 3.5).
6. Wash buffer (pH 3.5)
7. Silicon carbide.
8. Neutralizer.
9. 1.7 mL elution tube.
10. Elution buffer: 10 mM sodium phosphate, pH 12.5.

2.5 LC-MS Analysis

1. High-resolution, reversed-phase capillary LC system.
2. LTQ-Orbitrap Velos MS (Thermo Fisher Scientific, San Jose, CA).
3. Two Agilent 1200 nanoflow pumps (Agilent Technologies, Santa Clara, CA).
4. One Agilent 1200 capillary pump (Agilent Technologies, Santa Clara, CA).
5. PAL autosampler (Leap Technologies, Carrboro, NC).
6. (Software that allows for parallel event coordination).
7. Capillary reversed-phase columns: slurry pack 3 μm Jupiter C18 (Phenomenex, Torrance, CA) into 35 cm × 360 μm o. d. × 75 μm i.d. fused silica (Polymicro Technologies Inc., Phoenix, AZ).
8. Trapping columns: 3.6 μm Aeris WIDEPORE XB-C18 packed into a 4 cm length of 150 μm i.d. fused silica (*see* **Note 1**).

9. Mobile phase: 0.1% formic acid in water (A) and 0.1% formic acid acetonitrile (B) operated at constant flow of 300 nL/min.
10. LTQ-Orbitrap Velos mass spectrometer outfitted with a custom electrospray ionization (ESI) interface: electrospray emitters custom made by chemically etching 150 μm o.d. × 20 μm i.d. fused silica [21].
11. Extract_MSn from Bioworks Cluster 3.2 (Thermo Scientific).
12. Human protein database (UniProtKB, released 2010-05).

3 Methods

3.1 Urine Collection, Initial Processing, and Storage

1. Collect second morning void, midstream urine samples (50–100 mL) in sterile containers.
2. Centrifuge at 2000 × *g* for 20 min at room temperature within 1 h of collection (*see* **Note 2**).
3. Separate the supernatant from the pellet containing any particulate matter including cells and cell debris.
4. Adjust the supernatant pH to 7.0 with 1 M Tris–HCl.
5. Aliquot urine samples into 10 mL aliquots prior to freezing in order to ensure minimum impact of freeze-thaw cycles for multiple assays (*see* **Note 3**).
6. Store urine samples at −80 °C until further analysis.
7. Wash filtration unit with 10 mL of deionized water by centrifuging the tube at 3000 × *g* for 10 min to remove traces of glycerine.
8. For peptide extraction, separate urinary native peptides from larger molecular mass intact proteins by filtering the supernatant through the ultracentrifugal filtration tubes: add 10 mL of urine to the filter tube and centrifuge at 3000 × *g* for 20 min.
9. Transfer the filtrate to a 15 mL Falcon tube and store at −80 °C until time for analysis.

3.2 Sample Preparation

1. Use 4–2 mL aliquots for the evaluation of the SPE method and 4–2 mL aliquots for the mSPE method.
2. Pass 2 mL of 100% acetonitrile through the ProteoSpin column before equilibrating with 5 mL of 0.1% trifluoroacetic acid (TFA) at a rate of 1 drop(s).
3. Adjust the pH of the peptide samples to pH 3.0 with 50% TFA and load the peptides onto the cartridge.
4. Wash the cartridge with 5 mL of 0.1% TFA, and then elute with 1.5 mL 0.1% TFA in 70% acetonitrile mixture.

5. Repeat the elution step a second time, and bring the volume of the eluted peptides to 500 μL by SpeedVac concentration.
6. Extract the resulting peptide mixture with 1 mL ethyl acetate by mixing and vortexing for 1 min, and discard the upper organic layer.

3.3 Solid-Phase Extraction (SPE)

1. Repeat the previously described procedure using the Oasis HLB extraction cartridge for the initial peptide extraction.
2. Pass 2 mL of 100% acetonitrile through the column before equilibrating with 5 mL of 0.1% trifluoroacetic acid (TFA) at a rate of 1 drop/s.
3. Adjust the peptide samples to pH 3.0 with 50% TFA and then load the peptides onto the cartridge.
4. Wash the cartridge with 5 mL of 0.1% TFA and then elute with 1.5 mL 0.1% TFA in 70% acetonitrile.
5. Aliquot the peptide extract and store at −80 °C until time for further analysis.

3.4 Enrichment of Low-MW Endogenous Peptides (mSPE Method)

1. Perform an initial peptide extraction using the Oasis HLB extraction cartridge, followed by the ethyl acetate extraction method as mentioned previously.
2. After ethyl acetate extraction, add 40 μL of pH binding buffer (pH 3.5) to 200 μL of urine peptide sample extract to adjust the pH to pH 3.5.
3. Assemble the provided spin column with a collection tube and add 500 μL of column activation and wash buffer (both pH 3.5) to the column.
4. Process the column with silicon carbide and centrifuge for 2 min at 3300 × *g*.
5. Repeat the activation step a second time, load 500 μL of the pH-adjusted urine peptide extract from the Oasis HLB cartridge onto the column, and centrifuge for 2 min.
6. Pass the peptide extract through the column, add 500 μL of column activation and wash buffer to the column, and centrifuge the column for 2 min.
7. Discard the flow-through and repeat step 6 one more time.
8. Before eluting, add 4.65 μL of neutralizer to a fresh 1.7 mL elution tube and transfer the column to the elution tube.
9. Add 50 μL of elution buffer to the column, and centrifuge for 2 min to elute bound peptides.

3.5 LC-MS Analysis

1. Trap 5 μL sample injections, and wash on the trapping columns at 1.5 μL/min for 20 min prior to alignment with analytical columns.

2. Perform MS analysis on the LTQ-Orbitrap Velos mass spectrometer outfitted with the custom electrospray ionization (ESI) interface.
3. Maintain the heated capillary temperature and spray voltage at 350 °C and 2.2 kV, respectively.
4. Record full MS spectra at a resolution of 100 K (for ions at m/z 400) over the range of m/z 400–2000 with an automated gain control (AGC) value of 1e6.
5. Perform MS/MS in the data-dependent mode with an AGC target value of 3e4.
6. Select for the ten most abundant parent ions, excluding single charge states, for MS/MS using high-energy collisional dissociation (HCD) with a normalized collision energy setting of 40%. Use a dynamic exclusion time of 45 s.

3.6 Peptide and Protein Identifications

1. Convert LC-MS/MS raw data into ".dta" files using Extract_MSn from Bioworks Cluster 3.2.
2. Identify MS/MS spectra based on database searching against the human protein database (UniprotKB) using both SEQUEST (version 27, revision 12) and MS-Align+ algorithms [22] (*see* **Note 4**).
3. Apply filtering criteria using either MS-generating function score (MS-GF) <1E-10 or MS-Align+ false discovery rate (FDR) <0.05 to ensure the FDR of final peptide identifications is below 1% based on the decoy database searching strategy [23].

4 Notes

1. Two-column operation allows for columns to be "washed" (shortened gradients) and regenerates off-line without any cost to duty cycle.
2. Urine samples can be safely stored up to 1 h at room temperature and up to 12 h at 4 °C without significant protein degradation. Samples do not require the addition of protease inhibitors to improve sample integrity if stored at either 4 °C or −80 °C within 72 h of collection.
3. We suggest utilizing 10 mL urine so each aliquot only needs to be thawed once for a single experiment.
4. Figure 2 provides a comparison of representative chromatograms for SPE and mSPE.
5. MS-Align+ is an algorithm designed for intact protein identification and is capable of identifying high mass peptides from MS/MS data.

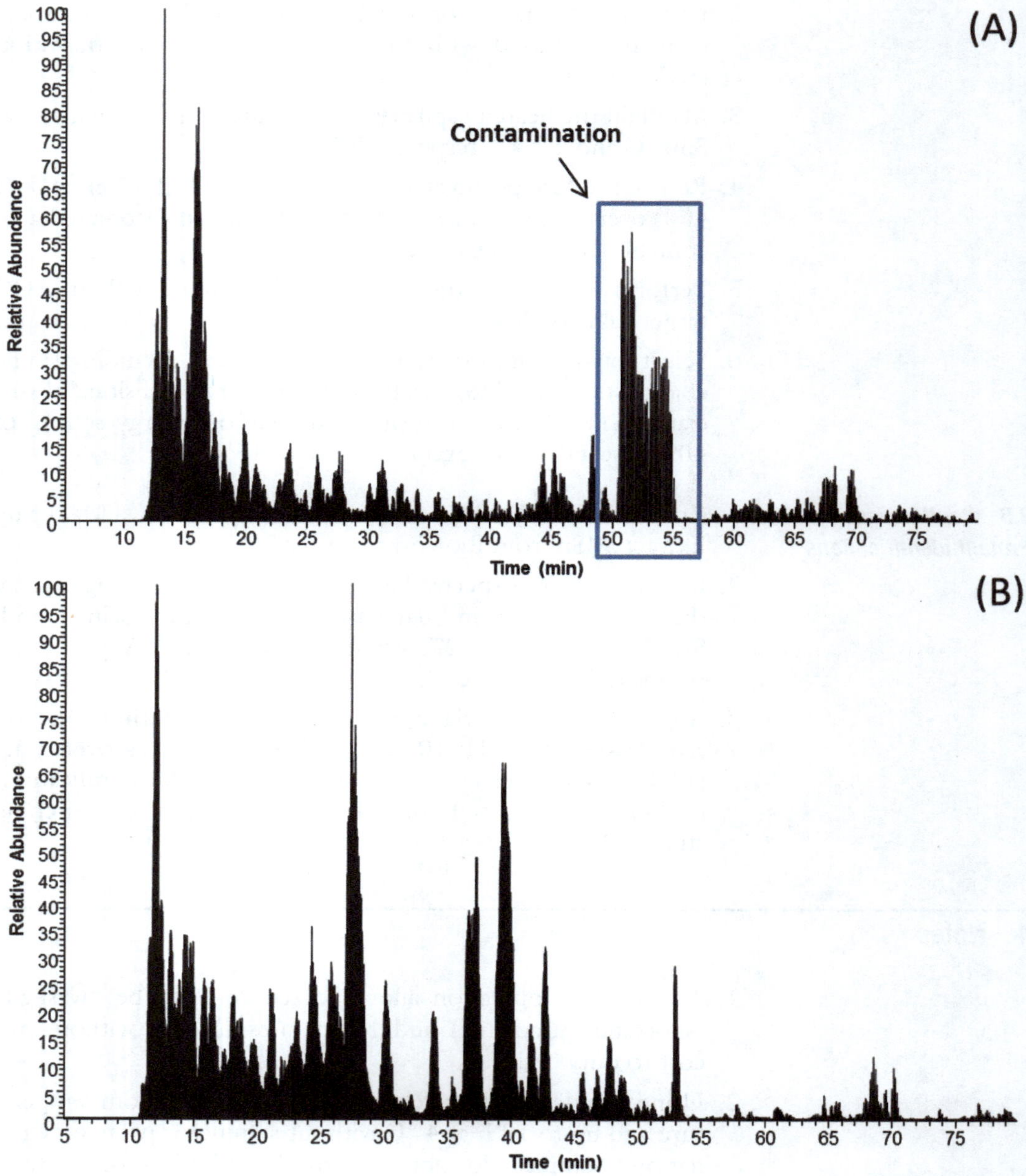

Fig. 2 Comparison of representative chromatograms for SPE and mSPE. (**a**) The LC chromatogram from the SPE method contains a large region of low intensity peaks presumably due to the ionization suppression effect from the contaminating species. (**b**) The LC chromatogram from the mSPE method displays well-distributed peaks throughout the LC span

Acknowledgments

This chapter is based on research performed as part of an American Recovery and Reinvestment (ARRA)-funded project under Award Number U01163594 (to P Heeger), from the National Institute of Allergy and Infectious Diseases. The work was carried out by members of the Clinical Trials in Organ Transplantation (CTOT) and Clinical Trials in Organ Transplantation in Children (CTOT-C) consortia. The experimental work described herein was performed in the Environmental Molecular Sciences Laboratory (EMSL), a US Department of Energy (DOE) national scientific user facility located at PNNL in Richland, Washington, and in the Sarwal Lab at Stanford University and California Pacific Medical Center Research Institute. PNNL is a multi-program national laboratory operated by Battelle Memorial Institute for the DOE under Contract DE-AC05-76RL01830.

References

1. Poste G (2011) Bring on the biomarkers. Nature 469(7329):156–157
2. Pisitkun T, Johnstone R, Knepper MA (2006) Discovery of urinary biomarkers. Mol Cell Proteomics 5(10):1760–1771
3. Sigdel TK, Gao X, Sarwal MM (2012) Protein and peptide biomarkers in organ transplantation. Biomark Med 6(3):259–271
4. Sigdel TK et al (2008) Optimizing protein recovery for urinary proteomics, a tool to monitor renal transplantation. Clin Transpl 22 (5):617–623
5. Sigdel TK, Lee S, Sarwal MM (2011) Profiling the proteome in renal transplantation. Proteomics Clin Appl 5(5–6):269–280
6. Sigdel TK, Sarwal MM (2012) Cell-free DNA as a measure of transplant injury. Clin Transpl 201–205
7. Sigdel TK et al (2013) A rapid noninvasive assay for the detection of renal transplant injury. Transplantation 96(1):97–101
8. Shimwell NJ et al (2013) Combined proteome and transcriptome analyses for the discovery of urinary biomarkers for urothelial carcinoma. Br J Cancer 108(9):1854–1861
9. Theodorescu D et al (2008) Discovery and validation of urinary biomarkers for prostate cancer. Proteomics Clin Appl 2(4):556–570
10. Albalat A, Mischak H, Mullen W (2011) Clinical application of urinary proteomics/peptidomics. Expert Rev Proteomics 8(5):615–629
11. Machtejevas E et al (2009) Profiling of endogenous peptides by multidimensional liquid chromatography: on-line automated sample cleanup for biomarker discovery in human urine. J Sep Sci 32(13):2223–2232
12. Schulz-Knappe P et al (2001) Peptidomics: the comprehensive analysis of peptides in complex biological mixtures. Comb Chem High Throughput Screen 4(2):207–217
13. Cutillas PR et al (2003) Detection and analysis of urinary peptides by on-line liquid chromatography and mass spectrometry: application to patients with renal Fanconi syndrome. Clin Sci (Lond) 104(5):483–490
14. Ling XB et al (2010) Integrative urinary peptidomics in renal transplantation identifies biomarkers for acute rejection. J Am Soc Nephrol 21(4):646–653
15. Wittke S et al (2005) Discovery of biomarkers in human urine and cerebrospinal fluid by capillary electrophoresis coupled to mass spectrometry: towards new diagnostic and therapeutic approaches. Electrophoresis 26 (7–8):1476–1487
16. Schaub S et al (2004) Urine protein profiling with surface-enhanced laser-desorption/ionization time-of-flight mass spectrometry. Kidney Int 65(1):323–332
17. Sigdel TK, Klassen RB, Sarwal MM (2009) Interpreting the proteome and peptidome in transplantation. Adv Clin Chem 47:139–169
18. Sigdel TK, Sarwal MM (2011) Recent advances in biomarker discovery in solid organ transplant by proteomics. Expert Rev Proteomics 8(6):705–715

19. Ling XB et al (2010) Urine peptidomics for clinical biomarker discovery. Adv Clin Chem 51:181–213
20. Hennion MC (1999) Solid-phase extraction: method development, sorbents, and coupling with liquid chromatography. J Chromatogr A 856(1–2):3–54
21. Kelly RT et al (2006) Chemically etched open tubular and monolithic emitters for nanoelectrospray ionization mass spectrometry. Anal Chem 78(22):7796–7801
22. Liu X et al (2012) Protein identification using top-down. Mol Cell Proteomics 11(6):M111 008524
23. Qian WJ et al (2005) Probability-based evaluation of peptide and protein identifications from tandem mass spectrometry and SEQUEST analysis: the human proteome. J Proteome Res 4(1):53–62

Methods in Molecular Biology (2018) 1788: 73–88
DOI 10.1007/7651_2017_89

Published online: 17 December 2017

Purification of Target Proteins from Native Tissues: CCT Complex from Bovine Testes and PP2Ac from Porcine Brains

Feng Guo, Nathan Wlodarchak, Patrick Menden, and Yongna Xing

Abstract

Proteomics is a powerful approach for systematic identification and quantification of the entire proteome of a biological system (cell, tissue, organ, biological fluid, or organism) at specific time points (http://www.nature.com). Extracting and purifying target proteins from native tissues are essential steps for many aspects of proteomic studies. In this chapter, we will introduce the experimental procedures to obtain soluble proteins from two different tissues: (1) the CCT (cpn-containing TCP-1) complex from bovine testes and (2) the protein phosphatase 2A (PP2A) catalytic subunit (PP2Ac or C) from porcine brains. With these two examples, we would like to provide some general guidelines for researchers on how to extract and purify target proteins from specific tissues and extend these approaches to other proteins of interest.

Keywords Bovine testes, CCT, FPLC, Porcine brains, PP2A, Protein purification, Proteomic

1 Introduction

With rapid advance in bioinformatics and research tools for system biology, proteomic analysis has become easy and efficient nowadays and is more commonly applied to modern biological research. Several proteomic databases provide valuable information for almost all proteins encoded in the human genome, such as PRIDE Archive (https://www.ebi.ac.uk/pride/archive/), ProteomicsDB (https://www.proteomicsdb.org/), Max-Planck Unified (MAPU) proteome database (http://www.mapuproteome.com/), etc. Recently, a map of human tissue proteome was reported, aggregating information from quantitative transcriptomics and protein profiling on tissue and organ levels [1]. The studies described in this report performed systematic analyses and provided comprehensive information on spatial localization of proteins in human tissues, which required extraction and purification of proteins from specific tissues [1]. In other specific studies, it is also necessary for researchers to obtain the protein samples from native tissues. For example, it is important to obtain native metalloproteins to examine the associated metal ions for characterization of enzymatic mechanisms [2]. New technologies in life sciences provide high-efficiency equipments for protein purification, such as AKTA (GE Healthcare) for fast protein liquid chromatography (FPLC). Researchers have also developed better strategies to purify proteins from native tissues than previous procedures developed for years. With years of experience on protein purification, here we would like to present two examples of such protocols and some general guidelines in this area (*see* **Note 5**). The purification of CCT complex from bovine testes applied a majority of common FPLC approaches. Thus, this example provides a general guidance to develop strategies for adopting different chromatography approaches to improve the yield and purification efficiency for distinct target proteins. In the second protocol, we took advantage of the strong binding affinity of PP2A regulatory subunit (A) and PP2Ac to extract and purify PP2Ac from diverse PP2A holoenzymes and regulation complexes in porcine brain [2]. In this case, the recombinant GST-PP2A-A was used to fish out the extracted PP2Ac from the crude sample directly as a high-affinity bait. As a result, the purification can be accomplished with a high purity in much fewer steps.

2 Materials

Prepare all solutions using ultrapure deionized water. Prepare and store all reagents at 4 °C (*see* **Note 6** (for ordering info)). All protein purification procedures are performed at 4 °C unless specified. Make sure to add fresh reducing reagents such as DTT (dithiothreitol) and/or beta-mercaptoethanol. Always save samples at each step to monitor the yield and purity of the target protein using SDS-PAGE.

2.1 Purify CCT Complex from Bovine Testes

1. Bovine testes (80 g) from −80 °C (*see* **Note 1**).
2. Lysis buffer [3]: 50 mM Hepes pH 7.4, 100 mM NaCl, 250 mM sucrose, 2 mM EDTA, 1 mM DTT, 1 mM PMSF, 10 μg/ml of leupeptin/aprotinin mix.
3. Cheesecloth or paper towel.
4. SQ Buffer A: 500 ml of 25 mM Tris pH 8, 100 mM NaCl, 3 mM DTT, 5% glycerol, 5 mM $MgCl_2$, 0.5 mM EDTA.
5. SQ Buffer B: 500 ml of 25 mM Tris pH 8, 1.0 M NaCl, 3 mM DTT, 5% glycerol, 5 mM $MgCl_2$, 0.5 mM EDTA.
6. Heparin Buffer C: 800 ml of 25 mM Tris pH 8, 200 mM NaCl, 3 mM DTT, 5% glycerol, 5 mM $MgCl_2$, 0.5 mM EDTA.
7. HIC Buffer D: 500 ml 25 mM Tris pH 8, 200 mM NaCl, 3 mM DTT, 5% glycerol, 1 M $(NH_4)_2SO_4$, 5 mM $MgCl_2$, 0.5 mM EDTA.
8. SEC Buffer E: same as Heparin Buffer C.
9. 12% SDS-PAGE.

2.2 Purify PP2Ac Subunit with Native Metals from Porcine Brains

1. Porcine brains (*see* **Note 1**).
2. Cheesecloth or paper towel.
3. Buffer 1: 10 mM Tris pH 8, 50 mM NaCl, 1 mM 2-ME, 100 μM EDTA.
4. Buffer 2: 10 mM Tris pH 8, 50 mM NaCl, 1 mM 2-ME, 10 μM EDTA.
5. Buffer 3: 95% ethanol, 1 mM PMSF, 10uM EDTA.
6. Buffer GSH: 50 mM Tris pH 8, 50 mM NaCl, 1 mM 2-ME, 5 mM reduced GSH.
7. Two GS4B (GE Healthcare Life Sciences, 17-5132-02) open columns: 2 ml resin/column.
8. 0.2 M NaOH.

2.3 Chromatography Columns and Reagents

1. Source 15Q 10/100 GL (GE Healthcare Life Sciences, 90-1004-65).
2. HiTrap Heparin HP (5 ml) (GE Healthcare Life Sciences, 17-0407-03).
3. HiTrap Phenyl HP (5 ml) (GE Healthcare Life Sciences, 17-5195-01).
4. Superdex 200, 10/300 GL (GE Healthcare Life Sciences, 17-5175-01).
5. Superose 6, 10/300 GL (GE Healthcare Life Sciences, 17-5172-01).
6. GS4B resin (GE Healthcare Life Sciences, 17-5132-02).
7. Econo-Pac® chromatography columns (Bio-Rad, 732-1010).
8. Superdex 75, 10/300 GL (GE Healthcare Life Sciences, 17-5174-01).

3 Methods

Based on the previous study on CCT complex [3] and PP2A [4], we developed and simplified the protein purification procedure to purify the CCT complex from bovine testes and PP2A from porcine brains. In general, the procedure will yield several milligrams of target protein in 2–3 days. The purification mainly relies on FPLC by the following steps:

1. Tissue preparation and cell lysis
2. $(NH_4)_2SO_4$ precipitation

3. Source 15Q (SQ) ion exchange chromatography
4. Additional FPLC methods, e.g., heparin column, hydrophobic interaction chromatography (HIC), or GST-pulldown
5. Final step: size exclusion chromatography (SEC), e.g., over Superdex 75, Superdex 200, or Superose 6

3.1 Purify Native CCT Complex from Bovine Testes

The eukaryotic and archaea chaperonin cylinder CCT (cpn-containing TCP-1) complex is a chaperonin providing favorable conditions for large protein folding in cytosol. It was also referred to as TCP-1 ring complex (TRiC) and classified as Group II chaperonin. The chaperone function of CCT complex utilizes ATP as energy source. Similar to its well-known prokaryotic homolog, GroEL from *E. coli*, the CCT complex resembles two donut-shaped rings stacked on top of one another to create a barrel. Each ring contains eight different but homologous subunits (50–60 kDa) [5–8]. Bovine testis was chosen to extract the CCT complex because of its high expression level in this tissue. In 2000, Raul G. Ferreyra and Judith Frydman published the protocol for the purification of CCT complex from bovine testes [3]. Ten years later, the same method was applied to purify the protein and determine the cyro-EM structure of the complex at 4.0 Å [8]. We modified this protocol to improve the efficiency by taking advantage of the AKTA protein purification system. Briefly, we shorten the sample preparation time by skipping the 1-h ultracentrifugation and the DE52 cellulose anion exchange chromatography but quickly moving to ammonium sulfate precipitation to reduce the sample volume for FPLC purification. We also introduced hydrophobic interaction chromatography (HIC) after Source 15Q and heparin column, because both columns require low salt for protein binding but HIC column needs a high concentration of ammonium sulfate to bind protein. According to the size of CCT complex (~ 800–960 kDa), the Superdex 200 column was chosen to remove the impurities smaller than the complex, while the Superose 6 column was applied to remove the large impurities or protein aggregates. Both size exclusion chromatography columns (Superdex 200 and Superose 6) can be applied as the final step because the sample was relatively clean after previous purification and the buffer can also be changed at the last step. Usually, one can finish the purification procedure in 2 days if managed well, e.g., equilibrate the next column while running SDS-PAGE gel to analyze the fractions from previous run. Such purification strategy and time management can also be applied to purify any other target proteins from specific tissues.

1. Lyse cells in tissue
   (a) Take 80–100 g testis from −80 °C, and soak the tissue in 100 ml ice-cold lysis buffer.
   (b) Blend briefly (30 s) to break down the tissue, and keep on ice.

   (c) Dounce 15 stokes in 40-ml Dounce tissue grinder.
   (d) Spin at 16,000 rpm for 30 min, 4 °C, and keep supernatant.
2. $(NH_4)_2SO_4$ (A.S.) precipitation
   (a) Filter the supernatant through three-layer paper towel.
   (b) Bring solution to final 1.3 M A.S. by adding ~17 g powder/100 ml, and stir until all powder is dissolved.
   (c) Spin at 16,000 rpm for 10 min, 4 °C, and keep supernatant.
   (d) Bring solution to 2.3 M A.S. by adding an additional ~13 g powder/100 ml.
   (e) Stir until all powder was dissolved; spin at 16,000 rpm for 10 min, 4 °C; and keep pellet.
   (f) Dissolve pellet in 10 ml Source 15Q Buffer A, spin at 16,000 rpm for 10 min at 4 °C, and then filter the solution (0.45 μm) (Fig. 1).
3. Ion exchange chromatography
   (a) On AKTA, use SQ Buffer A for Pump A and SQ Buffer B for Pump B, flow rate = 5 ml/min, and edit program with a 100 ml linear gradient of 0–50% Pump B buffer after 50 ml of the "washout unbound" step.
   (b) Equilibrate the Source 15Q (10/100) column with 30 ml of SQ Buffer A;

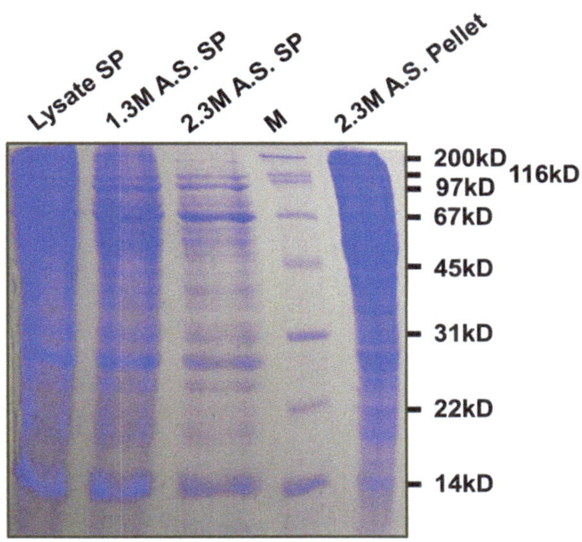

Fig. 1 SDS-PAGE analysis of samples from $(NH_4)_2SO_4$ precipitation. There is no clear band of CCT complex in crude sample at this step. The 2.3 M A.S. pellet was collected and proceeded to the next step

(c) Dilute sample from step 2.f to 40 ml with SQ Buffer A, load sample using Pump A, and dilute the last ~3 ml with an additional 20 ml of SQ Buffer A.

(d) Switch loading tip to SQ Buffer A after loading all the sample onto the column, continue the program described at step 3.a, and collect 5 ml/fraction.

(e) Pool fractions containing CCT together according to SDS-PAGE (Fig. 2).

4. HiTrap Heparin HP (5 ml) column

(a) On AKTA, use Heparin Buffer C for Pump A and SQ Buffer B for Pump B, flow rate = 3 ml/min, and edit

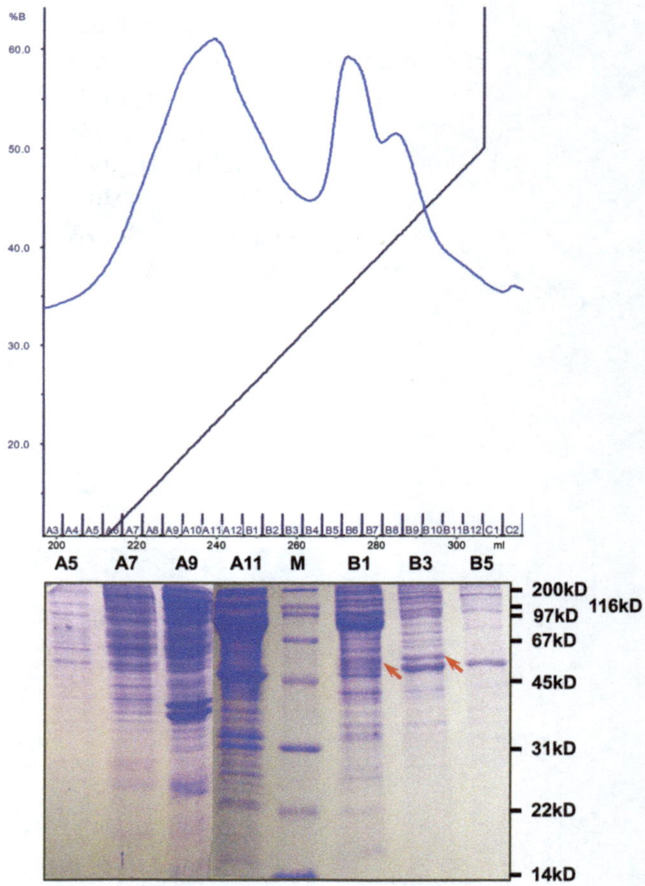

Fig. 2 Anion exchange chromatography over Source 15Q with a gradient of NaCl from 0.1 to 0.5 M (100 ml), followed by SDS-PAGE analysis of fraction A5-B5. Red arrows on the gel indicate CCT complex. Collect 5 ml/fraction. Pool fraction A5-B3 together (50 ml, ~150 mM NaCl), which can be loaded onto heparin column directly

program with a 100 ml linear gradient of 0–100% Pump B buffer after 50 ml of the "washout unbound" step.

(b) Equilibrate the HiTrap Heparin HP (5 ml) column with 20 ml of Heparin Buffer C (200 mM NaCl).

(c) Load sample (150 mM NaCl, dilute sample if >200 mM NaCl) from step 3.d using Pump A, and dilute the last ~3 ml with an additional 20 ml of Heparin Buffer C.

(d) Switch loading tip to Heparin Buffer C after loading all the sample onto the column, continue the program described at step 4.a, and collect 5 ml/fraction.

(e) Pool fractions containing CCT together according to SDS-PAGE (Fig. 3).

5. Hydrophobic interaction chromatography, HiTrap Phenyl HP (5 ml) column

(a) Test A.S. concentration for HIC column: from step 2 we learned that CCT will not precipitate at 1.3 M A.S., but will precipitate at 2.3 M A.S. To guarantee the protein remains stable in solution, we use 1.0 M A.S. as binding buffer for Pump A (*see* **Note 2**).

(b) On AKTA, use HIC Buffer D for Pump A and Heparin Buffer C for Pump B, flow rate = 3 ml/min, and edit program with a 100 ml linear gradient of 0–100% Pump B buffer after 50 ml of washout unbound.

(c) Equilibrate the HiTrap Phenyl HP (5 ml) column with 20 ml of HIC Buffer D (200 mM NaCl, 1.0 M A.S.).

(d) Mix sample from step 4.e (35 ml) with equal volume of 2.0 M A.S. solution and then filter it (0.45 μm).

(e) Load sample using Pump A, and dilute the last ~3 ml with an additional 20 ml of HIC Buffer D.

(f) Switch loading tip to HIC Buffer D after loading all the sample onto the column, continue the program described at step 5.b, and collect 5 ml/fraction.

(g) Pool fractions containing CCT together according to SDS-PAGE (Fig. 4).

6. Superdex 200

(a) Concentrate sample from step 5.e to 0.5–1.5 ml.

(b) Equilibrate the Superdex 200 column with 1.5 column volume (45 ml) of SEC Buffer E (Heparin Buffer C), flow rate = 0.5 ml/min.

(c) Inject concentrated sample through injection loop, and run the size exclusion chromatography program with SEC Buffer E.

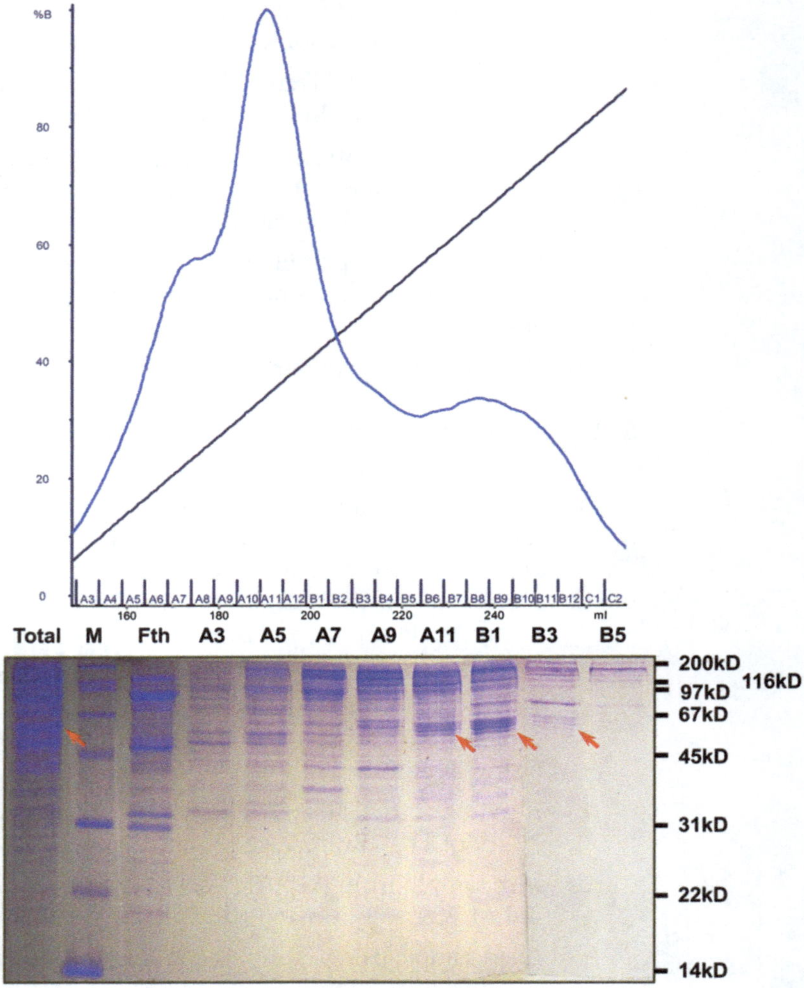

Fig. 3 HiTrap Heparin HP (5 ml) chromatography with a gradient of 0.2–1.0 M NaCl (100 ml), followed by SDS-PAGE analysis of fraction A3-B5. Red arrows on the gel indicate CCT complex. Collect 5 ml/fraction. Pool fraction A11-B5 together (total 35 ml)

(d) Pool fractions containing CCT together according to SDS-PAGE (Fig. 5).

7. Superose 6: same procedure as step 6 but using Superose 6 column

(a) Concentrate sample from step 6.d to 0.5–1.5 ml.

(b) Equilibrate the Superose 6 column with 1.5 column volume (45 ml) of SEC Buffer E (Heparin Buffer C), flow rate = 0.5 ml/min.

(c) Inject concentrated sample through injection loop, and run the size exclusion chromatography program with SEC Buffer E.

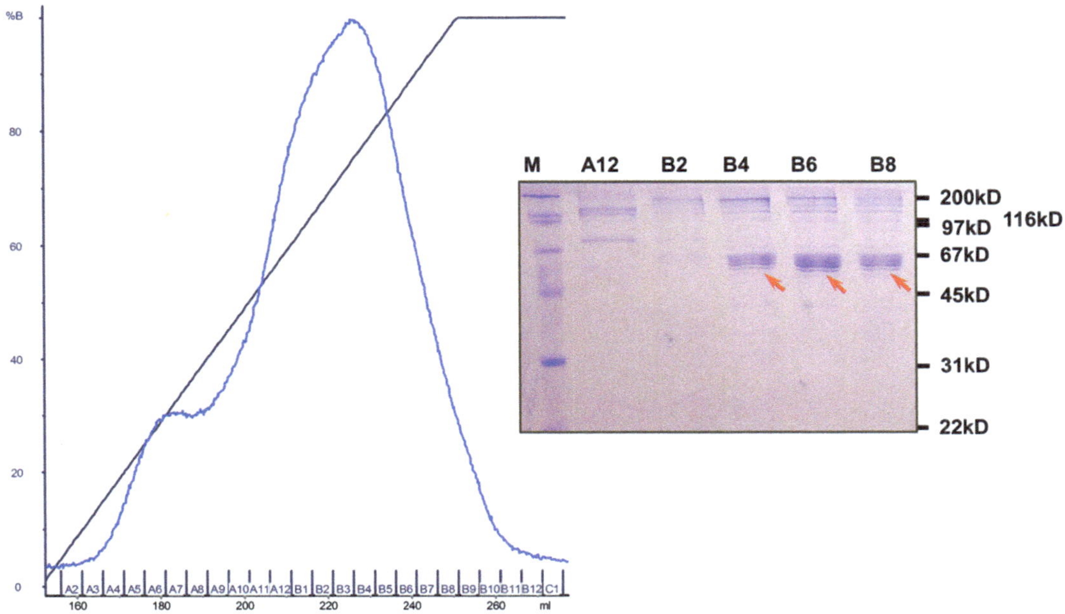

Fig. 4 Hydrophobic interaction chromatography in a gradient of 1.0-0 M A.S. (100 ml), followed by SDS-PAGE analysis of fraction A12-B8. Red arrows on the gel indicate CCT complex. Collect 5 ml/fraction. Pool fraction B4-B8 together (total 25 ml)

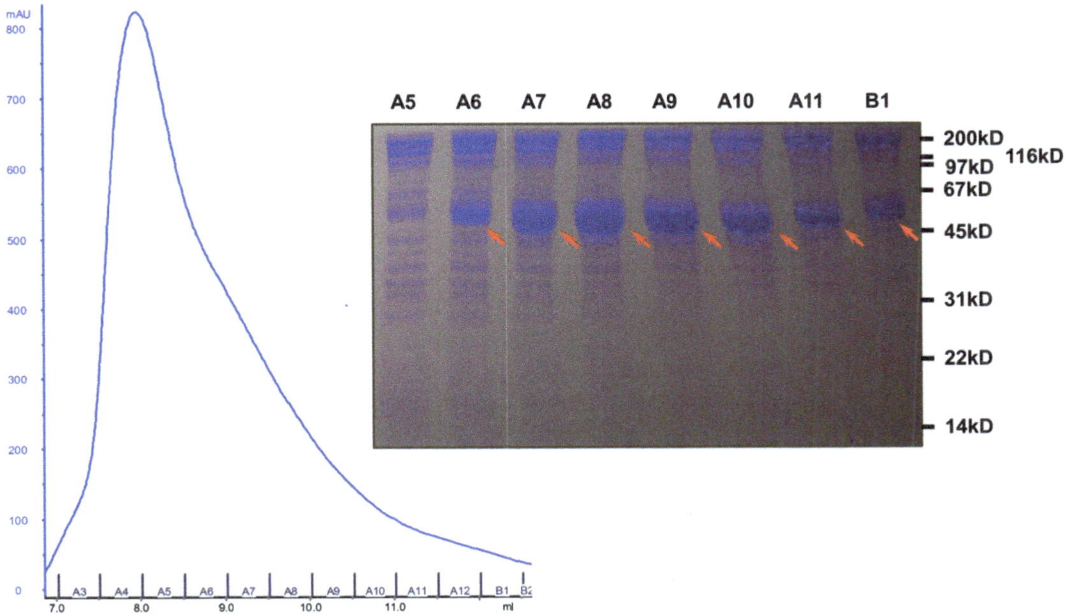

Fig. 5 Size exclusion (or gel filtration) chromatography over Superdex 200, followed by SDS-PAGE analysis of fraction A5-B1 (7–12 ml). *Red arrows* on the gel indicate CCT complex. Collect 0.5 ml/fraction. Pool fraction A6-B1 together (total 4 ml)

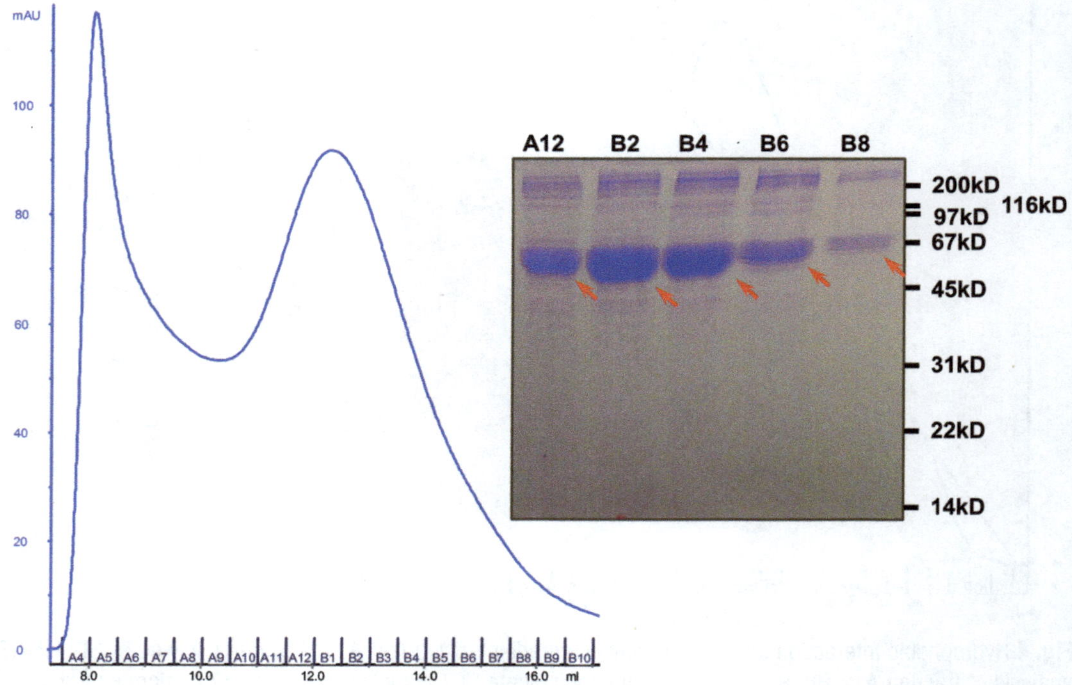

Fig. 6 Size exclusion (or gel filtration) chromatography over Superose 6. Fraction A12-B8 (11.5–16 ml) were examined on SDS-PAGE and visualized by Coomassie blue staining. *Red arrows* on the gel indicate CCT complex. Collect 0.5 ml/fraction. Pool fraction A12-B8 together (total 4.5 ml)

(d) Pool fractions containing CCT together according to SDS-PAGE.

(e) Determine the total amount of protein using Bradford assay or by extinction coefficient using absorbance at 280 nm (usually ~2–4 mg) (Fig. 6).

3.2 Purification of PP2Ac from Porcine Brains

Protein phosphatase 2A (PP2A) is a major serine/threonine phosphatase. It dephosphorylates a variety of specific substrates to regulate the cellular function of these target proteins. Deregulation of PP2A is closely related to different types of the human diseases, such as cancer, neurodegenerative disorders, and heart failure [9, 10]. The heterotrimeric holoenzyme contains a scaffold A subunit, a catalytic C subunit (PP2Ac), and a regulatory B subunit. Two divalent cations bind to the active site of PP2Ac, which are essential for the enzyme activity. Many divalent cations were reported to activate the enzyme in vitro, such as Mn^{2+}, Zn^{2+}, Mg^{2+}, Ni^{2+}, Co^{2+}, Fe^{2+}, etc. For many years, there is still a debate on what type of metal ions in the native PP2A [2, 11–14]. PP2A is enriched in different types of tissues and could be obtained from rabbit skeletal muscle (1998) [15], porcine brains (2004, 2014) [2, 16], bovine heart (2007) [4], etc. To examine the metal ion associated with PP2Ac, we developed and modified a protocol based on the previous studies. The native PP2A from porcine brains was subjected to inductively coupled plasma mass spectrometry (ICP-MS) analysis [2]. Briefly, the crude PP2A was purified by A.S. precipitation followed by ethanol precipitation. A second A.S. precipitation was performed to reduce the sample volume to ~100 ml. This sample was then applied onto GS4B column with immobilized GST-PP2A-A to pull down PP2Ac. The last step was buffer change for ICP-MS analysis using Superdex 75. This protocol can also be applied on other target proteins with strong binding partner.

1. Lysate preparation (*see* **Note 3**)
   (a) Thaw porcine brains (preferably) at 4 °C.
   (b) Rinse brains with deionized water to remove as much blood as possible.
   (c) Blend brains in 300–400 ml ice-cold Buffer 1.
   (d) Pass through cheesecloth to remove large pieces of bone/ fibrous membranes.
   (e) Add PMSF to 1 mM final concentration.
   (f) Dounce 20 stokes in a 40-ml Dounce tissue grinder, and collect the sample in a large beaker on ice.
   (g) Transfer supernatant to 40 ml tubes and spin at 16,000 rpm for 1 h at 4 °C, and collect supernatant.
   (h) Rinse pellets with Buffer 1 and spin rinse at 16,000 rpm for an additional 10 min at 4 °C, and combine supernatant from steps g and h.
2. Protein precipitation
   (a) First A.S. precipitation:
      (i) Precipitate solution at 2.9 M $(NH_4)_2SO_4$, by adding ~223 g powder to 500 ml supernatant at 4 °C, and stir to dissolve.
      (ii) Transfer to 40 ml tubes and spin at 16,000 rpm for 10 min at 4 °C.
      (iii) Remove supernatant and dissolve pellets in ~150 ml Buffer 2.
   (b) Ethanol precipitation:
      (i) *Quickly* add 5 volumes (~750 ml) of Buffer 3 at room temperature, and *quickly* spin for 5 min at 4000 rpm at 4 °C.
      (ii) Remove ethanol supernatant and dissolve the pellet in 500 ml Buffer 2 (use Dounce to resuspend); spin as in step 2.b.i and collect Buffer 2 supernatant.

(iii) Redissolve pellet in ~400 ml Buffer 2 (use Dounce), spin again as in step 2.b.i, collect and combine Buffer 2 supernatant from step 2.b.ii, and discard pellet.

(c) Second A.S. precipitation:

(i) Precipitate the combined Buffer 2 supernatant with 2.9 M $(NH_4)_2SO_4$ as in step 2.a, and spin solution as in step 2.a.ii to collect pellet.

(ii) Remove supernatant and dissolve pellets in 80 ml Buffer 2, spin as in step 2.a.ii, and collect supernatant.

(iii) Redissolve pellets in 20 ml Buffer 2, spin as in step 2.a.ii, and collect and combine supernatant with step 2.c.ii, for a total of ~100 ml.

3. GST-PP2A-A pulldown

(a) Pre-equilibrate two GS4B columns (2 ml/column) by Buffer 2, and then stack one on top of the other.

(b) Pass ~8 mg of GST-PP2A-A over the GS4B columns, and pass twice.

(c) After the second pass, wash the columns twice with 20 ml Buffer 2.

(d) Pass brain lysate from step 2.c.iii (100 ml) over the columns; pass twice and save the flow-through.

(i) After the second pass, wash the columns twice with 20 ml Buffer 2.

(ii) Elute each column with first 5 ml buffer GSH, followed by a second elution of an additional 15 ml of buffer GSH. Examine the sample on SDS-PAGE.

4. Buffer exchange using Superdex 75

(a) Concentrate the GST-PP2A-AC to 0.5–1.5 ml.

(b) Run the sample on Superdex 75 column, which is equilibrated with low conductivity diH_2O with 0.5 mM Tris pH 7.5 (*see* **Note 4**).

(c) Analyze fractions by SDS-PAGE.

(d) Collect and pool lowest conductivity fractions with target protein for ICP-MS (Fig. 7).

4 Notes

1. Bovine testes and porcine brains may be obtained from local slaughterhouse. One should ask the slaughterhouse to freeze them if they will not be used immediately. For bovine testes, remove the skin, vessels, and fat, cut the tissue into small pieces, seal 80–100 g per bag, and freeze in −80 °C. For porcine brains, seal each brain in a bag.

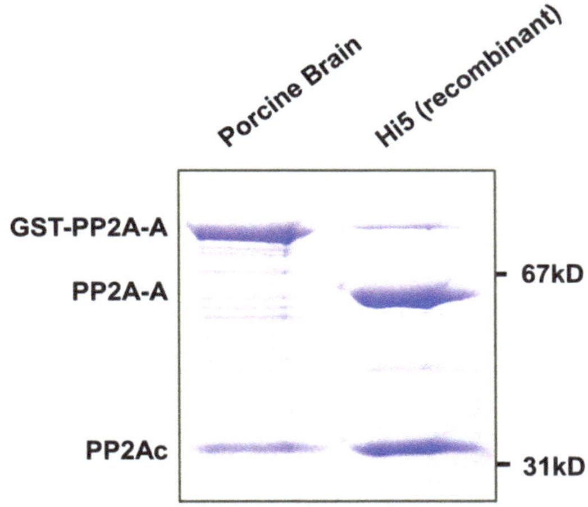

Fig. 7 SDS-PAGE analysis of the final sample of GST-PP2A-AC with C (PP2Ac) extracted from porcine brains. The recombinant PP2A-AC expressed from Hi5 insect cell was loaded on the same gel as the control

2. We recommend always determining the suitable A.S. concentration for binding buffer before running HIC, as proteins may precipitate in different condition. Generally, test if proteins precipitate in 0.5, 1, 1.5, and 2 M A.S. by mixing with an equal volume (200–500 μl) of 2× A.S. solution. The *highest final* A.S. concentration that *does not* precipitate protein will be used. Also, use 90% of the determined concentration to ensure the proteins will not precipitate on column. Depending on different protein behaviors, it is also a good idea to keep sample on ice for 5–10 min, after bringing the sample to desired A.S. concentration during a large scale purification. Some protein may still precipitate out after being kept on ice for a while, although it remains clear during the small test. Most of the proteins can be recovered by reducing the A.S. concentration because the A.S. precipitation is reversible.
3. Use 100 μM EDTA for lysis but 10 μM afterward to minimize the chelation of metal ion of PP2Ac. DTT may chelate metal ions and make the protein insoluble; thus, beta-mercaptoethanol is a better choice. Do not use TCEP.
4. Because ICP-MS analysis requires low NaCl concentration in the solution, we use only 0.5 mM Tris to stabilize the protein. Usually, 0.5–1.5 mg/ml protein sample can separate the salt peak nicely. Collect the fraction with low conductivity (<0.05 mS/cm) for best results.
5. General guidelines for purification of target proteins from tissue:
   (a) Before starting, one can check the human proteomic databases for the enriched tissues of target proteins. A recent

paper about tissue-based map of the human proteome published on *Science* will be helpful, which covers more than 90% of the putative protein-coding genes [1]. In this study, enriched proteins are mainly intracellular in the testis, mainly membrane-bound in the brain and kidney, and mainly secreted in the pancreas and liver [1]. Some fundamental information should be collected at this step, such as what type of protein you will work with and the cellular/subcellular location of target protein. Is your protein soluble, a membrane protein (insert or associate), or secreted to extracellular (body fluid)? Is it associated with exosomes or lipids? Is it in the cytoplasm or nucleus? Also, calculating the isoelectric point (pI) and hydrophobicity of the target protein based on its amino acid sequence would help to determine the FPLC method and procedure for protein purification.

(b) There are several concerns about protein solubility and stability in general:

- pH: most proteins are stable at pH 7–8.
- Salt concentration: low salt <50 mM, normal salt = 100–150 mM, high salt = 300 mM or higher. In most cases, normal salt concentrations are sufficient, but some proteins require high salt.
- Reducing condition: DTT and beta-mercaptoethanol are generally good to start with. TCEP is a strong reducing reagent that will precipitate protein or lose activity in some cases, such as PP2A.
- Temperature: most proteins are more stable at low temperature (on ice or 4 °C) than ambient temperature. Some proteins such as antibodies are relatively stable and thus can be handled at room temperature.
- Protease inhibitors: carefully select protease inhibitors that would not inhibit your target protein. For example, Na_3VO_4 is an inhibitor of protein tyrosine phosphatases, alkaline phosphatases, and a number of ATPases and thus should not be applied for these proteins. You can use it for kinase purification though. For a metal-binding protein, limited or no EDTA/EGTA should be used. Besides general inhibitors, one may also need specific protease inhibitors based on previous studies.
- DNase and RNase: these can digest large fragments of DNA or RNA released from the genome, thus improving the purity of target protein.

- Detergent: do you need detergent and what detergent should be used? It is critical to select proper detergent for membrane proteins, membrane-associated proteins, or lipid-related proteins.
- Metal ions: many enzymes require divalent cations such as Mg^{2+}, Mn^{2+}, Ca^{2+}, Zn^{2+}, Ni^{2+}, Fe^{2+}, Co^{2+}, etc. along with ATP or GTP to maintain or boost the activity.
- Water: always use deionized water. Some concern may be needed if local water quality is different, such as hard water with high mineral content. Some proteins are sensitive to these trace amounts of metal ions.
- Other special handling with posttranslational modifications: phosphorylation, acetylation, glycosylation, etc. Some chaperone protein may help to stabilize the target protein during purification.

(c) We recommend size exclusion chromatography at the final step to check the quality and conformation of target protein. There are several benefits to doing so: (1) change the sample to desired buffer; (2) check protein conformation, such as aggregation, polymerization, or dimerization; and (3) remove impurities that cannot be separated at previous steps.

(d) We also recommend confirming the target protein using Western blot or mass spectrometry (not present in this article).

(e) Troubleshooting: always monitor the protein during the purification procedure by SDS-PAGE, and save samples for each step. This is crucial especially when you are trying something new or working with novel proteins.

6. Reagents ordering information:

Hepes (Amresco, 0485)

Sucrose (Amresco, 0335)

DTT (Amresco, 0281)

Leupeptin/aprotinin (Thermo scientific, 78435/78432)

$(NH_4)_2SO_4$ (Amresco, 0191)

Glycerol (Fisher, BP229-4)

$MgCl_2$ (Amresco, 0281, 0288)

Bradford solution (Bio-Rad, 500-0006)

Tris (Amresco 0497)

NaCl (Sigma S7653-250G)

2-ME (β-mercaptoethanol) (JT Baker 4049-00)

EDTA (Amresco 0105-1KG)

Reduced GSH (Amresco 0399-50G)
95% ethanol (Barton Solvents)
PMSF (Amresco 0754-25G)
NaOH (Amresco, 0713)

Acknowledgment

This work was supported by ACS research scholar grant (Y. Xing), R01 GM096060-01 (Y. Xing), and T32 CA009135 (N. Wlodarchak). We would also like to thank Dr. Anming Xiong and Edward H. Wang for proofreading.

References

1. Uhlen M, Fagerberg L, Hallstrom BM, Lindskog C, Oksvold P, Mardinoglu A et al (2015) Proteomics. Tissue-based map of the human proteome. Science 1260419:347
2. Guo F, Stanevich V, Wlodarchak N, Sengupta R, Jiang L, Satyshur KA et al (2014) Structural basis of PP2A activation by PTPA, an ATP-dependent activation chaperone. Cell Res 24:190–203
3. Ferreyra RG, Frydman J (2000) Purification of the cytosolic chaperonin TRiC from bovine testis. Methods Mol Biol 140:153–160
4. Tran HT, Ferrar TS, Ulke-Lemee A, Moorhead GB (2007) Purification of PP2Ac from bovine heart. Methods Mol Biol 365:127–132
5. Frydman J, Nimmesgern E, Erdjument-Bromage H, Wall JS, Tempst P, Hartl FU (1992) Function in protein folding of TRiC, a cytosolic ring complex containing TCP-1 and structurally related subunits. EMBO J 11:4767–4778
6. Lewis VA, Hynes GM, Zheng D, Saibil H, Willison K (1992) T-complex polypeptide-1 is a subunit of a heteromeric particle in the eukaryotic cytosol. Nature 358:249–252
7. Gao Y, Thomas JO, Chow RL, Lee GH, Cowan NJ (1992) A cytoplasmic chaperonin that catalyzes beta-actin folding. Cell 69:1043–1050
8. Cong Y, Baker ML, Jakana J, Woolford D, Miller EJ, Reissmann S et al (2010) 4.0-A resolution cryo-EM structure of the mammalian chaperonin TRiC/CCT reveals its unique subunit arrangement. Proc Natl Acad Sci U S A 107:4967–4972
9. Virshup DM (2000) Protein phosphatase 2A: a panoply of enzymes. Curr Opin Cell Biol 12:180–185
10. Janssens V, Goris J (2001) Protein phosphatase 2A: a highly regulated family of serine/threonine phosphatases implicated in cell growth and signalling. Biochem J 353:417–439
11. Yu JS (1998) Activation of protein phosphatase 2A by the Fe2+/ascorbate system. J Biochem 124:225–230
12. Wozniak-Celmer E, Oldziej S, Ciarkowski J (2001) Theoretical models of catalytic domains of protein phosphatases 1 and 2A with Zn2+ and Mn2+ metal dications and putative bioligands in their catalytic centers. Acta Biochim Pol 48:35–52
13. Galadari S, Hago A, Patel M (2001) Effects of cations on ceramide-activated protein phosphatase 2A. Exp Mol Med 33:240–244
14. Swiatek W, Sugajska E, Lankiewicz L, Hemmings BA, Zolnierowicz S (2000) Biochemical characterization of recombinant subunits of type 2A protein phosphatase overexpressed in Pichia pastoris. Eur J Biochem 267:5209–5216
15. Cohen P, Alemany S, Hemmings BA, Resink TJ, Stralfors P, Tung HY (1988) Protein phosphatase-1 and protein phosphatase-2A from rabbit skeletal muscle. Methods Enzymol 159:390–408
16. Longin S, Jordens J, Martens E, Stevens I, Janssens V, Rondelez E et al (2004) An inactive protein phosphatase 2A population is associated with methylesterase and can be re-activated by the phosphotyrosyl phosphatase activator. Biochem J 380:111–119

Methods in Molecular Biology (2018) 1788: 89–111
DOI 10.1007/7651_2017_94

Published online: 18 November 2017

Integrative Analysis of Proteomics Data to Obtain Clinically Relevant Markers

Nathan Salomonis

Abstract

The analysis of proteomics data can be significantly challenging. Beyond the technical challenges of accurately identifying and quantifying peptides, identifying the most biologically coherent set of biomarkers can be a particularly daunting step. In this chapter, we will review a series of methods implemented in the software AltAnalyze that can be used to normalize proteomics peptide counts, identify a minimal set of the most distinguishing morbidity-associated biomarkers, and connect up these results to known pathways and interacting protein and regulatory networks. Here, we will apply this workflow to two examples that highlight different benefits of an integrated analysis workflow: (1) urine proteomics samples from patients with distinct kidney transplantation morbidities and (2) sudden infant death syndrome. By the end of this chapter, the reader should be able to apply a similar workflow to their own datasets to identify biologically significant protein markers and relevant networks.

Keywords Bioinformatics, Biomarker, Data analysis, Proteomics, Regulatory networks

1 Introduction to Bioinformatics Analysis

The advent of new technologies to accurately measure discrete molecular entities in mass (genes, proteins, RNAs, metabolites) has revolutionized the way in which researchers investigate cellular phenomena [1]. While technologies such as mass spectrometry, microarrays, and RNA-Seq (collectively referred to hereafter as "omics" technologies) have become commonplace in most research settings, the methods to analyze this data are still undergoing significant development [2, 3]. This "lag" in computation relative to assay development is due to the amazingly complex nature of biological systems and associated ability to model these systems and any reliable way. For example, gene regulation is dependent on multiple modalities of control, including epigenetic regulation, transcription, alternative splicing, alternative polyadenylation, translation, nonsense mediated decay, microRNA regulation, glycosylation, ubiquitination, phosphorylation, and proteolytic processing. Because each type of measurement technology only evaluates one or possibly two types of these modalities, the analysis of the data requires many assumptions in identifying the mode of gene/protein regulation.

In order to assemble a useful biological model from our experimental results, we must apply information from existing knowledge bases, such as growing gene and protein databases, that allow us to compare observed changes to prior observations. By doing so, we are able to effectively create new theoretical models that can be further validated through rigorous validation. In this way, bioinformatics serves to generate to new hypotheses and new tests rather than being an end to an experiment.

1.1 Importance of Knowledge Bases in Omics Analysis

A major goal of research in the twenty-first century is to expand upon the wealth of information in existing knowledge bases, by contributing to them and identifying new context-specific associations. New "omics" technologies provide one way to achieve this, by creating new molecular signatures of disease that can be integrated with existing knowledge bases and further refined to develop new testable disease hypotheses [4, 5]. Large-scale projects such as ENCODE and TCGA aim to expand these knowledge bases, by contributing new molecular signatures (e.g., ChIP-Seq, RNA-Seq, proteomics) to existing databases [6, 7]. With access to these large signature databases in combination with large databases of protein or RNA interactions (physical and regulatory), known biological pathways (WikiPathways, KEGG, PathwayCommons), and gene ontologies (GO, disease, phenotype), we can begin to develop new context-specific models of disease that can aid in the analysis of new datasets (Fig. 1) [8–13].

1.2 Software to Analyze and Comprehend User Data

Most laboratory trained researchers are not experts in the analysis of large "omics" datasets and have little to no advanced computational or bioinformatics skills. While more academic institutions are beginning to integrate such learning into their curriculums, it is particularly challenging for biologists to be experts in both the experimental and computational realms. In response, many research groups have built bioinformatics research tools that are more easily accessible by the non-informatics research community [14–16]. These tools include online web portals and downloadable computer applications to analyze simple gene lists, predicted interactions, or even raw omics data itself. Some tools, such as R and its large repository of life science-associated libraries in Bioconductor, provide programmatic as well as some graphical user interfaces that can be reasonably learned by the non-computational biologist [17]. A wide array of excellent commercial tools also exist as well for performing integrated bioinformatics analyses of omics data; however, cost and sharing of results become an issue when working across multiple research groups.

Here we discuss and step through the application of a software package called AltAnalyze, which when run will provide a wide array of analysis results, graphical outputs, and statistics depending on the type of data input [15, 18–25]. For gene microarray and RNA-Seq data, these include complex alternative splicing and single-cell analyses; however, for these and other data types, such

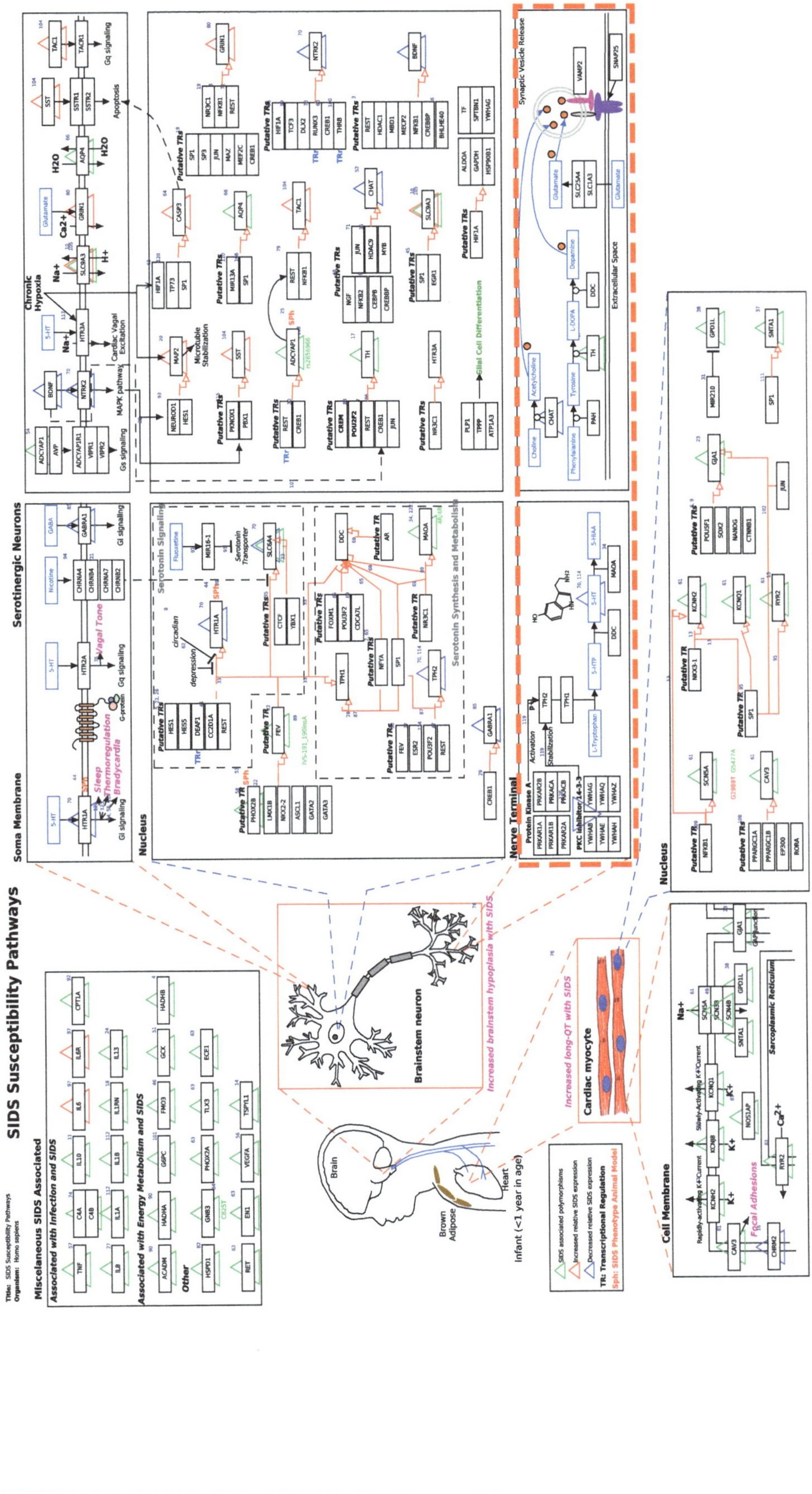

Fig. 1 Developing models of human disease from molecular omics data. A WikiPathways model of sudden infant death syndrome (SIDS) susceptibility, integrating cell systems, signaling pathways, genetics, proteomics, and animal model data. Such models can be derived from the scientific literature in combination with high-throughput molecular omics data providing a deeper understanding of molecular interactions and gene regulation. The *red* highlighted box indicates proteomics observations. http:/www.wikipathways.org/index.php/Pathway:WP706

as proteomics or metabolomics data, methods such as differential expression analysis, pathway, network, and expression visualization provide extremely useful means to evaluate the biological cohesiveness of user input datasets. Because this application is open-source and can be run through an intuitive graphical user interface or optionally via a command line (computer terminal window), it is accessible and even modifiable by researchers with various needs. From analyses with this and related tools, integrated molecular models can be created that combine new data with known genes, pathways, and interactions. By doing so new theoretical molecular models can be created, posted on the web, and subsequently used by other researchers to investigate and validate (Sect. 8).

1.3 AltAnalyze Installation

The AltAnalyze software can be downloaded for analysis on your own computer from http://www.altanalyze.org. The software is pre-built (compiled) so that it can immediately run on any newer Windows operating system, Mac OS X and certain Linux versions, without having to install any other software or download any other databases. For advanced users, source code is also available to allow the software to be run on other operating systems as a graphical user interface or as a command-line application. Various tutorials, frequently asked questions and answers, documentation and methods descriptions are available from this website, including additional details on installation (http://altanalyze.readthedocs.io). After downloading the software, unzip the compressed download (zip file or dmg) to your computer. The AltAnalyze software can be placed anywhere you pleased; however, it is recommended that you have at least 6GB of free hard disk space and at least 4GB of RAM. Once extracted, you can double-click on the AltAnalyze executable file (AltAnlayze.exe) in the AltAnalyze program folder and follow the prompts for installation of databases for your species of interest (just a few clicks needed). Once the species database has been installed, you will be ready to process the example data.

2 Disease Datasets for the Comprehensive Evaluation of Proteomics Data

Although there are multiple omics analysis technologies which can be examined in AltAnalyze, we will focus on its use with proteomics data, in particular two recently published disease datasets [24, 26]. As described in previous chapters, the initial identification of accurate peptide fragments is challenging with various discussed technical challenges. In this chapter we will begin with already obtained, protein fragment counts and proceed with the process of count normalization and quality control (QC). Most of the analysis steps we will be performing will be using AltAnalyze's default analysis workflow options, in order to process the input dataset in a single step. Nonetheless, we will describe the specific analyses that are performed in the background by AltAnalyze and what options are available for more complex analyses. In some of the below sections, we will further describe how a user can perform these functions outside of a full analysis workflow, focusing on a single step in the analysis, such as pathway enrichment, pathway visualization, network creation, expression visualization, Venn diagram comparison, principal component analysis, sample classification, merging files, and accession number translation. The two datasets we have chosen to apply these analyses to were generated with very distinct goals in mind but involve very similar methods of analysis.

2.1 Noninvasive Urine Biomarkers for Kidney Rejection

In this dataset, urine has been collected for individuals with and without kidney transplantation with known clinical outcomes [24]. These clinical outcomes include the acute transient rejection (AR) of the transplanted donor kidney, BK viremia (enhanced viral infection of the kidney following transplantation and immunosuppression), chronic allograft injury (CAI), and patients with a stable graft (no pathology). The log2 normalized expression data is available as an Excel spreadsheet as a publically accessible Synapse database file: https://www.synapse.org/#!Synapse:syn11246850.

2.2 Developing New Candidate Pathways Models of Sudden Infant Death Syndrome

This dataset measures the peptides present in the brain stems of presumably healthy infants that died as the result of either sudden infant death syndrome or other known causes [26]. Sudden infant death syndrome or SIDS occurs in a small percentage of healthy infants, under the age of 1 year, that suddenly and unexpectedly die for no other known reason [26, 27]. One possible explanation is the transient decreased presence of serotonin (or associated mis-regulation of other pathway components) in the brain stems of some infants that causes them to stop breathing [28–31]. Infants for this study [26] were selected who died of SIDS that were previously found to have decreased serotonin in the raphe nucleus of their brain stem and an absence of extrinsic risk factor for SIDS (e.g., sleep position). The raw spectral counts are available from an Excel spreadsheet in Synapse: https://www.synapse.org/#!Synapse:syn11246858.

3 Quality Control and Batch Effect Correction

Multiple relatively basic and advanced methods exist for assessing the quality of proteomics data, identifying batch or covariate-specific effects, and attempting to normalize these effects [32–34]. While several of these advanced methods can be applied using libraries for R available in Bioconductor, here we will focus on a few prepackaged and very useful methods available in the software AltAnalyze.

3.1 Importing the Dataset into AltAnalyze

In the case of already normalized data, such as the urine proteomics example dataset, all that is required is that the user save the results as a tab-delimited text file from Excel or similar software (Save As menu option, select the optional export file format) and that the file has a single annotation column and a single annotation row (already formatted as such). The expression data in this file is expressed in log2 format. To load this file in AltAnalyze, first open the AltAnalyze executable file, select **Homo Sapiens** under the "Select species" menu option and then **Other ID** under "Select vendor/data type" (Fig. 2). This text file has protein identifiers indicated using the **UniProt** database. To indicate the primary ID system is UniProt, select **UniProt** from the "Select platform" menu and then "Continue" to proceed with the analysis. This same workflow can be applied to the SIDS proteomics dataset, with the "Select platform" option set to **IPI** instead of UniProt.

Next, the user will be presented with two options: **Process Expression File** or **Additional Analyses**. Selecting this first option will allow you to process the tab-delimited expression file you have created using the standard AltAnalyze automated workflow. The second option will allow you to perform specific analyses present in the workflow and other options not available in the workflow, such as network analysis and identifier translation. Select the first option for now (press Continue), followed by selection of the input tab-delimited file location and results folder you wish to save outputs to when prompted. The results folder is often the same folder that contains the input file.

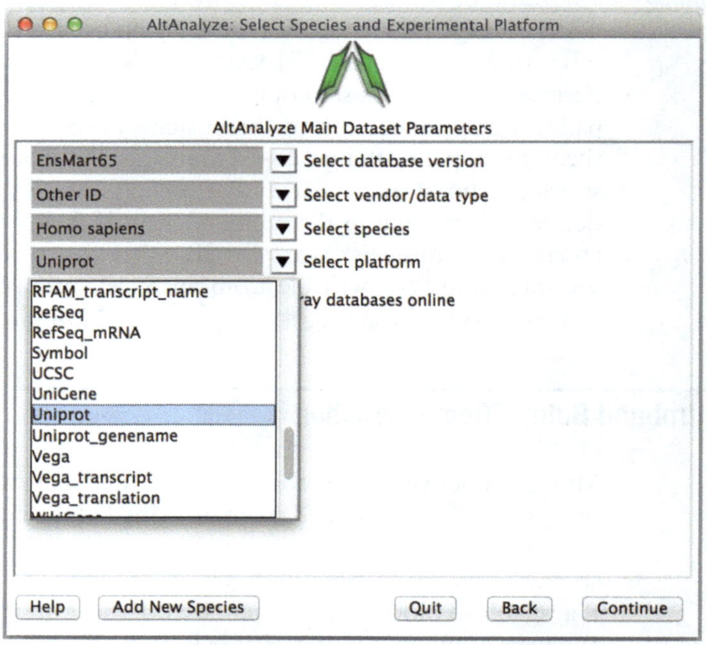

Fig. 2 Loading protein expression data in AltAnalyze. A screenshot of the main AltAnalyze software interface to select appropriate species (Homo sapiens) and dataset molecular identifiers. In AltAnalyze, specific supported omics platforms can be selected (e.g., RNA-Seq FASTQ files, BAM files, Affymetrix microarray, Agilent microarray) or the type of identifier present within an input normalized or un-normalized expression file. The identifiers should map to those listed within the input expression file (e.g., UniProt protein isoform accession numbers)

3.2 Dataset Normalization

There are two main normalization options for proteomics data in AltAnalyze: (1) quantile normalization and (2) batch effects correction. Quantile normalization adjusts all values for each sample in a dataset to have an identical range and distribution (shape). Hence, if some samples were found to already have low overall spectral counts, relative to the others, we would likely want to remove that sample prior to quantile normalization. For example, the sample "SIDS:82" has only 1380 total counts relative to "SIDS:79" with 4637 counts. If the dataset has been run as multiple independent batches of samples, batch correction may be required. This same type of correction can be appropriate if known confounding variable exists, such as gender differences, which can be classified as noncontinuous, discrete variables (e.g., male or female versus 1.3 and 2.3). The algorithm used is called Combat, which is the same method applied in the R library SVA [35].

Once the dataset has been selected, the following window, named **Expression Analysis Parameters**, provides all of the major analysis options in the software (Fig. 3). For the urine proteomics datasets, no additional normalization options need to be selected beyond the defaults. This is because the data has already been log2 normalized and is ready for analysis. The SIDS spectral count data however needs to be quantile normalized, requiring the "Normalize feature expression" option to be set from none to quantile. This data must also be log transformed, requiring changing of the option **log** in "Expression data format" to **non-log**. If batch effects are present in your data, you can also select the option "Remove batch effects (Combat)" as **yes**. For this option, you will need to later simply identify which samples are in which "batches." Here "batches" could be other discrete variables of interest.

3.3 Principal Component Analysis and Expression Visualization

One of the most important set of QC measures in the analysis of proteomics or other expression data is data visualization along the top principal components and expression clustering of significant and nonsignificant results. By default, AltAnalyze selects the option "Perform expression clustering and visual QC," which will output both principal component analysis (PCA) plots and hierarchically clustered proteins and samples as a graphical heatmap. Both of these analyses can be rerun with custom options using the **Additional Analyses** menu. For hierarchical clustering, the **Additional Analyses** menu option provides several powerful options to further explore your data, such as correlating your results to specific proteins, pathways, or other biologically important terms. These results will be saved to the folder "DataPlots" in the designated results folder. Ensuring that your samples are appropriately segregated according to the predicted biological groups is an important step in ensuring that the data has been normalized and processed correctly and that no unexpected confounding variables are present. Example outputs from the proteomics examples are shown in Fig. 4.

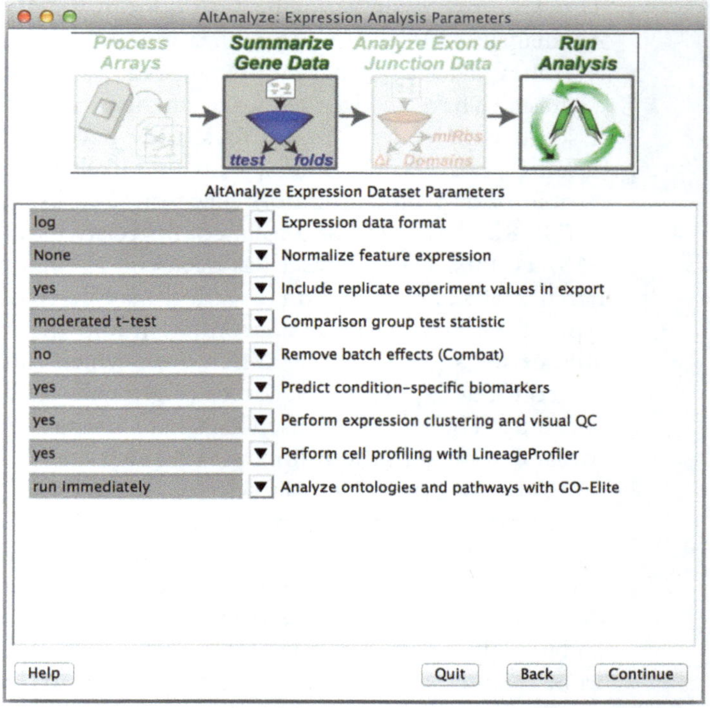

Fig. 3 Automated workflow options for normalizing and analyzing molecular omics data in AltAnalyze. AltAnalyze provides many options that enable advanced statistical analyses of user-provided expression datasets. These include the ability to select from comparison group testing procedures, normalization (e.g., quantile, batch effects correction), prediction of condition-specific biomarkers (MarkerFinder), identification of associated cell- or tissue-specific molecular profiles (LineageProfiler), and various gene-set/pathway analyses (GO-Elite) through a relatively simple graphical user interface

3.4 Cell-Level QC

The default selected menu option “Perform Cell Profiling with LineageProfiler” will perform an additional analysis that can be used to assess whether the appropriate cell types have been analyzed. This analysis will produce tables and plots in the results directory (ExpressionOutput and DataPlots folders) that will correlate the protein expression profiles to RNA microarray expression profiles, for a selected subset of highly informative cell-type-specific markers, for a large database of cell types and tissues. Although this analysis may not be informative, as your protein profiles will be compared to RNA expression levels, we have found that accurate tissue-level association will often be obtained for proteomics data, as is the case with the SIDS proteomics data (Fig. 4c). Hence, if a distinct cell/tissue prediction profile exists for one or more samples, any variables which could account for these differences should be seriously considered.

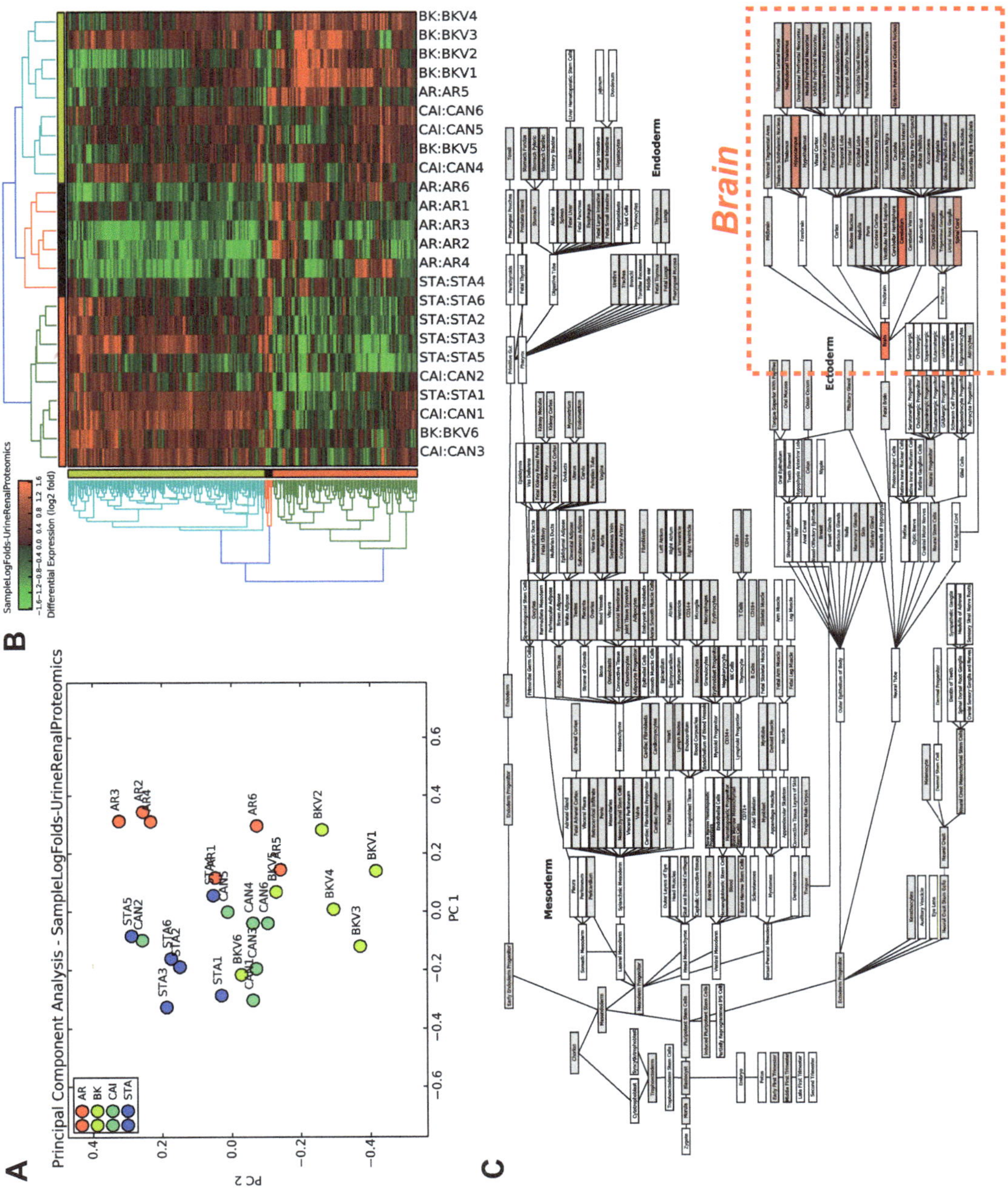
A
Principal Component Analysis - SampleLogFolds-UrineRenalProteomics
AR
BK
CAI
STA
PC 1
PC 2
B
SampleLogFolds-UrineRenalProteomics
Differential Expression (log2 fold)
BK:BKV4
BK:BKV3
BK:BKV2
BK:BKV1
AR:AR5
CAI:CAN6
CAI:CAN5
BK:BKV5
CAI:CAN4
AR:AR6
AR:AR1
AR:AR3
AR:AR2
AR:AR4
STA:STA4
STA:STA6
STA:STA2
STA:STA3
STA:STA5
CAI:CAN2
STA:STA1
CAI:CAN1
BK:BKV6
CAI:CAN3
C
Mesoderm
Endoderm
Ectoderm
Brain

4 Statistical Analysis of Proteomics Data

In the **Expression Analysis Parameters** window of AltAnalyze, there are numerous options present that allow us to evaluate the statistical significance or protein-level variation (following normalization), biological significance (pathway level), and even cellular significance (LineageProfiler). Each of these involves distinct statistical methods which we will briefly touch on in this section.

4.1 Differential Expression Analysis

By default, AltAnalyze compares pairs of user-annotated groups using two-sample test statistics and a comparison of all groups (analysis of variance or ANOVA). After these test statistics have been calculated, the values are adjusted for multiple hypothesis testing using the Benjamini–Hochberg (BH) adjustment. These distinct statistics are reported separately for each protein in the dataset (or other omics features being tested). If you do not have a strong statistical background, we recommend learning more about statistical hypothesis testing before rigorously evaluating your data, as major statistical considerations may be necessary in the design and analysis of your experiment.

There are two menus in which the statistics for differential expression analysis can be adjusted in AltAnalyze. The first is the "Comparison group test statistic" menu in the current window. Here, the user can select from a number of distinct pairwise group sample tests. The default statistical test, moderated *t*-test, which is a student *t*-test adjusted to account for the fact that the user is testing multiple features at once (described further below). For pairwise group comparisons, users can choose from unpaired and paired *t*-tests (assuming equal variance), rank sum, Mann–Whitney, and Kolmogorov–Smirnov. In addition to these unmoderated tests, AltAnalyze provides a moderated *t*-test option (unpaired, assuming equal variance), based on the limma empirical Bayes model. This model calculates an adjusted variance based on the variance of all genes analyzed for each comparison. Unlike a standard limma analysis, only the variances for the two groups being compared are used to compute the adjusted variance (s2 prior and df prior) for each

Fig. 4 Evaluating dataset quality control and global sample heterogeneity. Example automatically produced results from the test urine proteomics dataset obtained using the AltAnalyze default analysis options. (**a**) Principal component analysis (PCA) for differentially expressed proteins for samples corresponding to the four evaluated renal allograft patient groups (AR, BK, CAI, STA) (Sigdel TK et al. Mol Cell Proteomics. 2014). (**b**) Unsupervised hierarchical clustering for differentially expressed proteins for all pairwise group comparisons. Proteins are displayed as rows and samples as columns, with the expression of each protein normalized against the median log2 expression value. (**c**) Visualization of SIDS brain proteomic sample profiles classified along numerous cell-type and tissue-specific gene expression files from the AltAnalyze function LineageProfiler. *Red* reflects high and *gray* low tissue-specific *z*-scores from the evaluated sample group comparison as opposed to variance from all groups analyzed (when >2). The standard *t*-test assumes your samples have unequal variance between your sample groups, as does the moderated Welch *t*-test. The BH adjusted *p*-value will be referred to as the **adjp** in the AltAnalyze results files, while the non-adjusted *p*-value will be called the **rawp**. Typically, a Benjamini–Hochberg adjusted *p*-value of less than 0.05 is recommended as a threshold for further consideration; however, depending on the purpose of your analysis, your statistical threshold and selection of the may vary.

After selecting the comparison statistic to be used, if the analysis option "Analyze ontologies and pathways with GO-Elite" is set to the default "run immediately," you will be able to set additional statistical filters after selecting "Continue." This will bring up the **GO-Elite Parameters** menu, which we will later use to identify biologically relevant pathway-level associations [25] (Sect. 7). These options include which *p*-value cutoff to use via the "Maximum gene expression *t*-test *p*-value" entry field, whether to filter on the rawp or adjp in the "Filter based on the following *p*-value" menu, and which fold cutoff to filter upon in the "Minimum gene expression fold change" entry window. By default, the software will select the rawp; however, we recommend selection of adjp for the reasons described above. A fold change cutoff of 2 is typically reasonable, but for most proteomic studies, we are more considered with the *p*-value more than the fold cutoff. Hence, we can lower this value to 1.5, which will require a 50% difference (increase or decrease) in protein expression between the compared biological groups.

4.2 Statistical Parameters for Pathway Analysis

While still in the **GO-Elite Parameters** menu, the user can set which statistical filters to employ for pathway analysis as well as select which analyses should be performed (Fig. 5). AltAnalyze integrates the software GO-Elite for the purpose of pathway analysis. Pathway analysis is a general term applied to the statistical enrichment of different collection of gene sets that described different facets of known biology. These can be canonical signaling pathway genes, gene signatures obtained from experiments, transcription factor targets, cell-/tissue-specific markers, genes organized by disease phenotypes, or other similar categories. Because some of the categories are organized in the form of ontologies of related terms, redundancy in the genes associated with these terms is frequently present. GO-Elite addresses this redundancy by pruning down the ontology terms based on their relative enrichment scores when applied to your data. Since these methods do not differ too significantly, either should be sufficient for your analyses. However, we recommend that you select two rather than three genes for "Enter minimum number of changed genes" given that proteomics results produce typically fewer differentially regulated features than other omics technologies. However, selecting the option all (default) for

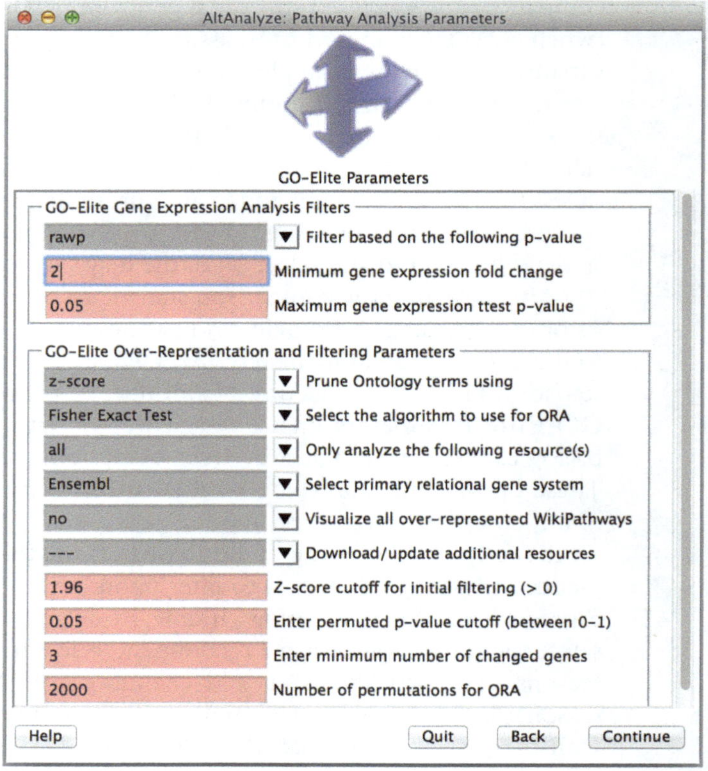

Fig. 5 Establishing differential expression, gene-set enrichment, and pathway visualization options. Users can define various options for their dataset analysis, such as statistical cutoffs using false discovery rate adjusted *p*-values (adjp) versus non-adjusted (rawp) and minimum differential expression cutoffs (fold and *p*-value). Differentially expressed genes or proteins from the associated comparison analyses are linked to over a dozen distinct gene-set and pathway databases to identify statistically enriched gene sets derived based on prior biological knowledge (e.g., Gene Ontology, WikiPathways, transcription factor targets). These options include the ability to automatically visualize gene comparison results directly within community-curated biological pathways from the WikiPathways database

"Only analyze the following resources" and yes for "Visualize all over-represented WikiPathways" will provide a greater breadth of predicted biological associations that will help understand the biological significance of your results. Once complete, select "Continue" to proceed to the next menu.

4.3 Establishing the Sample Groups and Comparisons

Following the previous menu, the user will be presented with **Group Annotation** window. By simply typing or pasting in the group names of each sample, the user will be able to proceed with specifying which groups to compare and how (Fig. 6). If you have hundreds of samples, additional options exist for importing these associations (see the AltAnalyze FAQ for details). After specifying the sample to group associations, select the experimental (e.g., disease) and reference (e.g., control) groups for comparison from

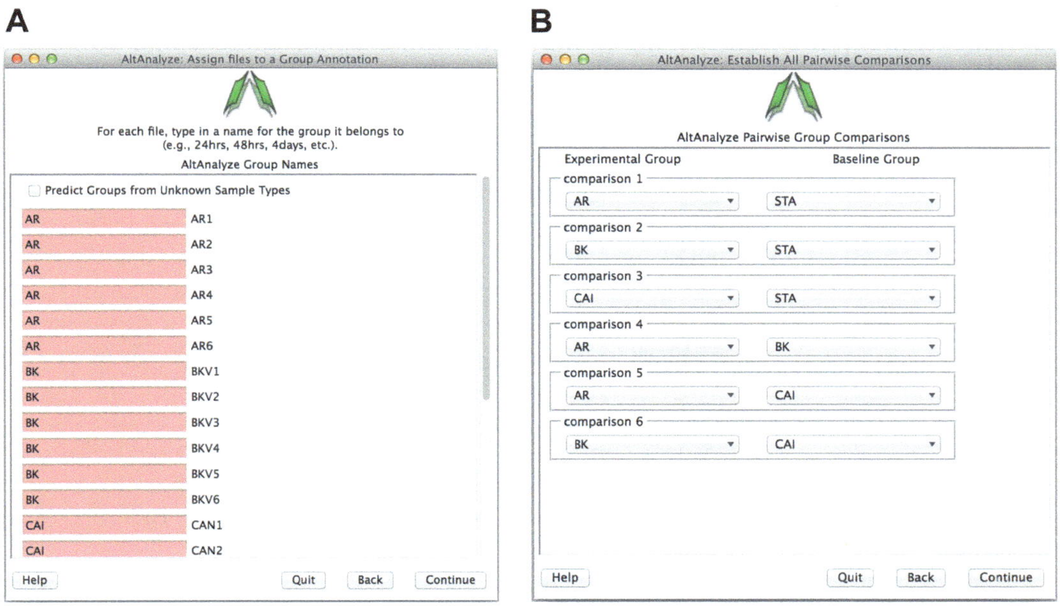

Fig. 6 Supervised comparison of known biological groups. AltAnalyze provides the capability to identify statistically significant molecular variance between known biological groups as well as define novel molecular subtypes through an unsupervised analysis called iterative clustering and guide-gene selection (ICGS) (Olsson A et al. Nature 2016). In panel **a**, the user-defined group annotations are entered prior to establishing possible pairwise group comparisons in panel **b**

the **Pairwise Comparison** window. These comparisons will later be used for the GO-Elite analyses and differential expression filtering options previously established. The protein results that are hierarchically clustered and output for PCA will also use any of the significantly differentially obtained proteins, for any of the designated comparisons. For the urine proteomics analysis, we recommend listing all possible pairwise comparisons between all indicated groups (AR, BK, CAI, STA).

4.4 Running the Analysis

Selecting Continue following selection of the pairwise group comparisons will immediately trigger AltAnalyze to run the workflow analysis. Depending on the number of samples, magnitude of results, and proteins analyzed, this analysis can take anywhere from minutes to hours. In the case of the two datasets suggested here, the analysis should take on the order of 5–20 min, most of which will be the GO-Elite pathway analyses and visualization, as nearly a dozen complex categories will be analyzed. The results will be stored in various directories which are described in the online documentation (http://www.altanalyze.org) and overviewed in Sect. 7 below.

5 Identifying Condition-Specific Biomarkers

One of the most informative analyses in AltAnalyze is biomarker detection. This analysis is performed using the MarkerFinder analysis, which is a component of the default analysis workflow run [20, 24]. Two main result files are produced from this analysis: (1) a heatmap illustrating the top 60 proteins associating with each annotated biological group and (2) a spreadsheet of all analyzed proteins assigned to one of these groups. This heatmap and associated data are saved to the DataPlots folder in a sub-directory named MarkerFinder (Fig. 7). In this example figure, Urine peptide expression profiles are shown for those proteins with the highest restricted expression to each kidney transplantation morbidity. The spreadsheet containing all proteins, assigned to a single biological group, can be found in the folder ExpressionOutput/MarkerFinder/AllGenes_correlations-ReplicateBased.txt. For each protein, a

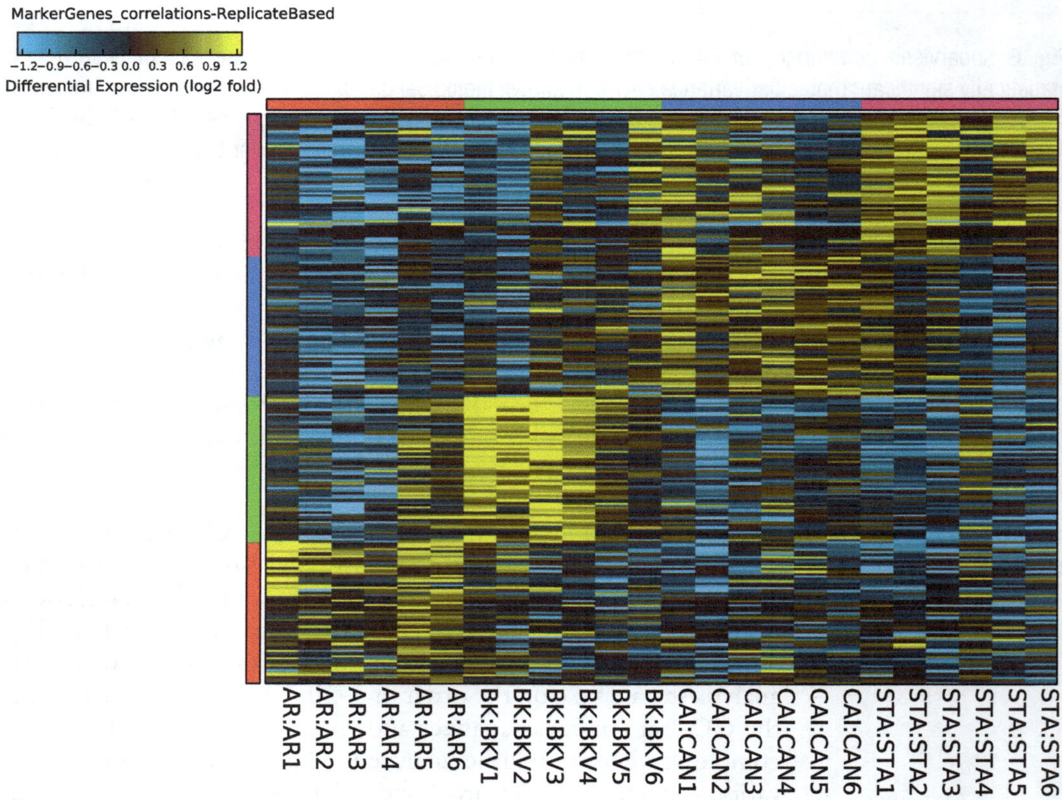

Fig. 7 Identifying condition-specific markers for distinct biological groups. In addition to the identification of differentially expressed genes, AltAnalyze enables the automatic prioritization of biological group-specific markers. The algorithm MarkerFinder provides these results in the form of genes or proteins (rows) and samples (columns), ordered according to the predicted specificity of those markers to each biological group

Pearson correlation coefficient is reported, which can be considered a similarity score to an ideal disease-specific profile (only expressed with that disease). Each protein is assigned to the morbidity it correlates to the most. By opening this file in Excel and sorting on the columns rho (Pearson correlation score) and tissue (annotated biological group), the user can obtain a prioritized list of biomarkers for each morbidity of interest. In this analysis of urine proteomics data, several biomarkers were selected for further downstream validation using an independent experimental assay.

6 Predicting the Source Cell Types of Proteins from Peripheral Body Fluids

The LineageProfiler analysis outlined in Sect. 4 of this chapter provides one possible means of determining which cell type(s) or tissue(s) may be represented in your analyzed sample. This is particularly useful for isolated tissue samples analyzed using proteomics, such as the brainstem SIDS samples. However, if you have peripheral fluids, you are more likely to identify associations to the peripheral fluid being analyzed (e.g., blood). In the case of urine, this is further complicated by a consistent lack of cells within a relatively sterile fluid. Hence, associated proteins are more likely to be derived from the kidney or other tissues in which significant cellular death is occurring.

When running the default AltAnalyze workflow, the GO-Elite pathway analysis will be performed using all available gene-set categories. This includes a set of cell- and tissue-specific biomarkers, typically ranging from 60 to 300 genes per tissue type. These are derived using a build process that examines hundreds of tissues and cell types across multiple transcriptome measurement platforms, for human and mouse (detailed description available from the AltAnalyze website). These unique profiles are obtained using a more structured version of the MarkerFinder algorithm described in Sect. 5. Evaluating these GO-Elite results can indicate whether specific cells or tissues are the origin of associated proteomics observations from urine or other peripheral fluids. These results are principally presented in three forms: (1) a heatmap illustrating the enrichment *z*-scores for the cellular categories, (2) a network of gene to cell/tissue types predicted to be enriched, and (3) a spreadsheet of the protein to cell associations presented in #2 with detailed statistics. Other tabular files are also produced in addition to these that include all protein to cellular associations (see the GO-Elite documentation at http://www.genmapp.org/go_elite). Such associations provide a means to identify the critical source cell types producing the excreted proteins, which can possibly be confirmed from other peripheral fluids or the source tissue itself.

7 Integrating Results Across Pathways and Gene Networks

With the completion of the AltAnalyze analysis, multiple results will be presented in the graphical user interface and on the user's computer. These include the following primary files, their locations, and software needed to explore:

1. Normalized proteomic expression profiles (ExpressionInput folder)—**Excel**
2. Differential expression analysis statistics and detailed annotations (ExpressionOutput/DATASET-[ExpName].txt)—**Excel**
3. Pathway analysis and visualization results (GO-Elite sub-directories)—**Excel** and **Adobe Acrobat**
4. Heatmaps, PCA plots, and LineageProfiler graphs (DataPlots folder)—**Adobe Acrobat**
5. Optimal morbidity-specific biomarkers (DataPlots/MarkerFinder and ExpressionOutput/MarkerFinder folders)—**Excel** and **Adobe Acrobat**

7.1 Evaluating the Initial Significance of Your Results

The DATASET file produced in #2 includes all statistical and annotation results for individual proteins. In this file is where it is most important to examine which known or novel peptides are regulated and to what extent. Prior knowledge regarding which proteins are expected to change and to what degree will likely be important for refining the filtering you perform initially in AltAnalyze. As such, it may be important to iterate your analysis using modified filtering options to increase the sensitivity for identifying previously validated observations. All significantly differentially expressed proteins will further appear in the SampleLogFold hierarchically clustered heatmaps in the DataPlots folder. Examination of this file, the OutlierLogFold, and the associated PCA plot will provide an additional means to evaluate the filters you employed and any potential outlier samples. For example, if one sample clearly is segregating by either of the top two PCA components, this sample may represent a biological or technical outlier that should be excluded from the analysis.

In the case of the SIDS proteomic analysis, only a few peptides will be considered differentially expressed using the employed default parameters, as a result of the small number of samples analyzed and heterogeneity in both controls and subjects. Because this is an initial discovery analysis, it is appropriate to decrease your significance thresholds to identify potential pathway-level observations that might shed more light on SIDS molecular pathophysiology. As such, we reanalyzed this data using a non-adjusted moderated *t*-test *p*-value threshold of 0.1 rather than 0.05, which returned a similar number of observations reported in the initial description of this data [26].

7.2 Understanding the Biological Significance of Your Results

Given that an inappropriate normalization, filtering, or other processing issues associated with your experiment can result in false-positive statistical associations, it becomes extremely important to validate your observations from a global biological perspective. This can be accomplished using the GO-Elite pathway analysis performed using the workflow steps outlined in previous sections of this chapter. As an example, we would expect brain-associated pathways and cellular markers to be enriched in our SIDS analyses, given that the biopsied nuclei were from a single brain region. Examination of the GO-Elite pruned results text file in a spreadsheet application identifies selective enrichment of the cell and tissue biomarker categories oligodendrocyte progenitor and brain, based on the differentially expressed proteins MBP, PLP1, and TPPP. Various other neuronal-specific biological processes (KEGG, Gene Ontology) and neuronal cellular components substantiate these observations, indicating that these differences are likely biologically significant. Further analysis of these differences at the transcriptional regulatory and pathway level identifies other proteomic difference between the two infant morbidity subsets that associate with previously observed pathophysiological differences, such as hypoxia response, brainstem gliosis, synaptic transmission deficits, and metabolic pathway differences that strongly point to informative and relevant biology [27].

Similar to the analysis of known pathways, the analysis of interacting proteins and regulatory network interactions can illustrate larger cohesive biological pathways that underlie protein-level observations. By default, this analysis is not performed in AltAnalyze, but can be easily run by the user from the **Additional Analyses** menu. As previously mentioned, this set of menus is available following the selection of the species, primary identifiers to be analyzed (initial AltAnalyze menu). Selecting the **Network Visualization** option from the **Additional Analyses** window will provide access to the tool NetPerspective [18] for the analysis of complex gene interactions, either within with a set of regulated genes or between genes within known biological categories. Selecting this menu will present a series of options for deriving a protein interaction network from various sources of data (Fig. 8a, b, d). Having run the workflow analysis, you will already have useful regulated protein sets from which to analyze. To do this:

1. Under **Option 2** select a regulated protein set corresponding to one of the morbidities of interest. For example, navigate to the folder GO-Elite/upregulated/ and select the file AR versus BK file.
2. For the menu "Select one or more interaction databases," select multiple available options (WikiPathways, KEGG, TFTargets, BioGRID). This will provide maximally informative interactions from pathways, from yeast-two-hybrid data, and from transcription factor targets [18].

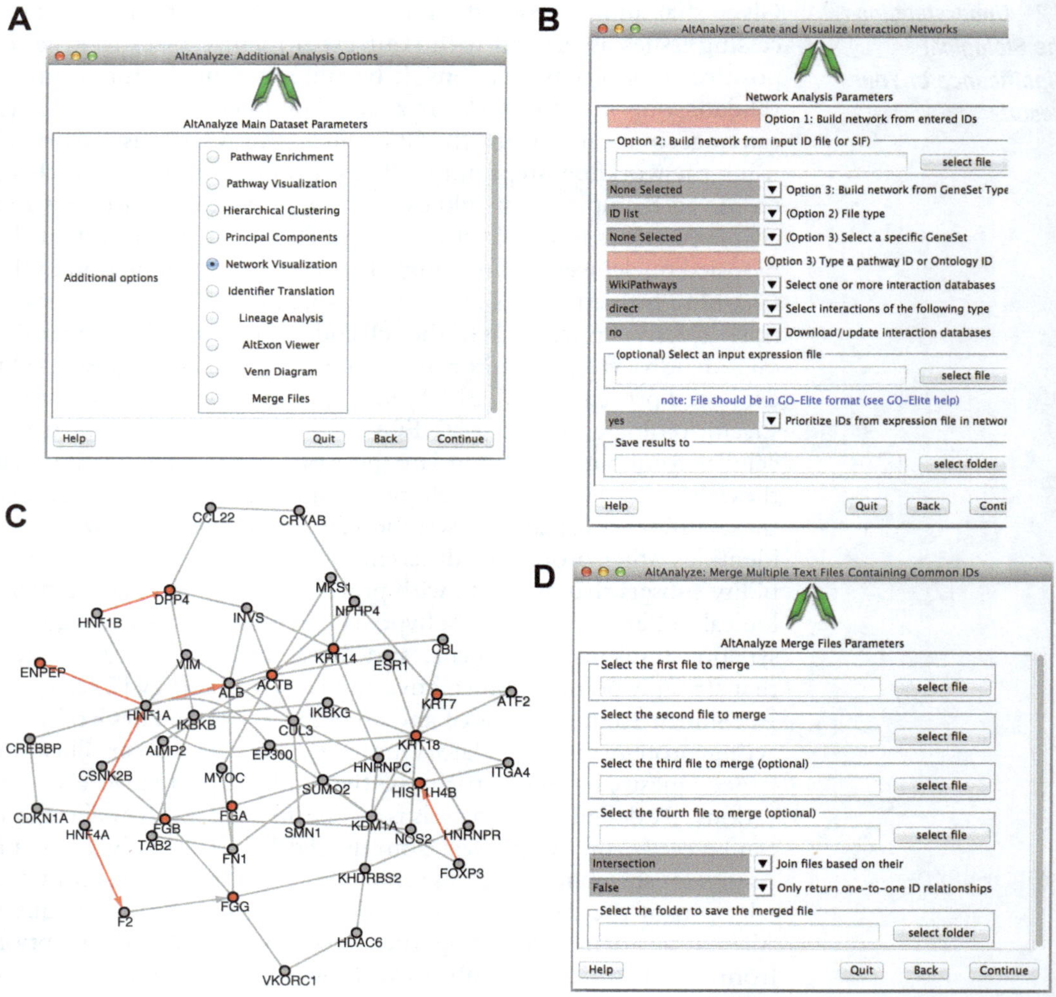

Fig. 8 Evaluating putative global protein and gene interaction networks derived from proteomics data. In addition to the automated workflows provided in AltAnalyze, users can perform accessory analyses for developing novel molecular regulatory or signaling hypotheses. (**a**) Using the Additional Analyses menu option, users selecting the analysis Network Visualization can query multiple genes or pathways of interest to obtain a molecular network representation of their data. (**b**) The associated software interface enables the selection of diverse protein–protein, protein–gene, protein–metabolite, and microRNA–gene interaction databases to evaluate direct or indirect interactions between a queried set of gene symbols (http:/altanalyze.readthedocs.io/en/latest/NetPerspective/). (**c**) An example output from the software NetPerspective is shown illustrating putative transcriptional regulatory interactions derived from prior annotated high-throughput chromatin immunoprecipitation analyses (*red arrows* = ChIP-Seq targets) and protein–protein interactions (BioGRID, WikiPathways, KEGG) for the AR kidney allograft group versus BK differentially expressed proteins. (**d**) Comparison of differing and common molecular profiles can be obtained using the Additional Analyses menu option Merge Files option in addition to the Venn diagram analysis tool

3. Under "Select interactions of the following type," select **shortest path**. This will look for any possible connections between the proteins in your regulated input file among the selected interaction databases. This analysis is particularly useful for small regulated gene sets, as it provides an exhaustive search.
4. For the option "Select an input expression file," select the same file as in #1.
5. Select a folder to "Save results to."
6. Select the Continue button to obtain all relevant protein interactions in the form of a colored network graph (Fig. 8c).

By comparing the different networks produced with distinct morbidities, one can begin to define the distinct proteomes and regulatory interactions defined by each disease regulated set.

7.3 Comparison of Distinct Results

Given that certain genes will be present in multiple morbidities relative to patients with stable grafts, it can be informative to repeat this process for genes shared among these different morbidities. To accomplish this prior to applying this tool to the urine proteomics data, we can select another option within the **Additional Analyses** window, called **Merge Files**, to identify a restricted set of genes for comparison. In this case, we wish to identify the common set of proteins regulated by any form of allograft injury. To do this:

1. From the **Merge Files** menu, select each of the renal transplantation morbidities versus stable from the folder GO-Elite/upregulated/ (AR, BK, CAI).
2. Leave the default option of "Join files based on their" to be equal to Intersection. Changing this to Union would return all proteins in each file, indicating any possible matches.
3. Select a folder to save the results to and then Continue.

This analysis will produce a file named MergedFiles.txt. This file can be directly run in NetPerspective as was the other GO-Elite input file previously, to identify interactions and proteins common to all forms of injury. The overlap of these different proteins can be further visualized in the **Additional Analyses** menu named Venn diagram using the same options present in the **Merge Files** menu (Fig. 7c).

We can compare the SIDS and urine proteomics datasets using these same options. Prior to do this, we need to translate the protein identifiers from the two different experiments to a common identifier system. Fortunately, this can be easily accomplished using the **Identifier Translation** option in the **Additional Analyses** menu. To do this you must:

1. Restart AltAnalyze and select a single proteomics input file.
2. Select the **Identifier Translation** menu and then select the file to translate (original dataset or GO-Elite regulated results file).
3. Select the identifier (ID) system present in the input file (first menu selection option).
4. Select a desired output ID system (e.g., Ensembl).
5. Select "Continue" to generate the file.

Once completed, a new file will be exported with the same name as the original file but containing a suffix that indicates the new ID system to the same location as the input file. The first column of this new file will contain the translated ID system IDs. Once this translation has been applied to both of the proteomics data files, these can be directly compared using the **Merge Files** function.

8 Creating a "Big Picture" from Omics Data

The above sections outline a complete workflow for the analysis of multiple proteomics datasets in tandem. The network and pathway results from these can be further used to create new reference pathway representations to be used by the greater research community. An example of this has been done with the SIDS proteomics dataset, to expand upon an existing theoretical model of SIDS genetic and molecular interactions using the open-access resource WikiPathways (Fig. 1) [27]. The example at http://www.wikipathways.org/index.php/Pathway:WP706 includes the proteomics observations obtained from the above illustrated pathway analysis. Such pathways can be further improved by an individual who signs up for a WikiPathways account. Once edited, these pathways become immediately available for GO-Elite pathway analysis in AltAnalyze and pathway visualization.

9 Conclusions and Future Directions: Going Beyond the Existing Tools

It is important to note that this chapter describes one possible set of workflows to analyze proteomics data, with many other tools providing valuable alternatives for the analysis and interpretation of such data. However, the described approaches provide many conceptual advantages for the integrated analysis of one or more datasets. These include the automated statistical analysis and visualization of data without the use of multiple analysis tools, robust and informative methods for pathway and cellular inference, and the creation of network models from which to derive new testable hypotheses. We anticipate multiple extensions to these individual tools in the future, as new methods emerge for increasing the accuracy and depth to which peptides are quantified.

References

1. Hasin Y, Seldin M, Lusis A (2017) Multi-omics approaches to disease. Genome Biol 18(1):83. https://doi.org/10.1186/s13059-017-1215-1. PubMed PMID: 28476144; PMCID: PMC5418815
2. Yan J, Risacher SL, Shen L, Saykin AJ (2017) Network approaches to systems biology analysis of complex disease: integrative methods for multi-omics data. Brief Bioinform. https://doi.org/10.1093/bib/bbx066. PubMed PMID: 28679163
3. Cisek K, Krochmal M, Klein J, Mischak H (2016) The application of multi-omics and systems biology to identify therapeutic targets in chronic kidney disease. Nephrol Dial Transplant 31(12):2003–2011. https://doi.org/10.1093/ndt/gfv364. PubMed PMID: 26487673
4. Sung J, Wang Y, Chandrasekaran S, Witten DM, Price ND (2012) Molecular signatures from omics data: from chaos to consensus. Biotechnol J 7(8):946–957. https://doi.org/10.1002/biot.201100305. PubMed PMID: 22528809; PMCID: PMC3418428
5. Subramanian A, Tamayo P, Mootha VK, Mukherjee S, Ebert BL, Gillette MA, Paulovich A, Pomeroy SL, Golub TR, Lander ES, Mesirov JP (2005) Gene set enrichment analysis: a knowledge-based approach for interpreting genome-wide expression profiles. Proc Natl Acad Sci U S A 102(43):15545–15550. https://doi.org/10.1073/pnas.0506580102. PubMed PMID: 16199517; PMCID: PMC1239896
6. Consortium EP (2012) An integrated encyclopedia of DNA elements in the human genome. Nature 489(7414):57–74. https://doi.org/10.1038/nature11247. PubMed PMID: 22955616; PMCID: PMC3439153
7. Cancer Genome Atlas Research Network (2008) Comprehensive genomic characterization defines human glioblastoma genes and core pathways. Nature 455(7216):1061–1068. https://doi.org/10.1038/nature07385. PubMed PMID: 18772890; PMCID: PMC2671642
8. Kohler S, Vasilevsky NA, Engelstad M, Foster E, McMurry J, Ayme S, Baynam G, Bello SM, Boerkoel CF, Boycott KM, Brudno M, Buske OJ, Chinnery PF, Cipriani V, Connell LE, Dawkins HJ, DeMare LE, Devereau AD, de Vries BB, Firth HV, Freson K, Greene D, Hamosh A, Helbig I, Hum C, Jahn JA, James R, Krause R, SJ FL, Lochmuller H, Lyon GJ, Ogishima S, Olry A, Ouwehand WH, Pontikos N, Rath A, Schaefer F, Scott RH, Segal M, Sergouniotis PI, Sever R, Smith CL, Straub V, Thompson R, Turner C, Turro E, Veltman MW, Vulliamy T, Yu J, von Ziegenweidt J, Zankl A, Zuchner S, Zemojtel T, Jacobsen JO, Groza T, Smedley D, Mungall CJ, Haendel M, Robinson PN (2017) The human phenotype ontology in 2017. Nucleic Acids Res 45(D1):D865–DD76. https://doi.org/10.1093/nar/gkw1039. PubMed PMID: 27899602; PMCID: PMC5210535
9. Schriml LM, Mitraka E (2015) The disease ontology: fostering interoperability between biological and clinical human disease-related data. Mamm Genome 26(9–10):584–589. https://doi.org/10.1007/s00335-015-9576-9. PubMed PMID: 26093607; PMCID: PMC4602048
10. Cerami EG, Gross BE, Demir E, Rodchenkov I, Babur O, Anwar N, Schultz N, Bader GD, Sander C (2011) Pathway commons, a web resource for biological pathway data. Nucleic Acids Res 39(Database issue):D685–D690. https://doi.org/10.1093/nar/gkq1039. PubMed PMID: 21071392; PMCID: PMC3013659
11. Pico AR, Kelder T, van Iersel MP, Hanspers K, Conklin BR, Evelo C (2008) WikiPathways: pathway editing for the people. PLoS Biol 6(7):e184. https://doi.org/10.1371/journal.pbio.0060184. PubMed PMID: 18651794; PMCID: PMC2475545
12. Ashburner M, Ball CA, Blake JA, Botstein D, Butler H, Cherry JM, Davis AP, Dolinski K, Dwight SS, Eppig JT, Harris MA, Hill DP, Issel-Tarver L, Kasarskis A, Lewis S, Matese JC, Richardson JE, Ringwald M, Rubin GM, Sherlock G (2000) Gene ontology: tool for the unification of biology. The Gene Ontology Consortium. Nat Genet 25(1):25–29. https://doi.org/10.1038/75556. PubMed PMID: 10802651; PMCID: PMC3037419
13. Ogata H, Goto S, Fujibuchi W, Kanehisa M (1998) Computation with the KEGG pathway

database. Biosystems 47(1–2):119–128. PubMed PMID: 9715755

14. Afgan E, Baker D, van den Beek M, Blankenberg D, Bouvier D, Cech M, Chilton J, Clements D, Coraor N, Eberhard C, Gruning B, Guerler A, Hillman-Jackson J, Von Kuster G, Rasche E, Soranzo N, Turaga N, Taylor J, Nekrutenko A, Goecks J (2016) The Galaxy platform for accessible, reproducible and collaborative biomedical analyses: 2016 update. Nucleic Acids Res 44(W1): W3–W10. https://doi.org/10.1093/nar/gkw343. PubMed PMID: 27137889; PMCID: PMC4987906
15. Emig D, Salomonis N, Baumbach J, Lengauer T, Conklin BR, Albrecht M (2010) AltAnalyze and DomainGraph: analyzing and visualizing exon expression data. Nucleic Acids Res 38(Web Server issue):W755–W762. https://doi.org/10.1093/nar/gkq405. PubMed PMID: 20513647; PMCID: PMC2896198
16. Kuehn H, Liberzon A, Reich M, Mesirov JP (2008) Using GenePattern for gene expression analysis. Curr Protoc Bioinformatics. Chapter 7:Unit 7 12. doi: https://doi.org/10.1002/0471250953.bi0712s22. PubMed PMID: 18551415; PMCID: PMC3893799
17. Gatto L, Christoforou A (2014) Using R and bioconductor for proteomics data analysis. Biochim Biophys Acta 1844(1 Pt A):42–51. https://doi.org/10.1016/j.bbapap.2013.04.032. PubMed PMID: 23692960
18. McLendon PM, Davis G, Gulick J, Singh SR, Xu N, Salomonis N, Molkentin JD, Robbins J (2017) An unbiased high-throughput screen to identify novel effectors that impact on cardiomyocyte aggregate levels. Circ Res 121(6):604–616. https://doi.org/10.1161/CIRCRESAHA.117.310945. PubMed PMID: 28655832; PMCID: PMC5581213
19. Rindler TN, Hinton RB, Salomonis N, Ware SM (2017) Molecular characterization of pediatric restrictive cardiomyopathy from integrative genomics. Sci Rep 7:39276. https://doi.org/10.1038/srep39276. PubMed PMID: 28098235; PMCID: PMC5241776
20. Olsson A, Venkatasubramanian M, Chaudhri VK, Aronow BJ, Salomonis N, Singh H, Grimes HL (2016) Single-cell analysis of mixed-lineage states leading to a binary cell fate choice. Nature 537(7622):698–702. https://doi.org/10.1038/nature19348. PubMed PMID: 27580035; PMCID: PMC5161694
21. Gao Y, Wang J, Zheng Y, Zhang J, Chen S, Zhao F (2016) Comprehensive identification of internal structure and alternative splicing events in circular RNAs. Nat Commun 7:12060. https://doi.org/10.1038/ncomms12060. PubMed PMID: 27350239; PMCID: PMC4931246
22. Soreq L, Guffanti A, Salomonis N, Simchovitz A, Israel Z, Bergman H, Soreq H (2014) Long non-coding RNA and alternative splicing modulations in Parkinson's leukocytes identified by RNA sequencing. PLoS Comput Biol 10(3):e1003517. https://doi.org/10.1371/journal.pcbi.1003517. PubMed PMID: 24651478; PMCID: PMC3961179
23. Gan Z, Wang J, Salomonis N, Stowe JC, Haddad GG, McCulloch AD, Altintas I, Zambon AC (2014) MAAMD: a workflow to standardize meta-analyses and comparison of affymetrix microarray data. BMC Bioinformatics 15:69. https://doi.org/10.1186/1471-2105-15-69. PubMed PMID: 24621103; PMCID: PMC3975178
24. Sigdel TK, Salomonis N, Nicora CD, Ryu S, He J, Dinh V, Orton DJ, Moore RJ, Hsieh SC, Dai H, Thien-Vu M, Xiao W, Smith RD, Qian WJ, Camp DG 2nd, Sarwal MM (2014) The identification of novel potential injury mechanisms and candidate biomarkers in renal allograft rejection by quantitative proteomics. Mol Cell Proteomics 13(2):621–631. https://doi.org/10.1074/mcp.M113.030577. PubMed PMID: 24335474; PMCID: PMC3916658
25. Zambon AC, Gaj S, Ho I, Hanspers K, Vranizan K, Evelo CT, Conklin BR, Pico AR, Salomonis N (2012) GO-Elite: a flexible solution for pathway and ontology over-representation. Bioinformatics 28(16):2209–2210. https://doi.org/10.1093/bioinformatics/bts366. PubMed PMID: 22743224; PMCID: PMC3413395
26. Broadbelt KG, Rivera KD, Paterson DS, Duncan JR, Trachtenberg FL, Paulo JA, Stapels MD, Borenstein NS, Belliveau RA, Haas EA, Stanley C, Krous HF, Steen H, Kinney HC (2012) Brainstem deficiency of the 14-3-3 regulator of serotonin synthesis: a proteomics analysis in the sudden infant death syndrome. Mol Cell Proteomics 11(1):M111 009530. https://doi.org/10.1074/mcp.M111.009530. PubMed PMID: 21976671; PMCID: PMC3270099
27. Salomonis N (2014) Systems-level perspective of sudden infant death syndrome. Pediatr Res 76(3):220–229. https://doi.org/10.1038/pr.

2014.90. PubMed PMID: 24964230; PMCID: PMC4193964

28. Bright FM, Byard RW, Vink R, Paterson DS (2017) Medullary serotonin neuron abnormalities in an Australian cohort of sudden infant death syndrome. J Neuropathol Exp Neurol 76 (10):864–873. https://doi.org/10.1093/jnen/nlx071. PubMed PMID: 28922849
29. Haynes RL, Frelinger AL 3rd, Giles EK, Goldstein RD, Tran H, Kozakewich HP, Haas EA, Gerrits AJ, Mena OJ, Trachtenberg FL, Paterson DS, Berry GT, Adeli K, Kinney HC, Michelson AD (2017) High serum serotonin in sudden infant death syndrome. Proc Natl Acad Sci U S A 114(29):7695–7700. https://doi.org/10.1073/pnas.1617374114. PubMed PMID: 28674018; PMCID: PMC5530643
30. MacFarlane PM, Mayer CA, Litvin DG (2016) Microglia modulate brainstem serotonergic expression following neonatal sustained hypoxia exposure: implications for sudden infant death syndrome. J Physiol 594(11):3079–3094. https://doi.org/10.1113/JP271845. PubMed PMID: 26659585; PMCID: PMC4887678
31. Duncan JR, Paterson DS, Hoffman JM, Mokler DJ, Borenstein NS, Belliveau RA, Krous HF, Haas EA, Stanley C, Nattie EE, Trachtenberg FL, Kinney HC (2010) Brainstem serotonergic deficiency in sudden infant death syndrome. JAMA 303(5):430–437. https://doi.org/10.1001/jama.2010.45. PubMed PMID: 20124538; PMCID: PMC3242415
32. Bittremieux W, Tabb DL, Impens F, Staes A, Timmerman E, Martens L, Laukens K (2017) Quality control in mass spectrometry-based proteomics. Mass Spectrom Rev. https://doi.org/10.1002/mas.21544. PubMed PMID: 28802010
33. Bittremieux W, Valkenborg D, Martens L, Laukens K (2017) Computational quality control tools for mass spectrometry proteomics. Proteomics 17(3–4). https://doi.org/10.1002/pmic.201600159. PubMed PMID: 27549080
34. Tabb DL (2013) Quality assessment for clinical proteomics. Clin Biochem 46(6):411–420. https://doi.org/10.1016/j.clinbiochem.2012.12.003. PubMed PMID: 23246537; PMCID: PMC3602213
35. Leek JT, Johnson WE, Parker HS, Jaffe AE, Storey JD (2012) The sva package for removing batch effects and other unwanted variation in high-throughput experiments. Bioinformatics 28(6):882–883. https://doi.org/10.1093/bioinformatics/bts034. PubMed PMID: 22257669; PMCID: PMC3307112

Methods in Molecular Biology (2018) 1788: 113–121
DOI 10.1007/7651_2017_74

Published online: 17 October 2017

Untargeted Screening of Urinary Peptides Using Offline Nano-Liquid Chromatography: MALDI-TOF/TOF Mass Spectrometry

François-Ludovic Sauvage, Sébastien Passeron, and Pierre Marquet

Abstract

In renal transplantation, the discovery of early urine biomarkers of graft lesions would be useful in helping physicians to improve patient care and minimize the use of invasive techniques such as biopsies. Over the last years, high-resolution mass spectrometry has been used extensively for the search of biomarkers in various biological fluids. Here we describe a procedure based on reverse-phase nano-HPLC, offline plate spotting, and MALDI-TOF and TOF/TOF applied in our laboratory for the search of natural peptides in urine samples from renal transplant patients.

Keywords: Biomarkers, Nano-liquid chromatography, Renal transplantation, Solid-phase extraction, Time-of-flight-mass spectrometry

1 Introduction

In renal transplantation, noninvasive, precocious urine biomarkers could be useful for both the diagnosis and prognosis of graft lesions, replacing the kidney graft biopsy, which is the current, although imperfect, gold standard [1–5]. Various analytical techniques have been developed for the untargeted detection and identification of potential biomarkers, particularly mass spectrometry [6–17]. For instance, using capillary electrophoresis coupled with mass spectrometry (CE-MS), Siwy et al. identified almost 1000 urinary peptides [11]. Other teams have used Matrix-Assisted Laser Desorption/Ionization (MALDI) mass spectrometry for the discovery of disease biomarkers [9, 13–14]. MALDI relies on the desorption by laser irradiation at a specified wavelength ($\lambda = 355$ nm) of crystallized spots, obtained after mixing the sample in liquid form (most frequently the effluent of an analytical column) with a specific matrix (for example α-cyano-4-hydroxycinnamic acid (CHCA) for peptidomic applications) able to transform photonic into electronic energy and facilitate ion transfer to the compounds of interest. Contrary to electrospray ionization (ESI), MALDI mostly generates mono-charged ions. Consequently, and also due to both the high mass resolution and precision inherent to the time-of-flight (TOF) technology, the peptide mass fingerprints obtained are easier to identify in spectra libraries than those issued from fragmentation of multi-charged ions as obtained with ESI.

Contrary to serum samples for which depletion of the major proteins is recommended, various sample extraction procedures can be applied to urine samples for proteomic and peptidomic analyses (e.g., solid-phase extraction [13–14], selective peptide extraction using magnetic beads [14], or only matrix desalting using PD-10 columns [15]). Recently, Sigdel et al. proposed to optimize sample preparation for urine peptidomic analysis by using modified solid-phase extraction (mSPE), defined as classical SPE followed by purification on a second column [16]. Court et al. proposed a standardized methodology for the urinary proteome, which is unfortunately not applicable to the analysis of natural peptides [12]. Indeed, procedures such as protein precipitation, sample denaturation, or enzymatic digestion may be detrimental when looking for natural, nontryptic peptides. In our laboratory, we first developed a solid-phase extraction procedure using C2 cartridges [13]. Furthermore, as this method was intended for the search of urinary biomarkers in renal transplantation, we assessed whether it was quantitative for a maximum of peptides commonly found in urine samples from kidney transplant recipients. For this, using liquid chromatography-tandem mass spectrometry [17], we monitored several peptides belonging to abundant proteins such as collagens and uromodulin, in order to evaluate the statistical performance of the extraction procedure. Finally, in order to have access to high-resolution mass spectrometry for the untargeted screening of urinary peptides as potential biomarkers in kidney graft recipients, we chose to use nano-liquid chromatography and MALDI-TOF(/TOF). Here we describe this method based on unspecific solid-phase extraction, reverse-phase nano-HPLC, off-line plate spotting, MALDI-TOF, and TOF/TOF.

2 Materials

Prepare all solutions using ultrapure water (prepared by purifying deionized water to attain a resistivity of 18 MΩ at 25 °C) and analytical grade reagents. Prepare and store all reagents at room temperature (unless indicated otherwise).

2.1 Reagents

2.1.1 Reagents for Solid-Phase Extraction

1. *Phosphoric acid solution (10%, by volume)*: Mix 450 mL of water and 50 μL of phosphoric acid (85%).
2. *Trifluoroacetic acid solution (1%, by volume)*: Mix 495 mL of water and 5 mL of trifluoroacetic acid (>99% pure).

3. *Trifluoroacetic acid solution in acetonitrile (3%, by volume)*: Mix 485 mL of acetonitrile and 15 mL of trifluoroacetic acid (>99% pure).
4. *Trifluoroacetic acid solution in acetonitrile (0.5%, by volume)*: Mix 497.5 mL of acetonitrile and 2.5 mL of trifluoroacetic acid (>99% pure).
5. *Trifluoroacetic acid solution (0.5%, by volume)*: Mix 497.5 mL of water and 2.5 mL of trifluoroacetic acid (>99% pure).
6. *Elution solution 1*: Mix 180 mL of the trifluoroacetic acid solution in acetonitrile (3%, by volume) and 20 mL of water.
7. *Elution solution 2*: Mix 100 mL of the trifluoroacetic acid solution in acetonitrile (0.5%, by volume) and 100 mL of trifluoroacetic acid solution (0.5%, by volume).
8. *Acetonitrile: water (50:50, by volume) solution*: Mix 100 mL of acetonitrile and 100 mL of water.
9. *Dilution solution*: Mix 99.9 mL of the acetonitrile–water (50:50, by volume) solution and 100 μL of formic acid (>98% pure) (see **Note 1**).

2.1.2 Reagents for Probot Spotting

10. *MALDI internal mass calibrator*: 630 nmol/L Glu-1-Fibrinopeptide B (Glu-Fib). Weigh 1 mg of Glu-Fib (Sigma, or equivalent) in a brown recipient and add 1000 mL of water. Mix until a perfect dissolution is observed. Aliquot and then store at −20 °C (see **Note 2**).
11. *CHCA dilution solution*: Mix 35 mL of Acetonitrile, 15 mL of water and 25 μL of trifluoroacetic acid (see **Note 3**).
12. *CHCA matrix solution*: In a 1.5 mL-brown Eppendorf, add 1.5 mL of CHCA dilution solution to 5 mg CHCA. Mix until a perfect dissolution is observed. Store at −20 °C until further use.
13. Prepare extemporaneously a 6.3 nmol/L Glu-Fib aqueous solution by mixing 5 μL of the 630 nmol/L Glu-Fib solution with 495 μL of water (see **Note 4**).
14. *Spotting solution*: Add 50 μL of the 6.3 nmol/L Glu-Fib aqueous solution to 950 μL of CHCA matrix solution (see **Note 5**).

2.1.3 Reagents for Nano-HPLC Analysis

15. *Mobile phase A (MP-A)*: Mix 980 mL of water, 20 mL of acetonitrile and 500 μL of trifluoroacetic acid.
16. *Mobile phase B (MP-B)*: Mix 500 mL of water, 500 mL of acetonitrile and 500 μL of trifluoroacetic acid.
17. *Mobile phase C (MP-C)*: Mix 200 mL of water, 800 mL of acetonitrile and 500 μL of trifluoroacetic acid.

18. *Pepmix 4 calibration solution*: Mix 10 μL of Pepmix 4 (LaserBioLabs, or equivalent) with 90 μL of CHCA Matrix solution. Vortex-mix and store at −20 °C until use.

2.2 Equipment

1. Sep-Pak Vac 3 (500 mg) tC_2 extraction cartridges.
2. C18 PepMap 100 (0.5 cm × 300 μm i.d.) trapping column.
3. Acclaim PepMap 100 (15 cm × 75 μm i.d.) analytical column.
4. Ultimate 3000 nano-high-pressure gradient pumping system or equivalent.
5. Probot plate-spotting system, or equivalent.
6. 4800 MALDI-TOF/TOF (ABSciex) mass spectrometer, or equivalent.

3 Methods

3.1 Sample Extraction

1. Install Sep-Pak tC_2 cartridges on a vacuum manifold.
2. Add 2.5 mL methanol and then open the tap. Let the drops flow with gravity (see **Note 6**).
3. Add 2.5 mL of the phosphoric acid solution. Let the drops flow with gravity (see **Note 7**).
4. Mix 2.5 mL of the phosphoric acid solution with a 2.5 mL thawed urine sample (see **Note 8**).
5. Apply the mixture to the cartridge. Let the drops flow with gravity.
6. Add 2 × 3 mL of the trifluoroacetic acid solution (1%, by volume). Let the drops flow with gravity.
7. Dry the cartridge under vacuum for at least 15 min (see **Note 9**).
8. Put a glass tube under each cartridge and elute with 3 mL of the elution solution 1. Let the drops flow with gravity (see **Note 10**).
9. Add 1.5 mL of the elution solution 2. Let the drops flow with gravity.
10. Dry the cartridge under vacuum for 1 min.
11. Evaporate the eluate at 40 °C with a low flow-rate of nitrogen (see **Note 11**).
12. Add 100 μL of the dilution solution. Vortex-mix for 30 s (see **Note 12**).
13. Transfer the solution to an injection vial (see **Note 13**).

3.2 Nano-Liquid Chromatography

1. Install a C18 PepMap 100 (0.5 cm × 300 μm i.d.) (trapping column) and an Acclaim PepMap 100 (15 cm × 75 μm i.d.) (elution column) on the chromatographic system.
2. Apply, on the trapping column, a constant flow-rate of 20 μL/min of MP-A.
3. Apply, on the elution column, the following gradient of MP-B and MP-C in MP-A, at ca. 30 °C and a constant flow-rate of 250 nL/min:
   (a) *0–5 min*: 2.5% MP-B, 2.5% MP-C.
   (b) *5–25 min*: linear increase from 2.5 to 12.5% MP-B, 2.5% MP-C.
   (c) *25–45 min*: linear increase from 12.5 to 20% MP-B, 2.5% MP-C.
   (d) *45–145 min*: linear increase from 20 to 40% MP-B, linear increase from 2.5 to 20% MP-C.
   (e) *145–161 min*: linear decrease from 40 to 5% MP-B, linear increase from 20 to 95% MP-C.
   (f) *161–167 min*: 5% B, 95% MP-C.
   (g) *167–168 min*: linear decrease from 5 to 2.5% MP-B, linear decrease from 95 to 2.5% MP-C.
   (h) *168–210 min*: equilibration with 2.5% MP-B, 2.5% MP-C.
4. Inject 2 μL sample extract.

3.3 Plate Spotting

1. Install 2 Opti-TOF LC/MALDI Inserts (123 × 81 mm) in the spotting system (see **Note 14**).
2. Mix the efflux of the elution column with the spotting solution delivered at a flow-rate of 500 nL/min.
3. Align the MALDI plates and the spotting needle using μCarrier, or equivalent software.
4. Start spotting 35 min after the extract injection, by fraction of 12 s.
5. Stop spotting 155 min after the injection, so that 600 spots per sample are obtained (see **Note 15**).

3.4 Time-of-Flight-Mass Spectrometry (TOF-MS)

3.4.1 TOF-MS Scan

1. TOF-MS calibration:
   (a) Spot 0.8 μL of a solution containing the Pepmix 4 calibration solution on the eight dedicated spots on the plate (see **Note 16**).
   (b) Check the calibration of the mass spectrometer by verifying the mass accuracy of the peptides, with a maximum error of 5 ppm.

2. TOF-MS analytical mode
   (a) Acquisition method:
      - Fixed Laser Intensity: 3200 (arbitrary units) (see **Note 17**),
      - Scan from 800 to 6000 amu (see **Note 18**).
   (b) Processing method:
      - Min S/N ratio: 25.
      - Internal calibration using Glu-Fib (m/z 1570.677), with a maximum error of 5 ppm (see **Note 19**).
   (c) Interpretation method:
      - Select a maximum of 100 precursors per spot.
      - Minimum S/N filter: 15.
      - Minimum chromatographic peak width: 2 spots.

3.4.2 TOF-MS/MS Scan

1. TOF-MS/MS calibration:
   (a) Use the spots prepared for MS calibration.
   (b) Check calibration by verifying the mass accuracy of the fragments obtained after fragmentation of ACTH [18–39] at an m/z of 2465.199 with a maximum error of 25 ppm.

3.5 Library Search

1. Using the Protein Pilot software, select "Identify Proteins" (A), "LC" (B) and "Add TOF/TOF Data" (C), and choose the data to be used for the library search (see Fig. 1),
2. Select the search algorithm to be used by clicking, in the "Process Using Window", either on "Paragon method" (D) or "Mascot method" (E) (see Fig. 1),
3. Check the search constraints by clicking on "Edit" (F) (see Fig. 1),
4. Select an appropriate directory and name for the result file to be saved (G) (see Fig. 1),
5. Perform library search by clicking on "Process" (H) (see Fig. 1).
6. Evaluate the level of confidence for the proposed peptides using both Paragon (Window A) and Mascot (Window B) algorithms (see Fig. 2).

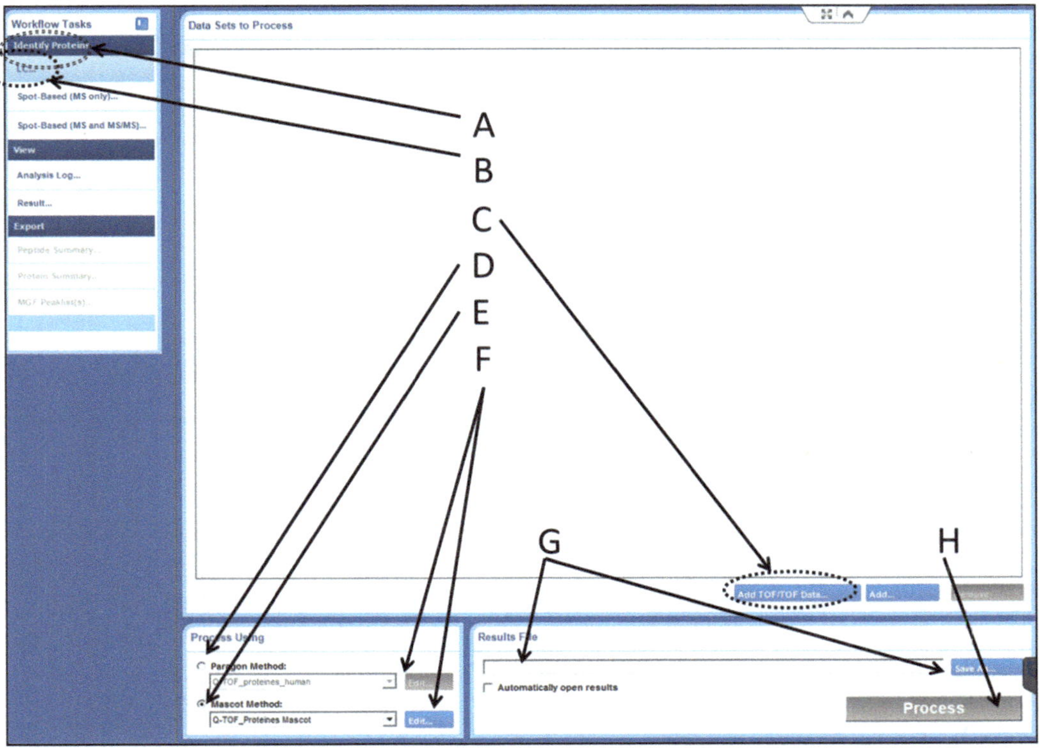

Fig. 1 (A) "Identify proteins" button, (B) "LC" button, (C) "Add TOF/TOF Data" button, (D) "Paragon® method" option, (E) "Mascot® method" option, (F) "Method edition," (G) "Results file" option, (H) "Process" button

4 Notes

1. Formic acid can be used either for MALDI or ESI ionization.
2. Smaller weights could be used depending on the precision of the balances used. Aliquot in small volumes (e.g., lower than 25 μL). Too many freeze-thaw cycles can affect the quality of the solution.
3. Wash the needle used for the spotting with this solution.
4. This solution can also be stored at −20 °C. However, the intensity of a peptide with m/z 1552.677 characterizing the loss of water on the Glu-Fib peptide increases after several freeze-thaw cycles.
5. Wait at least 1 min before opening the tap.
6. Close the tap after the end of water elution.
7. The thawed samples need to be centrifuged. Use the supernatant for the analysis.
8. A longer drying time allows evaporation of the remaining water after elution.

A

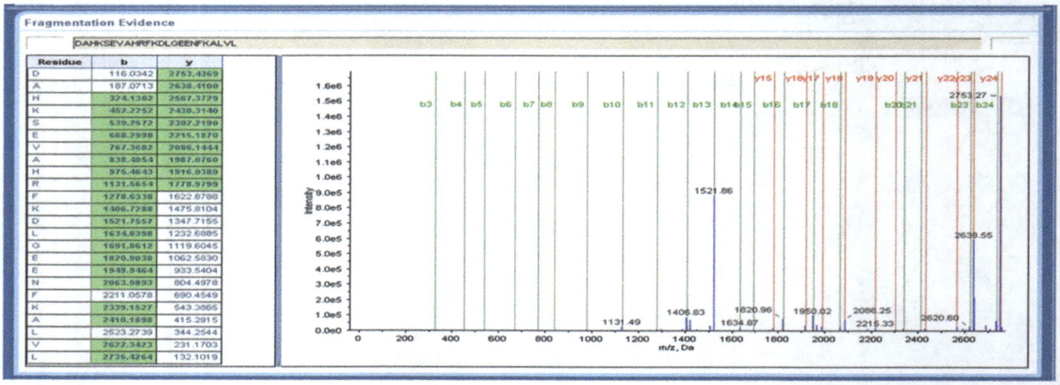

B

Monoisotopic mass of neutral peptide Mr(calc): 2752.4296
Ions Score: 150 Expect: 1.1e-011
Matches (Bold Red): 39/417 fragment ions using 56 most intense peaks

#	Immon.	a	a*	a0	b	b*	b0	d	Seq.	v	w	y	y*	y0	#
1	88.0393	88.0393		70.0287	116.0342		98.0237		D						24
2	44.0495	159.0764		141.0659	187.0713		169.0608		A	**2622.3787**		**2638.4100**	2621.3834	**2620.3994**	23
3	**110.0713**	296.1353		278.1248	**324.1302**		306.1197		H	2485.3198		**2567.3729**	2550.3463	2549.3623	22
4	101.1073	424.2303	407.2037	406.2197	**452.2252**	435.1987	434.2146		K	2357.2248	2356.2296	**2430.3140**	2413.2874	2412.3034	21
5	60.0444	511.2623	494.2358	493.2518	**539.2572**	522.2307	521.2467		S	2270.1928	2269.1975	**2302.2190**	2285.1925	2284.2084	20
6	102.0550	640.3049	623.2784	622.2943	**668.2998**	651.2733	650.2893		E	2141.1502	2140.1550	**2215.1870**	2198.1604	2197.1764	19
7	72.0808	739.3733	722.3468	721.3628	**767.3682**	750.3417	749.3577		V	2042.0818	2055.1022	**2086.1444**	2069.1178	2068.1338	18
8	44.0495	810.4104	793.3839	792.3999	**838.4054**	821.3788	820.3948		A	1971.0447		1987.0760	1970.0494	1969.0654	17
9	**110.0713**	947.4694	930.4428	929.4588	**975.4643**	958.4377	957.4537		H	1833.9858		**1916.0389**	1899.0123	1898.0283	16
10	**129.1135**	1103.5705	1086.5439	1085.5599	**1131.5654**	**1114.5388**	1113.5548	1018.5065	R	1677.8846	1676.8894	**1778.9799**	1761.9534	1760.9694	15
11	120.0808	1250.6389	1233.6123	1232.6283	**1278.6338**	1261.6072	1260.6232		F			1622.8788	1605.8523	1604.8683	14
12	101.1073	1378.7338	1361.7073	1360.7233	**1406.7288**	**1389.7022**	1388.7182	1321.6760	K			1475.8104	1458.7839	1457.7999	13
13	88.0393	1493.7608	1476.7342	1475.7502	**1521.7557**	1504.7292	**1503.7451**	1449.7710	D			1347.7155	1330.6889	1329.7049	12
14	86.0964	1606.8448	1589.8183	1588.8343	**1634.8398**	1617.8132	1616.8292	1564.7979	L			1232.6885	1215.6620	1214.6780	11
15	30.0338	1663.8663	1646.8398	1645.8557	**1691.8612**	1674.8347	1673.8507		G			1119.6045	1102.5779	1101.5939	10
16	102.0550	1792.9089	1775.8824	1774.8983	**1820.9038**	1803.8773	1802.8933	1734.9034	E			1062.5830	1045.5564	1044.5724	9
17	102.0550	1921.9515	1904.9250	1903.9409	**1949.9464**	1932.9199	1931.9358	1863.9460	E			933.5404	916.5138	915.5298	8
18	87.0553	2035.9944	2018.9679	2017.9839	**2063.9893**	2046.9628	2045.9788	1992.9886	N			804.4978	787.4713		7
19	120.0808	**2183.0628**	2166.0363	2165.0523	2211.0578	2194.0312	2193.0472		F			690.4549	673.4283		6
20	101.1073	2311.1578	2294.1313	2293.1472	**2339.1527**	2322.1262	2321.1422	2254.1000	K			543.3865	526.3599		5
21	44.0495	2382.1949	2365.1684	2364.1844	**2410.1898**	2393.1633	2392.1793		A			415.2915			4
22	86.0964	2495.2790	2478.2524	2477.2684	2523.2739	2506.2473	2505.2633	2453.2320	L			344.2544			3
23	72.0808	**2594.3474**	2577.3208	2576.3368	**2622.3423**	2605.3158	2604.3317	2580.3317	V			231.1703			2
24	86.0964								L			132.1019			1

Fig. 2 Typical results obtained using (**a**) Paragon® (confidence of 99%) and (**b**) Mascot® (score of 150) for the peptide sequence DAHKSEVAHRFKDLGEENFKALVL

9. As mentioned in **Note 6**, wait for at least 1 min before opening the tap. The color of the solid-phase changes and small bubbles can be observed in the discarded solvent.
10. Do not use higher temperatures to avoid possible degradation of the compounds.
11. Use acetonitrile in the dilution solution to allow the dissolution of lipophilic compounds.
12. Store injection vials at −80 °C before analysis. Thawed samples need to be centrifuged before injection to avoid overpressure on the columns.
13. Before use, wash plates with acetonitrile, water and acetonitrile again. Avoid dust by spraying air on plates.
14. As this kind of plate contains 1800 spots, three samples can be spotted on each one.

15. Let the spots dry before use.
16. Adapt the Laser Intensity on the intensity observed for the internal calibrator in each spot.
17. Natural peptides can indeed be longer than tryptic peptides.
18. Mass accuracy is checked for each spot using the internal calibrator.
19. Optimize fragmentation in order to obtain high enough intensity for both the lowest and highest mass fragments.

References

1. Pejcic M, Stojnev S, Stefanovic V (2010) Urinary proteomics—a tool for biomarker discovery. Ren Fail 32:259–268
2. Bestard O, Cruzado JM, la Franquesa M, Grinyo JM (2010) Biomarkers in renal transplantation. Curr Opin Organ Transplant 15:467–473
3. Alachkar N, Rabb H, Jaar BG (2011) Urinary biomarkers in acute kidney transplant dysfunction. Nephron Clin Pract 118:c173–c181
4. Rodríguez-Suárez E, Siwy J, Zürbig P, Mischak H (2014) Urine as a source for clinical proteome analysis: from discovery to clinical application. Biochim Biophys Acta 1(844):884–898
5. Ling XB, Mellins ED, Sylvester KG, Cohen HJ (2010) Urine peptidomics for clinical biomarker discovery. Adv Clin Chem 51:181–213
6. Metzger J, Schanstra JP, Mischak H (2009) Capillary electrophoresis-mass spectrometry in urinary proteome analysis: current applications and future developments. Anal Bioanal Chem 393:1431–1442
7. Mischak H, Coon JJ, Novak J, Weissinger EM, Schanstra JP, Dominiczak AF (2009) Capillary electrophoresis-mass spectrometry as a powerful tool in biomarker discovery and clinical diagnosis: an update of recent developments. Mass Spectrom Rev 28:703–724
8. Mischak H, Delles C, Klein J, Schanstra JP (2010) Urinary proteomics based on capillary electrophoresis—coupled mass spectrometry in kidney disease: discovery and validation of biomarkers, and clinical applications. Adv Chronic Kidney Dis 17:493–506
9. Ling XB, Sigdel TK, Lau K, Ying L, Lau I, Schilling J, Sarwal MM (2010) Integrative urinary peptidomics in renal transplantation identifies biomarkers for acute rejection. J Am Soc Nephrol 21:646–653
10. Mischak H, Schanstra JP (2011) CE-MS in biomarker discovery, validation, and clinical application. Proteomics Clin Appl 5:9–23
11. Siwy J, Mullen W, Golovko I, Franke J, Zurbig P (2011) Human urinary peptide database for multiple disease biomarker discovery. Proteomics Clin Appl 5:367–374
12. Court M, Selevsek N, Matondo M, Allory Y, Garin J, Masselon CD, Domon B (2011) Toward a standardized urine proteome analysis methodology. Proteomics 11:1160–1171
13. Benkali K, Marquet P, Rerolle JP, Le Meur Y, Gastinel L (2008) A new strategy for faster urinary biomarkers identification by Nano-LC-MALDI-TOF/TOF mass spectrometry. BMC Genomics 9:541
14. Quintana LF, Campistol JM, Alcolea MP, Banon-Maneus E, Sol-Gonzalez A, Cutillas PR (2009) Application of label-free quantitative peptidomics for the identification of urinary biomarkers of kidney chronic allograft dysfunction. Mol Cell Proteomics 8 (1):658–1673
15. Mischak H, Kolch W, Aivaliotis M, Bouyssié D, Court M, Dihazi H, Dihazi GH, Franke J, Garin J, Gonzalez de Peredo A, Iphöfer A, Jänsch L, Lacroix C, Makridakis M, Masselon C, Metzger J, Monsarrat B, Mrug M, Norling M, Novak J, Pich A, Pitt A, Bongcam-Rudloff E, Siwy J, Suzuki H, Thongboonkerd V, Wang LS, Zoidakis J, Zürbig P, Schanstra JP, Vlahou A (2010) Comprehensive human urine standards for comparability and standardization in clinical proteome analysis. Proteomics Clin Appl 4:464–478
16. Sigdel TK, Nicora CD, Hsieh SC, Dai H, Qian WJ, Camp DG II, Sarwal MM (2014) Optimization for peptide sample preparation for urine peptidomics. Clin Proteomics 11:7
17. Sauvage FL, Gastinel LN, Marquet P (2012) Untargeted screening of urinary peptides with liquid chromatography coupled to hybrid linear-ion trap tandem mass spectrometry. J Chromatogr A 1(259):138–147

Methods in Molecular Biology (2018) 1788: 123–129
DOI 10.1007/7651_2017_75

Published online: 23 March 2018

Identifying Clinically Relevant Proteins for Targeted Analysis in the Development of a Multiplexed Proteomic Biomarker Assay

Emily M. Mackay and Oliver F. Bathe

Abstract

In recent years, hundreds of candidate protein biomarkers have been identified using discovery-based proteomics. Despite the large number of candidate biomarkers, few proteins advance to clinical validation. Here, we describe a hypothesis driven approach to identify candidate biomarkers, previously characterized in the literature, with the highest probability of clinical applicability. A ranking method, the hypothesis directed biomarker ranking (HDBR) system, was developed to score candidate biomarkers based on seven criteria deemed important in the selection of clinically useful biomarkers. The HDBR system was initially applied to identify candidate biomarkers for the development of a diagnostic test for the early detection of colorectal cancer, but this system can be widely applied to identify biomarkers of relevance in different disease states, for diagnosis, prognostication, or any other specific purpose.

Keywords Candidate biomarkers, Diagnostic biomarkers, Oxford levels of evidence, Prognostic biomarkers, Proteomics, Therapeutic response biomarkers

1 Introduction

The field of cancer biomarker research for detection, prognostication, and therapeutic prediction has advanced rapidly in recent years due to improvements in analytical technologies. The emergence of "omic" platforms has resulted in the development of biomarkers based on multiple analytes. Gene and transcript biomarkers have been particularly prominent in recent years. However, the abundance of mRNA does not necessarily correlate with the amount of functional protein expressed, due to differences in post-transcriptional processing, translation and protein degradation [1, 2]. Therefore, protein markers may represent a more accurate reflection of the pathophysiological state, the "phenotype."

The large dynamic range of protein abundance, multiple isoforms, post-translational modifications, and variations in expression

Electronic supplementary material: The online version of this article (DOI: 10.1007/7651_2017_75) contains supplementary material, which is available to authorized users.

with time and cell type make characterization of the proteome challenging. With advances in mass spectrometry-based technologies, it is now possible to survey thousands of proteins in a sample, although obtaining a complete quantitative profile of the entire proteome is still not possible [3]. One significant challenge involves the wide range of protein abundance in biological samples, such as human plasma. Because the dynamic range of the proteome spans ten orders of magnitude, low abundance proteins (e.g., cytokines) may be masked by the presence of extremely abundant proteins (e.g., albumin) [4–6].

Despite technological advancements and the identification of hundreds of potential protein biomarkers, there are still very few biomarker assays validated for routine use in clinical settings. While identification by discovery phase studies has produced many potential biomarkers, subsequent validation using high-throughput assays for quantification of these proteins is lacking [7, 8]. Obstacles related to this include the high rate of false positive identifications during discovery phase experiments (a low signal-to-noise ratio in discovery experiments), the lack of biological relevance for some candidates, and the limited number of quantitative immunoassays available to verify differential abundance of candidate proteins. With the technical challenges and high costs associated with development of immunoassays, it is not feasible to design novel assays for testing large numbers of candidate biomarkers [9, 10].

The proteomic biomarker pipeline may be enhanced by using prior knowledge to target candidate proteins for analysis. That is, discovery assays for proteins that have been previously identified to be associated with a specific condition can be designed with the purpose of improving the signal-to-noise ratio. We have previously described a method by which candidate biomarkers can be identified through an initial literature search and subsequently ranked using a scoring system we called the Hypothesis Directed Biomarker Ranking (HDBR) system [11]. By applying specific selection criteria, one can prioritize candidates with the most probable value as a biomarker for the disease and application of interest. With this method, repetition of large-scale discovery experiments can be avoided.

Currently, a manual literature search is required to rank candidate proteins. In the future, it may be possible to automate the analysis, using standalone software or by leveraging software that is currently available. For example, literature search software to query text-based data can be used to replace a manual search. One example is the Agilent Literature Search Software, a meta-search tool for automatically querying multiple text-based search engines [12]. In addition, the multiattribute rankings output can be enhanced using an application such as LineUp [13], a scalable visualization tool that uses bar charts to depict the relative contribution of each ranking criterion. This would enable the user to explore the effects of changes and refinements in the parameters used to rank biomarker candidates.

2 Materials

No specific materials are necessary, except for connection to PubMed and/or EMBASE. A template for scoring and ranking proteins can be accessed at the following link: (Link to excel spreadsheet).

3 Methods

This protocol can be modified to be applicable to any disease (cancer, neurodegenerative, immunological, cardiac, etc.) and biomarker application (diagnosis, prognosis, etc.) of interest.

1. Conduct a comprehensive literature search on PubMed and/or EMBASE to identify candidate biomolecules relevant to the disease and application of interest (*see* **Note 1**).
2. Score the biomolecules identified in the initial literature search based on Oxford Levels of Evidence, specimen source, biomolecule type, assay throughput format, relevance to target disease, relevance of biological function, and number of NCBI PubMed citations. These seven criteria and their associated scores are outlined in Table 1.
3. Firstly, assess the candidate biomarker using the Oxford Center for Evidence-based Medicine (OCEBM) Levels of Evidence [14] which scores the quality of evidence available on a particular test based on the study design, validation of results in multiple centers, and use of appropriate reference standards. More specific information on how OCEBM levels of evidence are assigned is available at: http://www.cebm.net/oxford-centre-evidence-based-medicine-levels-evidence-march-2009/. A tool for the appraisal of diagnostic accuracy studies can be accessed at http://www.cebm.net/wp-content/uploads/2014/04/diagnostic-study-appraisal-worksheet.pdf. A tool for the appraisal of prognostic studies can be accessed at http://www.cebm.net/wp-content/uploads/2014/04/cebm-prognosis-worksheet.pdf.
4. Next, evaluate the specimen source. If a serum biomarker is desired for a specific diagnostic test, assign more points to biomarkers identified in the blood, over those identified in tissue or cell lines. Alternatively, if a tissue-based biomarker is being developed, then assign more points if the paper describes a protein detected in tissue.

Table 1
Example of the proteomic biomarker ranking system used to score protein biomarkers for the early diagnosis of colorectal cancer

		Score				
Key criteria	**Criteria for protein assay development**	**0**	**1**	**2**	**3**	**4**
Oxford levels of evidence[a]	Level 4		X			
	Level 3a and 3b			X		
	Level 2a, 2b and 2c				X	
	Level 1a, 1b and 1c					X
Specimen source[b]	Cell lines		X			
	Tissue			X		
	Blood				X	
Biomolecule evaluated	RNA		X			
	Protein			X		
Assay throughput format	High-throughput (multiplexed) assay with conflicting outcome on low-throughput confirmatory test	X				
	High-throughput (multiplexed) assay		X			
	High-throughput (multiplexed) assay with low throughput confirmatory test, or low throughput assay alone			X		
Relevance to target disease entity	Present in adenoma or carcinoma, but possibly also other conditions	X				
	Present in carcinoma only		X			
	Present in carcinoma and also in a related premalignant lesion			X		
Relevance to biological function	Hallmark of cancer unknown or absent	X				
	Hallmark of cancer identified		X			
Number of NCBI PubMed citations	<100	X				
	≥100		X			

[a]Level 5 evidence excluded
[b]Human

5. Evaluate biomolecule type. Since the goal is to identify candidate protein biomarkers, assign two points to molecules demonstrated to be associated with the disease state of interest at the protein level, and one point to molecules only identified to be associated with the disease state at the RNA level.

6. Score biomarkers based on the throughput format in which they were identified to be associated with disease state of interest. Assign a higher score to biomarkers assessed by high throughput assays and then confirmed with low throughput assays (or detected by low throughput assay alone) and a lower score to biomolecules evaluated using solely high throughput techniques (*see* **Note 2**).

7. Score the relevance of the candidate biomarkers in the context of the disease of interest and any precursor disease states (*see* **Note 3**).
8. Assess the functions of the biomarker candidates and their relevance in the disease of interest. The criteria used to score biomarkers for this step will need to be modified depending on the specific disease state (*see* **Note 4**).
9. Next, quantify the number of publications related to the biomarker candidates (*see* **Note 5**).
10. Finally, tally the scores and create the resulting rank list of biomarkers (*see* **Note 6**). Top ranked candidate biomarkers generated using the HDBR system represent ideal candidates for further investigation and experimental validation as biomarkers in the disease state of interest.
11. Identify and implement an experimental platform (immunohistochemistry, enzyme-linked immunosorbent assays (ELISA), multiple reaction monitoring (MRM) using mass spectrometry, etc.) that is appropriate for validation of top ranked candidate proteins in the disease and specimen source of interest (*see* **Note 7**).

4 Notes

1. To identify protein biomarkers for a serum-based diagnostic test for the early detection of colorectal cancer, the search terms used in the PubMed database included various combinations of "colorectal" "colon" "rectum" "adenoma" "carcinoma" "cancer" "biomarker" "diagnosis" "microarray" "protein" and "serum."
2. False positives from multiple hypothesis testing and erroneous statistical inferences are common at the discovery phase of research using high throughput assays due to the large number of variables measured and the statistical tests applied to each of these variables [15, 16]. Therefore, a higher score was assigned to biomolecules assessed by high throughput assays and then subsequently confirmed with low throughput assays.
3. For our example, where the biomarker test of interest was aimed at early diagnosis of colorectal cancer at the adenoma stage, the highest score was given to biomarkers present in adenomas and potentially in carcinomas. A lower score was assigned to biomarkers present in colon carcinomas only, and a score of zero was given to biomarkers that were identified to be present in colon adenomas and carcinomas while also present in other conditions, as these biomarkers would not be specific for the diagnosis of colorectal cancer.

4. As the HDBR system was designed to identify biomarkers relevant to cancer, proteins with known functions related to the hallmarks of cancer [17, 18] were of greatest interest. Therefore, proteins with a known pro-neoplastic function were given a higher score in the ranking system. Alternatively, if a biomarker for primary cardiovascular events were being developed, the relevance of proteins that had known effects on lipid metabolism or vascular physiology would be considered functionally relevant.
5. Points are assigned based on number of publications related to the biomarker candidates because it is assumed that the number of publications is a reflection of the degree of interest in that biomarker by the scientific community.
6. Using the results from our primary literature review, 151 candidate biomarkers were identified to be potentially useful for the early diagnosis of colorectal cancer using a serum based diagnostic test. The various criteria were utilized to derive a final HDBR score. Candidate biomarkers were ranked and the list of top ranked biomarkers generated by the HDBR system was available to be assessed experimentally. Overall, this system filters potential biomarkers ensuring that only the biomarkers with the most clinical potential are carried forward to validation.
7. Numerous experimental platforms are available for validation of candidate biomarkers. Platform selection depends on which specimen source (e.g., serum, tissue) is relevant in the biomarker assay of interest. For validation of biomarkers in serum samples, protein concentrations can be quantified in patient and control sera using methods such as ELISA or MRM (mass spectrometry). Additionally, two methods can be combined. For example, MRM can be used as a second independent validation method in the same sample population to confirm results from the ELISA validation step. If a tissue biomarker is desired, immunohistochemistry can be used to validate high expression of protein candidates in disease compared to control tissue.

Acknowledgment

The authors have declared no conflict of interest.

References

1. de Sousa Abreu R, Penalva LO, Marcotte EM, Vogel C (2009) Global signatures of protein and mRNA expression levels. Mol Biosyst 5:1512–1526
2. Vogel C, Marcotte EM (2012) Insights into the regulation of protein abundance from proteomic and transcriptomic analyses. Nat Rev Genet 13:227–232
3. Nilsson T, Mann M, Aebersold R, Yates JR et al (2010) Mass spectrometry in high-throughput proteomics: ready for the big time. Nat Methods 7:681–685

4. Anderson NL, Anderson NG (2002) The human plasma proteome: history, character, and diagnostic prospects. Mol Cell Proteomics 1:845–867
5. Anderson L, Hunter CL (2006) Quantitative mass spectrometric multiple reaction monitoring assays for major plasma proteins. Mol Cell Proteomics 5:573–588
6. Mallick P, Kuster B (2010) Proteomics: a pragmatic perspective. Nat Biotechnol 28:695–709
7. Anderson NL (2010) The clinical plasma proteome: a survey of clinical assays for proteins in plasma and serum. Clin Chem 56:177–185
8. Harlan R, Zhang H (2014) Targeted proteomics: a bridge between discovery and validation. Expert Rev Proteomics (6):657–661
9. Rifai N, Gillette MA, Carr SA (2006) Protein biomarker discovery and validation: the long and uncertain path to clinical utility. Nat Biotechnol 24:971–983
10. Paulovich AG, Whiteaker JR, Hoofnagle AN, Wang P (2008) The interface between biomarker discovery and clinical validation: the tar pit of the protein biomarker pipeline. Proteomics Clin Appl 2:1386–1402
11. Mackay EM, Koppel J, Das P, Woo J, Schriemer DC, Bathe OF (2015) A hypothesis-directed approach to the targeted development of a multiplexed proteomic biomarker assay for cancer. Cancer Inform 14:65–70
12. http://www.agilent.com/labs/research/litsearch.html
13. Gratzl S, Lex A, Gehlenborg N, Pfister H, Streit M (2013) LineUp: visual analysis of multi-attribute rankings—best paper award. IEEE Trans Vis Comput Graph (InfoVis '13) 19(12):2277–2286
14. OCEBM Levels of Evidence Working Group. "The Oxford 2011 levels of evidence." Oxford Centre for Evidence-Based Medicine. http://www.cebm.net/index.aspx?o=5653
15. Pounds SB (2006) Estimation and control of multiple testing error rates for microarray studies. Brief Bioinform 7:25–36
16. Tinker AV, Boussioutas A, Bowtell DDL (2006) The challenges of gene expression microarrays for the study of human cancer. Cancer Cell 9:333–339
17. Hanahan D, Weinberg RA (2000) The hallmarks of cancer. Cell 100:57–70
18. Hanahan D, Weinberg RA (2011) Hallmarks of cancer: the next generation. Cell 144:646–674

Methods in Molecular Biology (2018) 1788: 131–143
DOI 10.1007/7651_2017_86

Published online: 27 October 2017

In Situ Hybridization and Double Immunohistochemistry for the Detection of VEGF-A mRNA and CD34/Collagen IV Proteins in Renal Transplant Biopsies

Dejan Dobi and Zoltan G. Laszik

Abstract

Quantitative metrics on the tissue distribution of different cell phenotypes, extracellular matrix components, and signaling/cell cycle markers hold the promise for the advent of new-generation tissue-based predictive/prognostic biomarkers in clinical diagnostics. The workflow of this approach is composed of three major phases: (1) detection of multiple molecular targets on a single histologic section, (2) image acquisition, and (3) digital image processing and analysis. Here, we present the most prevalent current alternatives for step (1) and describe a three-plex staining and image acquisition platform that captures the spatial distribution of macromolecules from two different species.

Keywords: Fluorescent immunohistochemistry, In situ hybridization, Quantitative image analysis, Whole slide digital imaging

1 Introduction

In the case of protein targets, multiplexed immunohistochemistry is the method of acquiring data related to their spatial distribution and architectural relations [1]. While brightfield techniques have been reported, immunfluorescence is a far more common tool for methodical reasons [2, 3]. Two immunfluorescent strategies have emerged for the detection of multiple antigens on a single histologic section: (I) detection of all antigens consecutively or at the same time [1, 4] followed by one-step multispectral imaging and (II) dye cycling or sequential methods that are composed of repeated cycles of single or double immunofluorescent staining, imaging then the removal of the antibody-fluorophore complex [5]. Altough the results with both approaches are promising, they have some limitations that prevent them from immediately entering into the realm of routine clinical diagnostics. Multispectral image aquisition combined with whole slide scanning creates large data files, which makes image processing and analysis difficult. Dye cycling uses just a few primary antibodies per round of staining, making the iterative phase of the process labor intensive and time consuming. Even though it is expected that in the coming years these issues can be resolved [6], going beyond antibody-based analysis to RNA in situ hybridization (ISH) would allow us to understand disease pathomechanism on a more complex level, by combining histomorphology with proteomic and transcriptomic data. Since the advent of RNAscope platform [7], RNA ISH has not been limited to highly expressed genes and therefore is a feasible tool for putting gene expression data into morphological context. It uses 20 probe pairs that hybridize in tandem by pair to a 1KB region of the target RNA. This design, in combination with a two-step amplification method, leads to a high sensitivity and specificity. The detection system can be either chromogen-based or a fluorescent reporter and allows quantitative measurement of the signal based on dot counting [8]. In theory, the combination of single or multiplexed RNA ISH and multiplexed immunohistochemistry offers the possibilty to detect molecular targets from different species on a single histologic section.

Here, we present a two-step process: VEGF-A mRNA was detected with DAB chromogen with the RNAscope platform, then CD34, and collagen IV immunfluorescent stainings were performed on the same section. This sequence allowed the in situ quantification of three molecular targets without the need of multispectral imaging or dye cycling.

1.1 Biological Background

VEGF-A plays a dual role in kidney transplant [9]. On one hand, it is crucial in maintining renal intersitial microvascular integrity [10]; on the other hand—as a pro-angiogenic factor—it plays an important role in response to injury. Overexpression of VEGF can contribute to the pathogenesis of different kidney diseases (e.g., it can mediate glomerular hypertrophy and sclerosis in diabetes [11] or can contribute to the progression of injury in chronic allograft damage [12]). Therefore, VEGF-A expression pattern in renal tubuli and glomeruli, and its spatial relationship to microvasculature (CD34) and to the major component of basal lamina and mesangial matrix (collagen IV), is of interest.

2 Materials

2.1 In Situ Hybridization Phase

1. Materials supplied by ACD bio:
   (a) RNAscope VEGF-A probe, Catalog # 311351, Accession # NM 001025370. Probe region covers 1459 bp spanning nucleotide 1518–2977.
   (b) RNAscope negative probe DapB, Catalog # 310043, Accession # EF191515. Probe region covers 448 bp spanning nucleotide 414–862.

   (c) RNAscope positive probe HS PPIB, Catalog # 313901, Accession # NM 000942.4. Probe region covers 850 bp spanning nucleotide 139–989.
   (d) RNAscope 2.0 HD Detection Kit (Brown):
2. Pretreatment Kit (Catalog # 310020)
   (a) Pretreat 1
   (b) 10× Pretreat 2
   (c) Pretreat 3
3. 2.0 HD Detection Kit—*Brown* (Catalog # 310035)
   (a) 2.0Amp 1
   (b) 2.0Amp 2
   (c) 2.0Amp 3
   (d) 2.0Amp 4
   (e) 2.0Amp 5-*Brown*
   (f) 2.0Amp 6-*Brown*
   (g) DAB-A
   (h) DAB-B
4. Wash Buffer Kit (Catalog # 310091)
   (a) 50× Wash Buffer
5. The HybEZ System (110 VAC, Catalog # 310010) is a humidified oven and its accessories that prevent sections from drying out. Hybridization and incubation steps should be performed in this chamber (*see* **Note 1**)
   (a) Oven (110 VAC, Catalog # 310010)
   (b) Humidity control tray (with lid, Catalog #310012)
   (c) Slide rack (Catalog # 310014)
   (d) Humidifying paper pack (Catalog # 310015)

2.2 Immunofluorescent Staining Phase

1. Mouse CD34 (DAKO, Carpinteria, CA, USA: Catalog# M7165) [1:100]
2. Rabbit collagen IV (Abcam, Cambridge, MA, USA; Catalog # ab6586), [1:100]
3. Horse anti-mouse conjugated with FITC (Vectorlabs, Burlingame, CA, USA; Catalog # FI-2000), [1:100]
4. Goat anti-rabbit conjugated with Texas Red (Vectorlabs, Burlingame, CA, USA; Catalog # TI-1000), [1:100]
5. DAPI I. (Abbott Molecular, Abbott Park, IL, USA; Catalog # 06J49-001)

2.3 Miscellaneous Materials (Can Be Purchased from Any Major Laboratory Supplier in North America/Europe)

1. 100% ethanol (EtOH)
2. Xylene
3. Vertical slide rack
4. Staining dish
5. Clearing agent dish, xylene resistant
6. Cover glasses 24 × 50 mm
7. Ammonium hydroxide, 28–30%
8. Carboy (>3 L)
9. Water bath or incubator, capable of holding temperature at 40 ± 1 °C
10. Hot plate
11. Pipettors and tips, 1–1000 μL
12. Distilled water
13. Tubes (various sizes)
14. Fume hood
15. Graduated cylinder
16. Parafilm
17. Absorbent paper
18. 20% bleach
19. NaCl
20. KCl
21. Na_2HPO_4
22. KH_2PO_4

2.4 Image Acquisition Phase

1. Metafer slide scanning platform with VSlide and Metaviewer (Metasystems GmbH, Altlussheim, Germany)
2. Axio Imager Z2 microscope (Zeiss GmbH, Jena, Germany)
3. CoolCube 1 m digital camera (Metasystems GmbH, Altlussheim, Germany)
4. Metafer (v 3.10.2 or higher), VSlide, Metaviewer 2 softwares

3 Methods

Pre-ISH phase

3.1 Fixation, Paraffin Embedding, Cutting

1. Kidney biopsy core is obtained using 18G needles. After dividing the tissue, the part meant for light microscopy analysis is fixed in 10% neutral buffered formalin for 2 h at room temperature.
2. Fixed cores are paraffin embedded, sectioned at 2 μm and mounted to Superfrost Plus slides.

(The following steps are from the protocol provided by Advanced Cell Diagnostics for ISH.)

3. Keep the slides on hot plate for 1 h at 60 °C.

3.2 Deparaffinization

1. Put the slides into a vertical slide rack and submerge them into xylene for 2 × 5 min (use fresh xylene for the second round).
2. Put the slide rack into absolute ethanol for 2 × 5 min.
3. Air-dry the slides at room temperature.

ISH phase

3.3 ISH Pretreatment

1. Prepare 700 mL of 1× Pretreat 2 by adding 630 mL distilled water to 1 bottle (70 mL) 10× Pretreat 2 solution.
2. Place the beaker containing 1× Pretreat 2 on the hot plate. Cover the beaker with foil and turn the hot plate on maximum for 10–15 min.
3. Once 1× Pretreat 2 reaches boiling, set the hot plate knob to 104 °C to maintain uniform boiling.
4. Dribble 5 drops of Pretreat 1 over each slide to cover the entire section (*see* **Note 2**).
5. Incubate slides for 10 min at room temperature.
6. Wash the slides in distilled water by moving the vertical slide rack up and down five times in a staining dish.
7. Incubate the slides in mildly boiling (100–104 °C) 1× Pretreat 2 for 15 min.
8. Transfer the hot slide rack from the 1× Pretreat 2 to the staining dish containing distilled water. Do not let the slides cool in Pretreat 2.
9. Wash the slides in distilled water by moving the rack up and down five times, then change the distilled water, and repeat the process.
10. Put the slide rack into absolute alcohol and wash the slides by moving the rack up and down five times.
11. Air-dry the slides.
12. Draw a barrier with a hydrophobic barrier pen around the section. Let the barrier dry completely.
13. Add five drops of Pretreatment 3 to entirely cover each section, remove the humidity control tray from the oven, place the slides into the oven, and insert the tray back. Incubate the slides for 30 min at 40 °C (*see* **Note 3**).
14. Remove the slides from the oven and submerge them into distilled water.

3.4 Hybridization

1. Prepare 3 L of 1× Wash Buffer by adding 2.94 L distilled water and 1 bottle (60 mL) of 50× Wash Buffer to a large carboy. Mix well.
2. Add four drops of VEGF-A/RTU/PPIB RNA probe to each section, and place them back into the oven for two hours at 40 °C (*see* **Note 4**).
3. Submerge the slides into 1× Wash Buffer twice (use fresh Buffer for the second time).

3.5 Amplification

1. Dribble 4 drops of Amplification 1 and incubate the slides in the humidified oven for 30 min at 40 °C.
2. Submerge the slides into 1× Wash Buffer twice (use fresh Buffer for the second time).
3. Dribble 4 drops of Amplification 2 and incubate the slides in the humidified oven for 30 min at 40 °C.
4. Submerge the slides into 1× Wash Buffer twice (use fresh Buffer for the second time).
5. Dribble 4 drops of Amplification 3 and incubate the slides in the humidified oven for 30 min at 40 °C.
6. Submerge the slides into 1× Wash Buffer twice (use fresh Buffer for the second time).
7. Dribble 4 drops of Amplification 4 and incubate the slides in the humidified oven for 30 min at 40 °C.
8. Submerge the slides into 1× Wash Buffer twice (use fresh Buffer for the second time).
9. Dribble 4 drops of Amplification 4 and incubate the slides in the humidified oven for 30 min at 40 °C.
10. Submerge the slides into 1× Wash Buffer twice (use fresh Buffer for the second time).
11. Dribble 4 drops of Amplification 5 Brown and incubate the slides in the humidified oven for 30 min at 40 °C.
12. Submerge the slides into 1× Wash Buffer twice (use fresh Buffer for the second time).
13. Dribble 4 drops of Amplification 6 Brown and incubate the slides in the humidified oven for 30 min at 40 °C.
14. Submerge the slides into 1× Wash Buffer twice (use fresh Buffer for the second time).

3.6 Detection

1. Pipette 120 μL DAB-A and DAB-B 1:1 to each slide for 10 min.
2. Wash the slides in distilled water for 1 min and then repeat the process with fresh distilled water.

3.7 Immunofluorescent Staining Phase

1. Prepare 500 mL of phosphate-buffered saline (PBS):
2. Add 8 g of NaCl, 0.1 g of KCl, 0.72 g of Na_2HPO_4, and 0.12 g of KH_2PO_4 to 400 mL distilled water. Mix well. Add distilled water to 500 mL.
3. Add 3 drops of DAPI I to 1 mL PBS.
4. Cover the slides with normal horse and goat serum for 30 min at room temperature.
5. Wash the slide in PBS for 1 min, and repeat the process with fresh PBS.
6. Cover each section with 70 μL of the primary antibody cocktail, collagen IV, and CD34 (*see* Sect. 2), for 60 min at room temperature.
7. Wash the slide in PBS for 1 min, and repeat the process with fresh PBS.
8. Cover each section with 70 μL of the secondary antibody cocktail, horse anti-mouse conjugated with FITC, and goat anti-rabbit conjugated with Texas Red (*see* Sect. 2), for 30 min at room temperature.
9. Wash the slide in PBS for 1 min, and repeat the process with fresh PBS.
10. Cover the slides with 70 μL of diluted DAPI I.

3.8 Image Acquisition Phase (See Note 5)

1. Start Metafer on the Metafer Slide Scanning Platform PC.
2. Set Mode to Metacyte.
3. Set up a reflector cube for the ISH DAB signal that is not assigned to any fluorochrome and couple a color transmission filter with it, on Configure\Metafer Parameters\Microscope Tab, Display channels 7–12: checked.
4. Set Metafer to save training files:
   (a) Configure\Metafer Parameters\Search tab\Training Images section.
   (b) Store training images during search: checked.
   (c) Training file format: Std + TXT&TIF.
5. Set Metafer not to use compression while exporting Training Images:
   (a) Configure\Metafer Parameters\Output tab\Image export section.
   (b) Use Compression for TRN image TIFF export: unchecked.
6. Determine the exposure times for each slide and each color channel (i.e., DAPI, FITC, Texas Red) by right clicking on the live image of a representative area of the section and adjusting the exposure time until no red areas are seen.

7. Set up the Focus Parameters.

   The optimal focus parameter set that results in a well-focused image is dependent on section quality and thickness; therefore a specific set of parameters—that would fit every scenario—can not be given. Here, we would like to emphasize those features that should be adjusted and the principles based on which these adjustments can be performed.

   (a) Set the No. of focus planes and the plane distance so that the planes span the entire thickness of the section.

   (b) As these focus features will be applied as Fine Focus Parameter Set (see under 8a), Grid Width/Grid Margin/Neighborhood features will not affect the focus process; therefore default settings can be left unchanged.

   (c) We found that unchecking "Predict Focus Start Position" leads to more, well-focused field of views.

   (d) Focusing Mode, 0; Contrast Mode,1.

   (e) Values for Max. Grid Gap Percentage/Grid Interpolation Mode/Min. Object Area/Min. Foc. Contrast/Global/ Min. Foc. Contrast/ZMinMax/Max. Height Second Peak/Max. Saturation Area/Min./Max. Total Area can be left unchanged.

8. Set up a classifier for each slide.

   (a) Set your Focus Parameter Set as "Fine Focus."

9. Adjust the Capture tab.

   (a) Set the color channels with the appropriate exposure times for FITC, Texas Red, and DAPI. It is important to choose the brightest signal as the "Counterstain Channel." Set the exposure time for the "ISH DAB" channel at the lowest possible value.

   (b) Expsure mode for each channel: fixed.

   (c) Saturation area for each channel: 400.

   (d) No. of focus planes for each channel: 1.

   (e) Camera gain: 4.5.

10. Adjust the Exposure tab.

    (a) No. of exposures for each channel: 1.

    (b) Autofocus for each channel: checked.

    (c) Focus plane mode: single focus.

    (d) Correct for inhomogenous illumination: checked.

    (e) Extend Image to Capture Margins: checked.

11. Adjust the Image processing tab.

    (a) For DAPI channel, set SBHistoMax processing method in Operations for Raw Images section.

12. Image quality, Cell selection, Cell Processing, Features, and Gallery Tabs:
    (a) These tabs are helpful primarily in optimizing ISH signal detection; therefore default settings can be left unchanged.
13. Other tabs.
    (a) Every checkbox should be unchecked except for Store Overview Image, Store Training Images, and Unidirectional XY Positioning.
14. Set up scan.
    (a) Click Setup on main screen.
    (b) Set the location of your data files.
    (c) Activate the necessary slide positions.
    (d) Name the slides.
    (e) Assign the appropriate classifiers for the appropriate slide positions.
    (f) Choose "Manual" for scan area.
    (g) Count: set its value to an arbitrarily high figure (3000000).
15. Start scan.
    (a) Select Search.
    (b) Select scan area: define the smallest sourrounding rectangle of the biopsy; follow the instructions on the screen, and find the rectangle's bottom right and upper left corner.
    (c) The microscope will find the center of the scan area that you have to focus on. If no tissue is located in the center, move the stage into a position where a representative area of the section can be seen. Focus on live image.
    (d) Repeat these steps for all slides.
    (e) Once all slides are completed, you will be prompted to "Prepare the microscope" and start scanning.

3.9 Image Processing and Analysis Phase

The whole image processing and analysis algorithm is beyond the scope of this chapter. Here, we describe an image stitching and processing workflow that results in .tif whole slide images or area of interests from whole slide images that can be further processed and analyzed in ImageJ (v1.43 or higher) software [13]. For further details regarding image processing and analysis, we refer to an e-handbook written by Peter Bankhead [14].

1. Image stitching: Once Metafer finishes the scanning process, it saves and stores the pictures taken of the individual field of

views as separate images. VSlide by Metasystems is capable of stitching these tiles and can create a whole slide image.

(a) Open VSlide.

(b) Set up the stitching Parameter Set.

(c) Configure\VSlide Parameter Set.

(d) Input tab → DAPI, Texas Red, FITC, and "ISH DAB" channel with a relative weight of 100%.

(e) Output tab → Color Image sub-tab.

(f) Create Color Image: checked, select all input colors, File format: .vsi, Compression: RAW, Use Multiresolution: checked, Use Tiled Tiff Mode: unchecked, Color Display Mode: Mixed colors.

2. Select Offline Mode.

(a) Input Path: the location where your training file was stored by Metafer.

(b) Output path: the folder where you would like your whole slide images to be stored.

(c) Select Parameter Set.

3. Select your slides from the right "Available slides" box.

4. Move them to the "Selected slides" box by "Add."

5. Open Metaviewer 2.

(a) Open the .vsi file.

(b) Draw a rectangle that encompasses the whole section or the area of interest.

(c) Export the chosen area as .tif.

(d) Separate; pixel-to-pixel aligned color channels can be extracted as 8 bit gray scale images.

4 Summary

Here, we described a combined in situ hybridization/immunfluorescent staining method that allows the detection of three different targets from two molecular species (CD34, collagen IV proteins, VEGF mRNA). The method does not require multispectral imaging or dye cycling, which makes the image acquisition faster, while the pixel-to-pixel alignment of the separate color channels is maintained. First, VEGF mRNA is detected with DAB chromogen (ACD RNAscope platform), followed by double immunofluorescence for CD34 and collagen IV with DAPI. Second, all four channels are scanned by a fluorescent slide scanner, simultaneously. Third, the acquired image is further processed and is feasible to be further analyzed in ImageJ or in another open source software (Fig. 1).

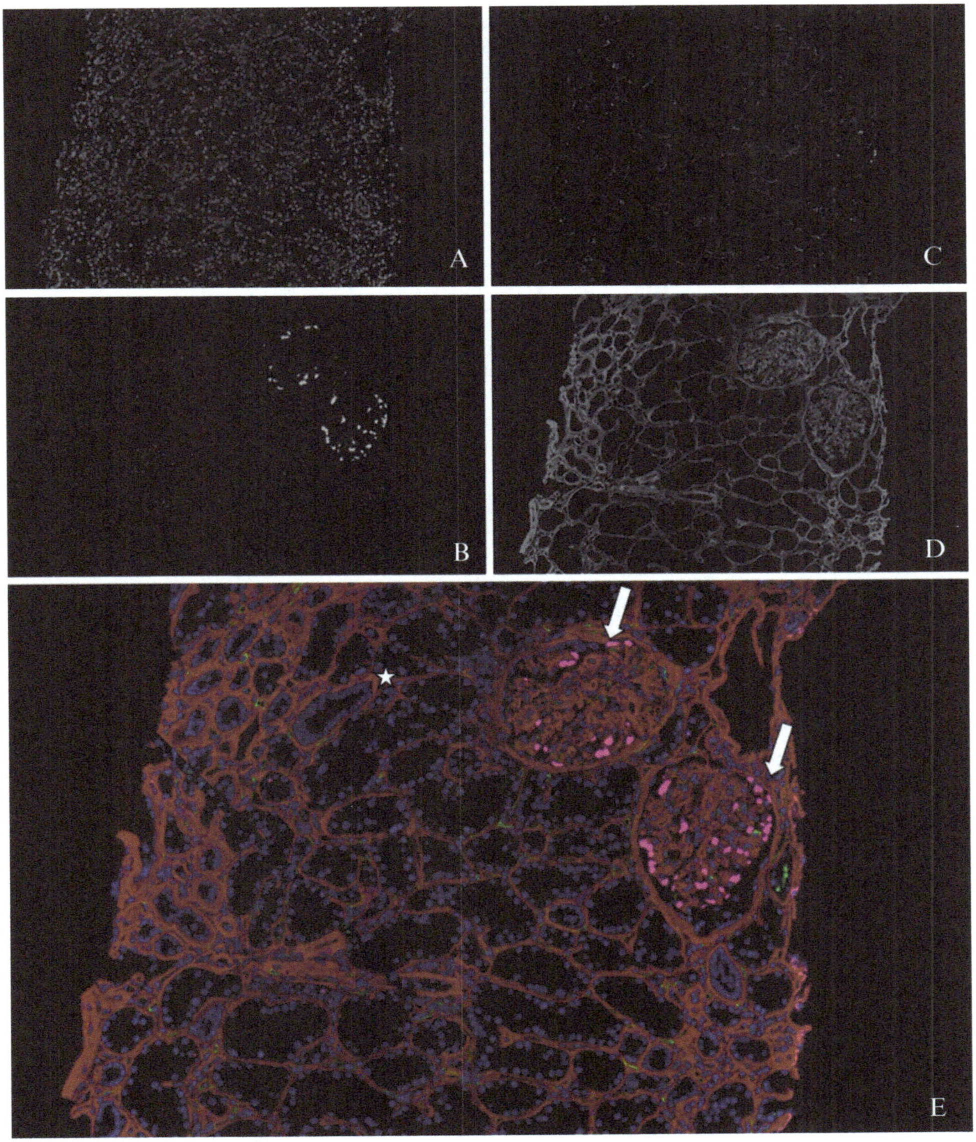

Fig. 1 Representative area from the whole slide image of a kidney biopsy. Separate color channels show DAPI (**a**), VEGF mRNA (**b**), CD34 (**c**), collagen IV (**d**). On the composite image (**e**) pixel-to-pixel alignment of the separate channels reveals the high level of VEGF mRNA expression of podocytes (*arrow*) and its lower level in the distal tubular epithelial cells (*asterisk*)

5 Notes

1. Although we used the ACD oven for our hybridization steps, any oven is feasible if it can keep the tissue sections humid.
2. Some authors [15] have eliminated the Pretreatment 1 step, as this is required only for horseradish peroxidase (HRP) detection

methods and implemented a slightly modified Pretreatment 3 step with diluted Pretreatment Solution 3 (1:15 dilution in PBS) and an incubation time of 30 min.

3. It is worth switching on the oven 30 min before Sect. 3.3, **step 13**.
4. Preheat the probes in oven at 40 °C and let them cool down at room temperature before using. This will prevent precipitation.
5. Under **step 3.8**, we highlighted those settings that differ from the default settings, which were determined during the installation and set up of the Metafer Slide Scanning Platform. For the best image acquisition results, further fine-tuning of the global Metafer Settings, Metacyte Classifiers, or Focus Parameter Set may be needed.
6. Glomerular CD34 signal distribution and intensity after ISH slightly differ from the CD34 staining pattern if no previous ISH was done on the section (Fig. 2). To overcome this

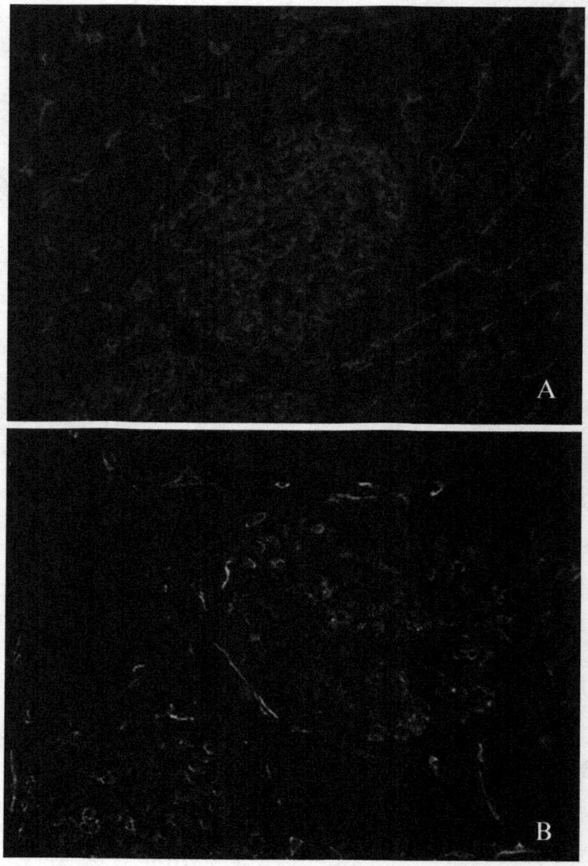

Fig. 2 Glomerular CD34 staining pattern and intensity differ if prior in situ hybridization (ISH) is performed [(**a**) CD34 staining without ISH, (**b**) CD34 staining after ISH] (see **Note 6**)

discrepancy, we suggest further fine-tuning of the primary/secondary antibody dilution or the implementation of **Note 2**.

References

1. Stack EC, Wang C, Roman KA, Hoyt CC (2014) Multiplexed immunohistochemistry, imaging, and quantitation: a review, with an assessment of tyramide signal amplification, multispectral imaging and multiplex analysis. Methods 70(1):46–58
2. van der Loos CM (2010) J Histotechnol 33:31–40
3. Glass G, Papin JA, Mandell JW (2009) J Histochem Cytochem 57:899–905
4. Tsurui H et al (2000) Seven-color fluorescence imaging of tissue samples based on Fourier spectroscopy and singular value decomposition. J Histochem Cytochem 48:653–662
5. Wählby C, Erlandsson F, Bengtsson E, Zetterberg A (2002) Sequential immunofluorescence staining and image analysis for detection of large numbers of antigens in individual cell nuclei. Cytometry 47(1):32–41
6. Angelo M, Bendall SC, Finck R, Hale MB, Hitzman C, Borowsky AD, Levenson RM, Lowe JB, Liu SD, Zhao S, Natkunam Y, Nolan GP (2014) Multiplexed ion beam imaging of human breast tumors. Nat Med 20 (4):436–442
7. Wang F, Flanagan J, Su N, Wang LC, Bui S, Nielson A, Wu X, Vo HT, Ma XJ, Luo YJ (2012) RNAscope a novel in situ RNA analysis platform for formalin-fixed, paraffin-embedded tissues. Mol Diagn 14(1):22–29
8. Wang Z, Portier BP, Gruver AM, Bui S, Wang H, Su N, Vo HT, Ma XJ, Luo Y, Budd GT, Tubbs RR (2013) Automated quantitative RNA in situ hybridization for resolution of equivocal and heterogeneous ERBB2 (HER2) status in invasive breast carcinoma. J Mol Diagn 15(2):210–219
9. Mayer G (2011) Capillary rarefaction, hypoxia, VEGF and angiogenesis in chronic renal disease. Nephrol Dial Transplant 26 (4):1132–1137
10. Choi YJ, Chakraborty S, Nguyen V et al (2000) Peritubular capillary loss is associated with chronic tubulointerstitial injury in human kidney: altered expression of vascular endothelial growth factor. Hum Pathol 31:1491–1497
11. Reinders MEJ, Rabelink TJ, Briscoe DM (2006) Angiogenesis and endothelial cell repair in renal disease and allograft rejection. J Am Soc Nephrol 17:932–942
12. Malmström NK, Kallio EA, Rintala JM et al (2008) Vascular endothelial growth factor in chronic rat allograft nephropathy. Transpl Immunol 19:136–144
13. Schneider CA, Rasband WS, Eliceiri KW (2012) NIH image to ImageJ: 25 years of image analysis. Nat Methods 9:671–675
14. Bankhead P (2014) Analyzing fluorescence microscopy images with ImageJ. GitBook. [online] Available at: https://www.gitbook.com/book/petebankhead/imagej-intro/details
15. Stempel AJ, Morgans CW, Stout JT, Appukuttan B (2014) Simultaneous visualization and cell-specific confirmation of RNA and protein in the mouse retina. Mol Vis 20:1366–1373

Methods in Molecular Biology (2018) 1788: 145–156
DOI 10.1007/7651_2017_93

Published online: 08 November 2017

LC-SRM-Based Targeted Quantification of Urinary Protein Biomarkers

Yuqian Gao, Hui Wang, Carrie D. Nicora, Tujin Shi, Richard D. Smith, Tara K. Sigdel, Minnie M. Sarwal, David G. Camp II, and Wei-Jun Qian

Abstract

Liquid chromatography (LC)-selected reaction monitoring (SRM) is a powerful protein quantification technique in terms of sensitivity, reproducibility, and multiplexing capability. LC-SRM can accurately measure the concentrations of surrogate proteotypic peptides for targeted proteins in complex biological samples by using their stable heavy isotope-labeled counterparts as internal standards. Herein, we describe a step-by-step protocol of the application of LC-SRM to quantify candidate protein biomarkers in human urine.

Keywords Targeted quantification, LC-SRM, Urine, Biomarker, Stable heavy isotope-labeled peptide, Skyline

1 Introduction

Liquid chromatography coupled with selected reaction monitoring (LC-SRM) has become an increasingly popular technology for accurate quantification of targeted proteins in complex biological samples [1–4]. LC-SRM has recently been recognized as an alternative to immunoassays for protein biomarker verification [5–8]. For developing protein-based clinical tests, voided urine is often a good source, with an advantage over other clinical samples because it is noninvasive and easily accessible in large quantities. However, the verification of protein biomarkers in patient urine is challenging with LC-SRM because the urinary protein concentration is low and varies significantly at the intraindividual and interindividual levels [9, 10]. To address this issue in our LC-SRM workflow of measuring candidate urinary biomarkers, we use a low molecular weight protein cutoff filter to concentrate the proteins from patient urine and either the total urinary protein mass or the urinary creatinine concentration to normalize the urinary protein concentration.

In this chapter, we describe the detailed procedure for the application of LC-SRM to accurately quantify candidate urinary protein biomarkers that were selected based upon our previous discovery data and/or literature reports [10]. The detailed LC-SRM workflow includes urine sample preparation, SRM assay development, LC-SRM quantification, and SRM data analysis.

2 Materials

2.1 Urine Sample Collection and Storage

Urine sample collection is based on recommendations from the Human Kidney and Urine Proteome Project (HKUPP).

1. Collect 50–100 mL of morning void midstream urine in sterile containers.
2. Centrifuge at 2000 × *g* for 20 min at room temperature within 1 h of collection.
3. Separate the supernatant from any particulate matter (including cells and cell debris).
4. Measure the urinary creatinine level by standard colorimetric assay.
5. Adjust the pH of the supernatant to be 7.0.
6. Aliquot the sample into 10 mL aliquots, and store at −80 °C until further analysis.

2.2 Heavy Isotope-Labeled Synthetic Peptides

Crude heavy synthetic peptides labeled with $^{13}C/^{15}N$ on the C-terminal lysine (K) and arginine (R) (Thermo Fisher Scientific, San Jose, CA).

2.3 Protein Extraction, Digestion, and Cleanup Components

1. Amicon Ultra centrifugal filtration units (10 kDa molecular weight cutoff, Millipore, Bedford, MS).
2. Ultracentrifugal filtration chamber exchanging buffer: 50 mM NH_4HCO_3, pH 8.0.
3. Bicinchoninic acid (BCA) protein assay kit (Pierce Biotechnology Inc., Rockford, IL).
4. Solid high-purity urea (Sigma, St. Louis, MO) for denaturing the proteins.
5. Reducing reagent: 500 mM dithiothreitol (DTT) in water.
6. Thermomixer (Eppendorf North America, Hauppauge, NY).
7. Alkylation reagent: 1 M iodoacetamide (IAA) in water.
8. Digestion buffer: 0.5 M triethylammonium bicarbonate buffer (TEAB).
9. Trypsin solution: sequencing grade modified porcine trypsin (Promega, Madison, WI) freshly dissolved in digestion buffer to a final concentration of 1 μg/μL.
10. 1 mL solid-phase extraction (SPE) C18 column (Supelco, Bellefonte, PA).

11. SPE conditioning solution: 100% methanol and 0.1% trifluoroacetic acid (TFA) in water.
12. SPE washing solution: 0.1% TFA in 5% acetonitrile/95% water (v/v).
13. SPE eluting solution: 0.1% TFA in 80% acetonitrile/20% water (v/v).

2.4 LC-SRM Components

1. LC instrumentation: nanoACQUITY UPLC® system equipped with 5 μL injection loop, 100 μm × 100 mm BEH 1.7 μm C18 column.
2. LC solvents: mobile phase A, 0.1% formic acid (FA) in water; mobile phase B and weak needle wash, 0.1% FA in 90% acetonitrile/10% water (v/v); strong needle wash, 100% acetonitrile; and seal wash, 20% methanol/80% water (v/v).
3. MS instrumentation: Thermo Scientific TSQ Vantage triple-stage quadrupole mass spectrometer.

3 Methods

Typically there are five steps in the LC-SRM workflow (see Fig. 1): (1) SRM assay development, (2) urine sample processing, (3) addition of synthetic heavy isotope-labeled peptides into protein digests, (4) LC-SRM analysis, and (5) data analysis using Skyline software [11].

3.1 SRM Assay Development

Before performing urine sample analysis, SRM assays for the biomarker proteins of interest need to be developed, which include surrogate peptide selection, purchasing synthetic heavy isotope-labeled peptides, peptide transition selection, and achieving optimal collision energy for each transition.

3.1.1 Surrogate Peptide Selection for Biomarker Proteins of Interest

The selection of surrogate peptides is a key step for sensitive and accurate quantification of target proteins. There are numerous tryptic peptides for a single protein; the selection step aims to select peptides with the best MS response to represent each protein. The number of observations from the MS/MS shotgun proteomics data repository or a theoretical prediction by protein sequences is typically used to facilitate the selection process. The most frequently used MS/MS data repositories are PeptideAtlas [12] and GPM [13] (*see* **Note 1**):

1. Present the information about the identified peptide sequences and their number of observations in a table similar to Table 1.
2. Calculate the length of peptide sequence in terms of number of amino acids.
3. Blast the sequences in the MS-Homology website (http://prospector.ucsf.edu/prospector/cgi-bin/msform.cgi?

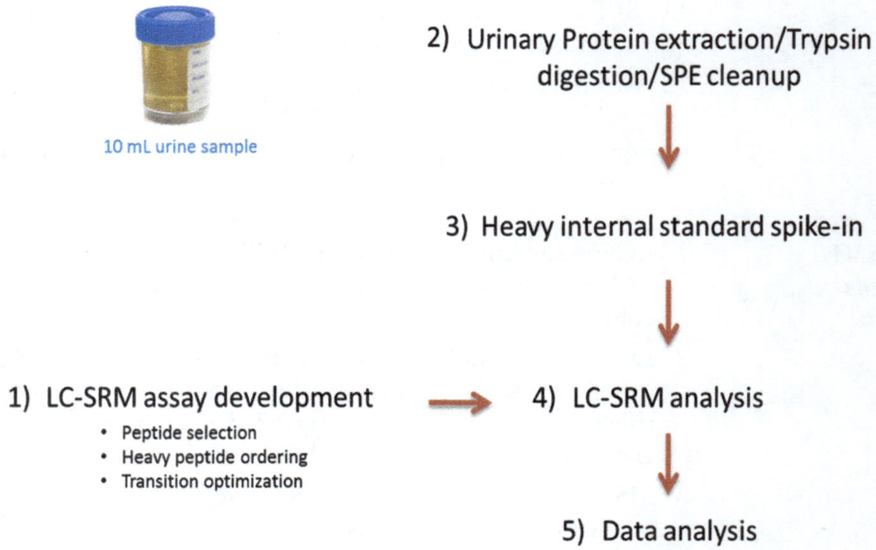

Fig. 1 Workflow of LC-SRM quantification of candidate urinary protein biomarkers

form=mshomology) for any shared sequences with other proteins in homo sapiens database.

4. Examine the peptide usefulness based on the length of amino acids within a 6–25 range and without any other complicating factors, such as containing posttranslational modifications (PTMs) or methionine and covering the signal peptide region (*see* Table 1).
5. Rank the peptides by their number of observations in PeptideAtlas [12] and GPM [13] (*see* **Note 2**):
   - PeptideAtlas: http://www.peptideatlas.org
   - GPM: http://gpmdb.thegpm.org/index.html
6. Select the three best peptides based on the number of observations; the larger the number of observations, the better the candidate for SRM study.
7. When there is no MS/MS data available, base the peptide ranking on prediction scores from CONSeQuence [14] and ESPPredictor [15]; the larger the score, the better the candidate for SRM study:
   - CONSeQuence: http://king.smith.man.ac.uk/CONSeQuence
   - ESPPredictor: http://www.broadinstitute.org/cancer/software/genepattern/modules/ESPPredictor.html

3.1.2 Synthetic Heavy Isotope-Labeled Peptides

1. The crude synthetic heavy isotope-labeled peptides are purchased based on the selection above with the following criteria:
   - Isotopically label the C-terminus with heavy arginine ($[^{12}C_6, ^{15}N_4]$-arginine) or lysine ($[^{12}C_6, ^{15}N_2]$-lysine).

Table 1
Peptide selections of urine protein biomarker candidate of glucose-6-phosphate isomerase (G6PI)

Clean sequence	# Obs in PeptideAtlas	# Obs in GPM	Peptide length	Unique peptide	Comments
TLAQLNPESSLFIIASK	**26**	**80166**	**17**	**Yes**	**Potential surrogate peptide**
HFVALSTNTTK	**5**	**46316**	**11**	**Yes**	**Potential surrogate peptide**
VWYVSNIDGTHIAK	**3**	**44475**	**14**	**Yes**	**Potential surrogate peptide**
ILLANFLAQTEALMR	29	101445	15	Yes	Containing M
TFTTQETITNAETAKEWFLQAAK	4	13384	23	Yes	Missed cleavage
EWFLQAAK	3	9735	8	Yes	Relatively lower # obs
INYTEGR	3	7307	7	Yes	Relatively lower # obs
TFTTQETITNAETAK	2	58239	15	Yes	Relatively lower # obs
KIEPELDGSAQVTSHDASTNGLINFIK	2	45152	27	Yes	# of amino acids >25
VDHQTGPIVWGEPGTNGQHAFYQLIHQGTK	2	37597	30	Yes	# of amino acids >25
LTPFMLGALVAMYEHK	2	11890	16	Yes	Containing M
VKEFGIDPQNMFEFWDWVGGR	2	4067	21	Yes	Containing M
SNTPILVDGKDVMPEVNK	1	26371	18	Yes	Containing M

The three rows in boldface indicate the selected three best peptides
Note: "#Obs" stands for number of observations

- Protect all cysteines by carbamidomethylation (CAM).

2. Upon receiving the crude heavy peptides, store at −20 °C for further use.

3.1.3 Transition Optimization of Selected Peptides

In a large-scale study, there are often hundreds of surrogate peptides used for targeted protein quantification. To get the best transitions and their optimal collision energies (CEs), the optimization of each individual surrogate peptide by direct infusion is time-consuming. The methods we use here to select the best transitions are based on Orbitrap HCD MS/MS data of the synthetic crude heavy peptides [16].

1. Prepare a stock solution of crude heavy peptides in a 1.5 mL Eppendorf safe-lock tube at the concentration of 1000 fmol/μL in 0.1% TFA in water as solvent.
2. Aliquot the stock solution into 100 μL aliquots.
3. Take one aliquot for assay development, and store the rest at −80 °C for future use.
4. Prepare a solution of crude heavy peptides at 500 fmol/μL in 30 μL 0.1% FA in water.

5. Load 5 μL of the above 500 fmol/μL heavy peptide solution onto a LC-MS/MS system, and obtain HCD MS/MS data (see **Note 3**).
6. Analyze the data by MSGF+, and import the analysis results into Skyline to build a peptide spectral library (https://skyline.gs.washington.edu/labkey/wiki/home/software/Skyline/page.view?name=building_spectral_libraries for instructions).
7. Select the top 5 ranked y-ion transitions for each peptide.
8. Save the Skyline file for further analysis (*see* **Note 4**).

3.2 Urine Sample Processing (See Note 5)

1. Wash the Amicon 15 mL ultracentrifugal filtration device with 10 mL of 50 mM NH_4HCO_3 (pH 8.0), and spin at 4000 × *g* for 10 min to remove the trace of glycerine.
2. Thaw the frozen urine samples on ice.
3. Load a 10 mL aliquot of the urine samples to the filter chamber, and centrifuge the sample at 4000 × *g* at 10 °C for 20 min to separate small MW peptides and other pigments (< 10 kDa) from the larger proteins.
4. Buffer exchange the sample twice by adding 10 mL of 50 mM NH_4HCO_3 (pH 8.0) to the filter chamber and centrifuging as described in **step 1**. Pipet the final retentate in the filter chamber to a fresh and labeled 1.5 mL Eppendorf tube.
5. Adjust the final volume of the retentate to 400 μL with 50 mM NH_4HCO_3 (pH 8.0).
6. Determine the protein concentration by BCA protein assay.
7. Add powdered urea into the tube to a final concentration of 8 M for protein denaturation.
8. Add 500 mM DTT solution to a final concentration of 10 mM for reduction. Sonicate the sample briefly, and incubate at 37 °C for 1 h with constant shaking in a ThermoMixer (*see* **Note 6**).
9. Add 1 M IAA to a final concentration of 40 mM, and incubate at 37 °C for 1 h in the dark with constant shaking for alkylation (*see* **Note 6**).
10. Dilute sample by ten times with digestion buffer, add trypsin solution at protein/trypsin ratio of 50:1 (w/w), incubate at 37 °C for 3 h, and then add TFA solution to a final concentration of 0.1% to stop the reaction.
11. Precondition 1 mL SPE C18 columns by slowly passing 3 mL methanol and then 4 mL SPE conditioning buffer through the column.
12. Load each tryptic digest onto separate SPE C18 columns; pass each sample through, and wash each column with 4 mL of SPE washing buffer.

13. Elute the peptides from each SPE C18 column with 1 mL of SPE eluting buffer, and dry each sample under a reduced vacuum using a SpeedVac.
14. Redissolve the sample with 100 μL water.
15. Determine the peptide concentration by BCA protein assay.
16. Store samples at −80 °C freezer until further use.

3.3 Addition of Heavy Peptide Internal Standards

1. Mix all the heavy isotope-labeled peptides in one 1 mL stock solution at a concentration of 1000 fmol/μL of each peptide using 0.1% TFA in water as solvent.
2. Calculate the volumes needed for LC-SRM analysis:

$$\text{Volume of urine protein digest needed } (V_{\text{digest}}) = V_{\text{LC-SRM}} \times C_{\text{digest, LC-SRM}} \div C_{\text{digest, original}}$$

$$\text{Volume of heavy peptide stock needed } (V_{\text{heavy peptide}}) = V_{\text{LC-SRM}} \times C_{\text{heavy, LC-SRM}} \div C_{\text{heavy peptide stock}}$$

$$\text{Volume of deionized water needed } (V_{\text{water}}) = V_{\text{LC-SRM}} - V_{\text{digest}} - V_{\text{heavy peptide}}$$

where $V_{\text{LC-SRM}}$ is the volume of solution needed for final LC-SRM analysis (μL), which is typically 20 μL for several injections; $C_{\text{digest,LC-SRM}}$ is the peptide concentration of final sample (μg/μL) in the final LC-SRM solution; $C_{\text{digest original}}$ is the peptide concentration (μg/μL) for the original urine protein digest for a given patient sample; $C_{\text{heavy, LC-SRM}}$ is the molar concentration of heavy internal standard peptides (fmol/μL) in the final LC-SRM solutions; and $C_{\text{heavy peptide stock}}$ is the molar concentration of heavy internal standard peptide stock solutions (fmol/μL). Typically, the peptide concentration of the final LC-SRM solutions is 1 μg/μL for 1 μL injection, and the heavy peptide concentration is 100 fmol/μL for crude heavy peptides.
3. Add the deionized water and urine protein digest sequentially into a Waters glass vial, and shake at 800 × *g* for 6 min.
4. Add heavy peptide stock into the same glass vial, and shake at 800 × *g* for 6 min.
5. Centrifuge the glass vial at 4000 × *g* for 2 min to eliminate any air bubbles. The sample is ready for LC-SRM analysis.

3.4 LC-SRM Analysis

The LC-SRM analysis is performed in two steps. The first step is to finalize the LC-SRM assay using the transition list obtained from Orbitrap HCD MS/MS, and the second step is to analyze the urine samples from individual subjects with the finalized method.

3.4.1 LC-SRM Setup

1. LC-SRM is performed with a nanoACQUITY UPLC system coupled online to a TSQ Vantage triple quadrupole mass spectrometer.
2. Pack the reversed-phase capillary column, ACQUITY UPLC BEH C18 column, with 1.7 μm particles, 100 mm length × 100 μm i.d.
3. Degas mobile phases online using a vacuum degasser.
4. Maintain the LC column temperature at 42 °C.
5. Use the following LC gradient:

Time (min)	Flow (μL/min)	%B
0	0.5	0.5
11	0.5	0.5
11.5	0.4	0.5
**At 13.3 min, end the injection*		
13.5	0.4	10
17	0.4	15
38	0.4	25
49	0.4	38.5
50	0.4	95
55	0.5	95
60	0.5	95
61	0.5	0.5
**At 70 min, end the method*		

6. Operate the TSQ Vantage at 1.5 mTorr, and maintain the ESI voltage at 2400 V in positive polarity with a 20 μm i.d. emitter tip. Etch the emitter tips following the previously described methods [17].
7. Scan the transitions with 0.002 m/z scan width and 0.7 m/z peak widths (FWHM) for both Q1 and Q3.
8. In nonscheduled LC-SRM method, set the scan/dwell time for each transition to 10 ms, while in scheduled LC-SRM method, set the total cycle time to 1.1 s.
9. Inject 1 μL (e.g., 1 μg) of protein digest on the LC column.

3.4.2 Finalizing the LC-SRM Method

1. Analyze one typical urine sample iteratively using nonscheduled LC-SRM with the transitions obtained from Sect. 3.1.3. Each nonscheduled method should contain a maximum of 100 transitions (including both light and heavy peptide transitions). For example, if there are 35 proteins, 105 peptides, and 1050 transitions, it will take 11 nonscheduled LC-SRM methods.

2. Import the results into the Skyline file.
3. Select the top three transitions with the highest intensity and lowest interferences.
4. Draw the peak boundaries of each peptide to achieve retention time (RT) for each peptide.
5. Export the scheduled LC-SRM method with the average RT of all three transitions.

3.4.3 Analyze the Individual Urine Samples

1. Randomize all the urine samples and perform the scheduled LC-SRM method.
2. Run one blank sample (buffer A) between adjacent urine samples to minimize cross-sample contamination.

3.5 Data Analysis

Since both the concentrations of heavy internal standards and the loading amounts are the same across all samples, the peak area ratios between endogenous transitions and heavy internal standard transitions will represent the molar ratios between the amounts of endogenous and heavy internal standard peptides. The peak area ratios can be calculated using Skyline software [11], especially for large-scale studies.

1. Import the LC-SRM datasets into the Skyline file.
2. Manually examine the peak boundaries of all the peptides for each individual dataset.
3. Evaluate whether the peptide is detected and which transition to use for the peak area ratio calculation. The detection of endogenous peptides is mainly examined by their signal-to-noise (S/N) ratios and the agreement between the relative intensities of the three transitions of the endogenous peptides and that of the heavy isotope-labeled internal standard. As you can see in Fig. 2, the S/N ratios of all three transitions of endogenous peptide A are more than 10, and the relative abundance of all three transitions of the endogenous peptide A is in good agreement with that of its internal standard, so we can confirm that the endogenous peptide A is clearly detected in the urine sample. We can either use the best transition, the transition with highest SRM response and lowest noise-to-signal ratio, in this case, the Blue transition, or the average peak area ratios of the three transitions to quantify the protein. In comparison, the endogenous peptide B is not detected because the signals of all three transitions of endogenous peptide B are close to their noises. In the case of peptide C, a significant level of interferences is observed. The S/N ratio of the Blue transition of endogenous peptide C is more than 10, but the S/N ratios of the other two transitions, both Purple and Red, are close to 1. Furthermore, the relative intensity between these three transitions of endogenous peptide C is not comparable with that

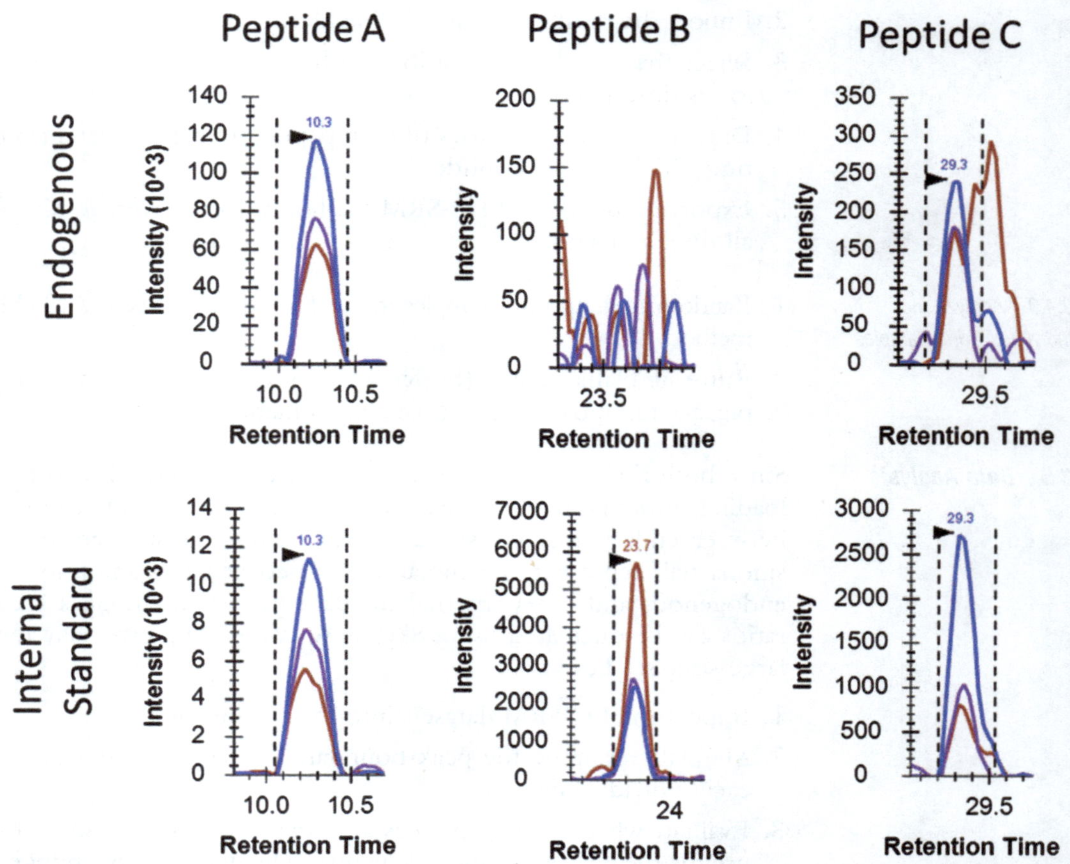

Fig. 2 Extracted ion chromatograms (XICs) of three peptides (A, B, C) in a urine sample. The dotted lines demonstrate the peak boundaries, while the arrows indicate the retention times. Peptide A shows confident detection and quantification, while peptide B lacks of clear signals, and peptide C suffers from matrix interferences

of the internal standard. In that case, the endogenous peptide C is still considered detected, but only the best transition, the Blue transition, can be used for the protein quantification.

4. Export the results with "Peptide Sequence," "Precursor Mz," "Product Mz," "Dataset Name," and "Area Ratio." (The "Area Ratio" is the peak area ratio for each individual transition between endogenous and internal standard peptides.)

5. Calculate the concentration of protein in urine using the following equation. The equation is based on the assumption that there is the same peptide recovery across all of the urinary proteins, including targeted proteins (*see* **Note 7**):

$$\text{Target protein concentration } (\text{ng}/_{\text{mL}}) = L/_{H} \times C_{\text{heavy, LC-SRM}} \div C_{\text{digest, LC-SRM}} \times C_{\text{protein}} \times 400\,\mu\text{L} \times \text{MW} \times 10^{-6}\frac{\text{ng}}{\text{fg}} \div 10\text{ mL}$$

where $L/_{H}$ is the peak area ratio of endogenous (light) and heavy internal standard peptides; $C_{\text{heavy, LC-SRM}}$ is the molar concentration of heavy internal standard peptides (fmol/μL) in the final LC-SRM solutions; MW is the molecular weight of targeted protein (Da or g/mol); $C_{\text{digest,LC-SRM}}$ is the mass concentration of total protein digest (μg/μL) in the final LC-SRM solutions; and C_{protein} is mass concentration of total protein (μg/μL) in the 400 μL concentrated retentate from 10 mL of original urine.

6. Normalize the protein concentration by either total urinary protein mass or urinary creatinine concentration:

$$\text{Normalized target protein concentration } \left(\text{ng}/_{\mu\text{g}}\right) = \text{target protein concentration } (\text{ng}/_{\text{mL}}) \div \text{urinary protein mass } (\text{mg}/_{\text{L}})$$

or

$$\text{Normalized target protein concentration } \left(\text{ng}/_{\mu\text{g}}\right) = \text{target protein concentration } (\text{ng}/_{\text{mL}}) \div \text{urinary creatinine concentration } (\text{mg}/_{\text{L}})$$

4 Notes

1. In the PeptideAtlas MS/MS data repository, look for "Human Urine" since we are working with urine samples.
2. In general, the number of observations in PeptideAtlas provides more accurate information than those in GPM.
3. The detailed LC-MS/MS operation was described in our recent paper [16].
4. The optimal collision energy (CE) for each y-ion transition was determined using Skyline software [11].
5. Unless otherwise stated, all solutions should be prepared in deionized water with a resistivity of 18.2 MΩ·cm.
6. DTT and iodoacetamide solution should be made fresh for each operation of digestion.
7. The internal standard peptides are in crude quality, so the protein concentration obtained is only relative concentration instead of absolute concentration.

Acknowledgments

Portions of the research were supported by NIH grant P41GM103493 and R01DK083447. The experimental work described herein was performed in the Environmental Molecular Sciences Laboratory, Pacific Northwest National Laboratory, a national scientific user facility sponsored by the DOE under contract DE-AC05-76RL0 1830.

References

1. Lange V, Picotti P, Domon B, Aebersold R (2008) Selected reaction monitoring for quantitative proteomics: a tutorial. Mol Syst Biol 4:222
2. Picotti P, Rinner O, Stallmach R, Dautel F, Farrah T, Domon B, Wenschuh H, Aebersold R (2010) High-throughput generation of selected reaction-monitoring assays for proteins and proteomes. Nat Methods 7(1):4
3. Cox J, Mann M (2011) Quantitative, high-resolution proteomics for data-driven systems biology. Annu Rev Biochem 80:27
4. Shi T, Su D, Liu T, Tang K, Camp DG II, Qian WJ, Smith RD (2012) Advancing the sensitivity of selected reaction monitoring-based targeted quantitative proteomics. Proteomics 12 (8):19
5. Rifai N, Gillette MA, Carr SA (2006) Protein biomarker discovery and validation: the long and uncertain path to clinical utility. Nat Biotechnol 24(8):13
6. Whiteaker JR, Zhao L, Anderson L, Paulovich AG (2010) An automated and multiplexed method for high throughput peptide immunoaffinity enrichment and multiple reaction monitoring mass spectrometry-based quantification of protein biomarkers. Mol Cell Proteomics 9(1):13
7. Tang HY, Beer LA, Barnhart KT, Speicher DW (2011) Rapid verification of candidate serological biomarkers using gel-based, label-free multiple reaction monitoring. J Proteomie Res 10 (9):13
8. Whiteaker JR, Lin C, Kennedy J, Hou L, Trute M, Sokal I, Yan P, Schoenherr RM, Zhao L, Voytovich UJ, Kelly-Spratt KS, Krasnoselsky A, Gafken PR, Hogan JM, Jones LA, Wang P, Amon L, Chodosh LA, Nelson PS, McIntosh MW, Kemp CJ, Paulovich AG (2011) A targeted proteomics-based pipeline for verification of biomarkers in plasma. Nat Biotechnol 29(7):10
9. Selevsek N, Matondo M, Sanchez Carbayo M, Aebersold R, Domon B (2011) Systematic quantification of peptides/proteins in urine using selected reaction monitoring. Proteomics 11(6):13
10. Sigdel TK, Salomonis N, Nicora CD, Ryu S, He J, Dinh V, Orton DJ, Moore RJ, Hsieh SC, Dai H, Thien-Vu M, Xiao W, Smith RD, Qian WJ, Camp DG 2nd, Sarwal MM (2014) The identification of novel potential injury mechanisms and candidate biomarkers in renal allograft rejection by quantitative proteomics. Mol Cell Proteomics 13(2):11
11. MacLean B, Tomazela DM, Shulman N, Chambers M, Finney GL, Frewen B, Kern R, Tabb DL, Liebler DC, MacCoss MJ (2010) Skyline: an open source document editor for creating and analyzing targeted proteomics experiments. Bioinformatics 26(7):3
12. EW Deutsch HL, Aebersold R (2008) PeptideAtlas: a resource for target selection for emerging targeted proteomics workflows. EMBO Rep 9 (5):6
13. R Craig JC, Beavis RC (2004) Open source system for analyzing, validating, and storing protein identification data. J Proteomie Res 3 (6):9
14. Eyers CE, Lawless C, Wedge DC, Lau KW, Gaskell SJ, Hubbard SJ (2011) CONSeQuence: prediction of reference peptides for absolute quantitative proteomics using consensus machine learning approaches. Mol Cell Proteomics 10(11):M110.003384
15. Fusaro VA, Mani DR, Mesirov JP, Carr SA (2009) Prediction of high-responding peptides for targeted protein assays by mass spectrometry. Nat Biotechnol 27:9
16. Wu C, Shi T, Brown JN, He J, Gao Y, Fillmore TL, Shukla AK, Moore RJ, Camp DG II, Rodland KD, Qian WJ, Liu T, Smith RD (2014) Expediting SRM assay development for large-scale targeted proteomics experiments. J Proteome Res 13(10):9
17. Kelly RT, Page JS, Luo Q, Moore RJ, Orton DJ, Tang K, Smith RD (2006) Chemically etched open tubular and monolithic emitters for nanoelectrospray ionization mass spectrometry. Anal Chem 78(22):6

Methods in Molecular Biology (2018) 1788: 157–164
DOI 10.1007/7651_2017_91

Published online: 02 December 2017

Fractionation of Soluble Proteins Using DEAE-Sepharose, SP-Sepharose, and Phenyl Sepharose Chromatographies for Proteomics

Parhom Towfighi, Jacob Shaw, and Tara Sigdel

Abstract

In order to simplify a complex mixture of soluble proteins from tissues, a protocol to fractionate samples prior to two-dimensional (2D) gel electrophoresis has been developed. These methods involve the use of DEAE-Sepharose, SP-Sepharose, and phenyl Sepharose chromatographic columns and the fractionation of the protein mixtures based on differential anionic, cationic, and hydrophobic properties of the proteins, respectively. Fractionation of the soluble proteins with DEAE-Sepharose can result in an increase in the number of detectable 2D gel spots. These gel spots are amenable to protein identification by using in-gel trypsin digestions, matrix-assisted laser desorption/ionization time-of-flight mass spectrometry, and peptide mass fingerprinting. The DEAE-Sepharose column fractionation acts to partition soluble proteins from cell extracts. Similarly, a SP-Sepharose column can fractionate soluble proteins and increase the number of detectable gel spots. Lastly, fractionation of cell extract with a phenyl Sepharose column can also result in an increase in the number of detectable 2D gel spots. This chapter describes an easy, inexpensive way to fractionate soluble proteins and a way to better profile proteomes.

Keywords DEAE-Sepharose, Fractionation, Phenyl Sepharose, SP-Sepharose

1 Introduction

Progress in genomics has not only dramatically revolutionized our understanding of the number and nature of genes but has also opened the door for a global and comprehensive study of cellular components and metabolic pathways. Efforts have been made to analyze the levels of gene expression by using microarrays and DNA chip methods. Specifically, DNA array data has been used to profile the levels of mRNA within a particular cell type. However, gene expression is very complex, and the direct correlation of mRNA levels to the protein content is not often possible [1].

Proteomics, which traditionally involved the use of two-dimensional (2D) gels to display large sets of proteins, has become increasingly popular as it integrated the use of mass spectrometry (MS) and database searching to identify the proteins corresponding to 2D gel spots. However, because of the heterogeneity of samples and the large differences in relative protein concentrations, the detection and identification of all of the proteins in a proteome is still a challenge. Recently, there have been efforts to develop better identification and quantification methods for the large number of proteins present in 2D gels. The development of easy, yet effective, sample fractionation methods prior to isoelectric focusing (IEF) on immobilized pH gradient (IPG) strips has become a focus area [2–7].

Different approaches have been taken to improve the resolution and the level of detection in 2D gels. Narrow-range IPG strips and larger-sized gels have been used to improve the resolution and separation of gel spots [8]. To address the issue of protein detection levels, fractionation strategies such as cellular localization and anion exchange chromatography [9] have been used to simplify cell extracts before electrophoresis. Herein, we describe a method to fractionate cell extracts before 2D gel electrophoresis (2DGE) by using different chromatographic columns, which separate proteins in terms of their ionic and hydrophobic properties.

The advantages of these approaches are (1) the columns are inexpensive and easy to use and adequately partition the tested extracts of soluble proteins, (2) there is no requirement for high-performance liquid chromatography or other automated liquid chromatographic systems, and (3) at least three times more of the proteome can be profiled using these strategies. In addition, these techniques can be readily modified to further increase our ability to probe a large majority, if not all, of certain proteomes. For example, the use of different IPG strips that cover a greater pH range (this method only uses pH 4–7 strips) will allow for the detection of proteins with pI values greater than 7 or less than 4. Increased fractionations, by using either more salt elutions or combining two or more of the columns, will lead to the detection of more proteins. Silver staining, instead of using Coomassie blue, will allow for the detection of proteins that are present in lower concentrations. The use of membrane protein fractionation and solubilization strategies [10–12] will allow for the profiling of those important proteins. Lastly, the use of longer IPG strips and larger SDS-PAGE gels will improve the resolution of gel spots. The implementation of these modifications should allow for the profiling of the entire proteomes.

2 Materials

2.1 Preparation of Soluble Protein Extracts from Tissue

1. A TissueLyser/crusher (this can be any instrument that is able to simultaneously disrupt and homogenize sample through a combination of turbulence and mechanical shearing), e.g., TissueRuptor from Qiagen (Cat No./ID: 9001271) (see **Note 1**).
2. Reagents to subsequently extract total protein from the homogenized tissue, such as AllPrep DNA/RNA/Protein Mini Kit

(Qiagen, Qproteome Mammalian Protein Prep Kit, Cat No./ID: 379012, or AllPrep DNA/RNA/Protein Mini Kit, Cat No./ID: 80004) (see **Note 2**).

3. Protein concentration assay: Pierce™ Coomassie Plus (Bradford) Assay Kit (Thermo Fisher Scientific, Cat. No. 23236) (see **Note 3**).

2.2 Fractionation with DEAE-Sepharose

1. D-(+)-Glucose (Sigma, St. Louis, MO).
2. Bovine serum albumin (Sigma, St. Louis, MO).

1. Diethylaminoethyl (DEAE)-Sepharose (Amersham Biosciences, Piscataway, NJ).
2. 1.0 × 10 cm Flex-column.
3. Column buffer A: 25 mM Tris–HCl, pH 7.5, containing 0.5% Tween 20 and 10% glycerol.
4. 25 mM Tris–HCl, pH 7.5, containing 10% glycerol and 0.15 M NaCl.
5. 25 mM Tris–HCl, pH 7.5, containing 10% glycerol and 1.0 M NaCl.

2.3 Fractionation with SP-Sepharose

1. Sulfopropyl (SP)-Sepharose (Amersham Biosciences, Piscataway, NJ).
2. 0.7 × 10 cm Flex-column.
3. 50 mM Tris–HCl, pH 7.0.
4. 50 mM Tris–HCl, pH 7.0, containing 1 M NaCl.

2.4 Fractionation with Phenyl Sepharose

1. Phenyl Sepharose (Amersham Biosciences, Piscataway, NJ).
2. 0.7 × 10 cm Flex-column.
3. 50 mM Tris–HCl, pH 7.5, containing 1.0 M $(NH_4)_2SO_4$.
4. 2.0 M $(NH_4)_2SO_4$.
5. 50 mM Tris–HCl, pH 7.5.

2.5 Acetone Precipitation and Sample Cleanup for 2D Gels

1. Prechilled acetone at −20 °C.
2. Refrigerated microfuge.
3. IEF sample buffer/rehydration buffer: 8 M urea, 50 mM dithiothreitol, 4% CHAPS, 0.2% carrier ampholytes (Bio-Rad, Hercules, CA), and 0.0002% bromophenol blue.
4. ReadyPrep 2-D Cleanup Kit (Bio-Rad, Hercules, CA).

2.6 Isoelectric Focusing and Two-Dimensional SDS-PAGE Gel Electrophoresis

1. 11 cm immobilized IPG pH 4–7 gradient strips (Bio-Rad, Hercules, CA).
2. Equilibration buffer I: 6 M urea, 2% SDS, 0.05 M Tris–HCl, pH 8.8, 20% glycerol, and 2% dithiothreitol.

3. Equilibration buffer II: 6 M urea, 2% SDS, 0.05 M Tris–HCl, pH 8.8, 20% glycerol, and 2.5% iodoacetamide.
4. Protean IEF Cell (Bio-Rad).
5. Criterion precast gels: Tris–HCl, 8–16% resolving, 4% stacking (Bio-Rad, Hercules, CA).
6. Coomassie Blue Protein Assay Dye (Bio-Rad, Hercules, CA).

2.7 Imaging and Image Analysis

1. VersaDoc Imaging System (Bio-Rad).
2. PDQuest 2-D Analysis Software 7.1.1 (Bio-Rad).

2.8 In-Gel Trypsin Digestions, MALDI-MS Analyses, and Database Searching

1. 25 mM NH_4HCO_3/50% acetonitrile solution.
2. SpeedVac (Savant Instruments, Holbrook, NY).
3. 10 mM dithiothreitol in 25 mM NH_4HCO_3.
4. 55 mM iodoacetamide.
5. 12.5 ng/μL trypsin in 25 mM NH_4HCO_3.
6. 50% acetonitrile/5% formic acid.
7. 10 mg/mL α-cyano-4-hydroxycinnamic acid in 50% acetonitrile/0.05% trifluoroacetic acid.
8. Matrix-assisted laser desorption/ionization (MALDI) target.
9. Bruker Reflex III MALDI-MS instrument.

3 Methods

3.1 Preparation of Soluble Protein Extracts from Tissue

Based on the Method Pertaining to TissueRuptor (Qiagen)

1. Familiarize with the operating instruction of TissueRuptor (see **Note 4**).
2. Place the sample material into a suitably sized vessel, and add the appropriate volume of lysis buffer (see **Note 5**).
3. Place the tip of the disposable probe into the tube containing the sample and lysis buffer. To avoid damage to the TissueRuptor and probe during sample disruption, make sure that the tip of the probe is submerged in the buffer. Important: Use a minimum of 180 μL lysis buffer in a suitably sized vessel.
4. Disrupt the sample for at least 30 s at full speed. The homogenization time can be extended until no debris is visible. Move the tip of the probe within the sample vessel during processing to increase the efficiency of disruption and homogenization (see **Note 6**).
5. After sample disruption, purify protein selecting an appropriate total protein extraction kit, such as AllPrep DNA/RNA/Protein

Mini Kit (Qiagen, Qproteome Mammalian Protein Prep Kit, Cat No./ID: 379012, or AllPrep DNA/RNA/Protein Mini Kit, Cat No./ID: 80004). Use manufacturer's manual for a proper isolation of total protein.

3.2 Fractionation with DEAE-Sepharose

1. Prepare the DEAE-Sepharose column by packing 5 mL of DEAE-Sepharose resin into a 1.0 × 10 cm Flex-column.
2. Wash the resin bed with five volumes of water and about ten volumes of column buffer A.
3. Load the cell extract, containing not more than 100 mg of total protein, onto the column, and collect the flow-through fraction.
4. Wash the column with two column volumes of column buffer A.
5. Elute bound proteins with 6 mL of 25 mM Tris–HCl, pH 7.5, containing 10% glycerol and 0.15 M NaCl, and then with 6 mL of 25 mM Tris–HCl, pH 7.5, containing 10% glycerol and 1.0 M NaCl.
6. Collect the resulting fractions to be used in subsequent steps.

3.3 Fractionation with SP-Sepharose

1. Prepare the SP-Sepharose column by packing 3 mL of SP-Sepharose resin into a 0.7 × 10 cm Flex-column.
2. Wash the resin bed with five column volumes of water and ten column volumes of 50 mM Tris–HCl, pH 7.0.
3. Load a cell extract containing no more than 50 mg of total protein onto the column, and collect the flow-through.
4. Wash the column with two column volumes of 50 mM Tris–HCl, pH 7.0.
5. Elute bound proteins from the column with 6 mL of 50 mM Tris–HCl, pH 7.0, containing 1 M NaCl.

3.4 Fractionation with Phenyl Sepharose

1. Prepare a phenyl Sepharose column by packing 3 mL of phenyl Sepharose resin into a 0.7 × 10 cm Flex-column.
2. Wash the resin bed with five column volumes of water and ten column volumes of 50 mM Tris–HCl, pH 7.5, containing 1.0 M $(NH_4)_2SO_4$.
3. Precipitate a fraction of proteins from a cell extract containing no more than 50 mg of total protein by making the solution 2.0 M in $(NH_4)_2SO_4$.
4. Separate the precipitate by centrifugation.
5. Load the supernatant onto the equilibrated phenyl Sepharose column, and discard the flow-through.

6. Wash the column with two column volumes of 50 mM Tris–HCl, pH 7.5, containing 1.0 M $(NH_4)_2SO_4$.
7. Elute bound proteins from the column with 6 mL of 50 mM Tris–HCl, pH 7.5.

3.5 Acetone Precipitation and Sample Cleanup for 2D Gels

1. Mix the sample (not less than 0.1 mg/mL) to be precipitated with an equal volume of prechilled acetone at −20 °C and vortex the tube.
2. Incubate the mixture on ice for 30 min and centrifuge at top speed in a refrigerated microfuge for 30 min.
3. Dry and resuspend the pellet very thoroughly in IEF sample buffer.
4. Use a 500 g sample in the gel cleanup procedure with the Ready-Prep 2-D Cleanup Kit.

3.6 Isoelectric Focusing and Two-Dimensional SDS-PAGE Gel Electrophoresis

1. Dissolve protein samples (100–300 μg) in rehydration/sample buffer.
2. Condition the IPG strips by treating them in equilibration buffers I and II.
3. Conduct isoelectric focusing on the 11 cm IPG strips (pH 4–7) by using the Protean IEF Cell (Bio-Rad).
4. After focusing, store the IPG strips at −80 °C until use in subsequent steps.
5. Run the second dimension SDS-PAGE gel using Criterion Precast Gels and stain the gels with Coomassie blue.

3.7 Imaging and Image Analysis

1. Collect images of gels by using the VersaDoc Imaging System and analyze by using the PDQuest 2-D Analysis Software 7.1.1.
2. Ascertain the number of protein spots by using the Spot Detection Wizard in the PDQuest software package.

3.8 In-Gel Trypsin Digestions, MALDI-MS Analyses, and Database Searching

1. Excise and vortex protein spots with 100 μL of 25 mM NH_4HCO_3/ 50% acetonitrile solutions for 10 min.
2. Repeat this step at least twice before drying the gels completely by using the SpeedVac.
3. Treat the dried gel pieces with 25 μL of 10 mM dithiothreitol in 25 mM NH_4HCO_3 and incubate at 56 °C for 1 h.
4. Remove the supernatant, add 25 μL of 55 mM iodoacetamide onto the gels, and incubate the mixture at room temperature for 45 min in the dark.
5. Discard the supernatants, treat the dried gel pieces with 25 μL of 12.5 ng/μL trypsin in 25 mM NH_4HCO_3, and incubate the mixtures overnight at 37 °C.
6. Collect the supernatants from the trypsin-digested mixtures in separate tubes.

7. Extract the peptides twice by treating the gel pieces with 30 μL 50% acetonitrile/5% formic acid and centrifuging the samples for 5 min.
8. Dry the extract to a 10 μL volume by using the SpeedVac.
9. Mix a 2 μL sample of the peptide extract with 2 μL of 10 mg/mL α-cyano-4-hydroxycinnamic acid in 50% acetonitrile/0.05% trifluoroacetic acid, and load 2 μL of this mixture onto a matrix-assisted laser desorption/ionization (MALDI) target.
10. Employ the Bruker Reflex III MALDI-MS instrument for MALDI-time-of-flight (TOF)-MS analysis of trypsin-digested peptides.
11. Obtain mass spectra by setting the instrument in reflectron mode with a reflectron detector voltage of 1.45 kV. Collect spectra in the m/z range of 1000–2600 Da.
12. Use peptide mass fingerprints obtained by MALDI-TOF analysis to search databases in SWISS-PROT (http://us.expasy.org) and NCBI and GenBank (www.ncbi.nlm.nih.gov), and assign using ProFound (http://prowl.rockefeller.edu) and MASCOT (www.matrixscience.com).

4 Notes

There are multiple methods of tissue disruption and homogenization. Researchers can choose a method that is suitable for their need in terms of size and multiplexing ability.

1. Tissue extraction kits are available from different vendors, and researchers should make a choice based on their availability and based on the quality/price.
2. Multiple kits exist and multiple methods exist to quantify total protein in samples. Researchers can use their discretion to use the method of their preference.
3. Make sure to be aware of the subsequent total protein extraction kit protocol. After storage in RNAlater RNA Stabilization Reagent, tissues become slightly harder than fresh or thawed tissues. However, disruption and homogenization of this stabilized tissue is usually not a problem. Fibrous tissue samples may not be completely homogenized after processing with the TissueRuptor; small amounts of debris do not affect the RNA purification procedure, and the tissue sample will usually be fully digested after treatment with proteinase K in the RNA purification procedure. Disruption in lysis buffer may lead to foaming. Use a suitably sized vessel for disruption. If foaming occurs, either let the sample stand at room temperature (15–25 °C) for a few minutes or centrifuge the sample briefly before continuing with the purification protocol.

Disruption for more than 15 s may generate heat, depending on the disruption vessel used.

4. For efficient disruption of tissue samples, the size of the sample must not be greater than half the diameter of the TissueRuptor probe. Cut the sample into smaller pieces, if necessary.
5. For purification of total nucleic acids, keep the disruption time as short as possible. Longer disruption times may lead to shearing of genomic DNA.
6. Sample material can be homogenized for up to 5 min without adversely affecting the quality of the RNA.

References

1. Gygi SP et al (1999) Correlation between protein and mRNA abundance in yeast. Mol Cell Biol 19(3):1720–1730
2. Righetti PG et al (2005) Prefractionation techniques in proteome analysis: the mining tools of the third millennium. Electrophoresis 26 (2):297–319
3. Righetti PG et al (2003) Prefractionation techniques in proteome analysis. Proteomics 3(8):1397–1407
4. Lilley KS, Razzaq A, Dupree P (2002) Two-dimensional gel electrophoresis: recent advances in sample preparation, detection and quantitation. Curr Opin Chem Biol 6(1):46–50
5. Gorg A et al (2002) Sample prefractionation with Sephadex isoelectric focusing prior to narrow pH range two-dimensional gels. Proteomics 2(12):1652–1657
6. Bae SH et al (2003) Strategies for the enrichment and identification of basic proteins in proteome projects. Proteomics 3(5):569–579
7. Issaq HJ et al (2002) Methods for fractionation, separation and profiling of proteins and peptides. Electrophoresis 23(17):3048–3061
8. Hoving S, Voshol H, van Oostrum J (2000) Towards high performance two-dimensional gel electrophoresis using ultrazoom gels. Electrophoresis 21(13):2617–2621
9. Pasquali C, Fialka I, Huber LA (1999) Subcellular fractionation, electromigration analysis and mapping of organelles. J Chromatogr B Biomed Sci Appl 722(1–2):89–102
10. Molloy MP et al (2002) Profiling the alkaline membrane proteome of Caulobacter crescentus with two-dimensional electrophoresis and mass spectrometry. Proteomics 2(7):899–910
11. Santoni V, Molloy M, Rabilloud T (2000) Membrane proteins and proteomics: un amour impossible? Electrophoresis 21 (6):1054–1070
12. Molloy MP (2000) Two-dimensional electrophoresis of membrane proteins using immobilized pH gradients. Anal Biochem 280(1):1–10

Methods in Molecular Biology (2018) 1788: 165–175
DOI 10.1007/7651_2017_76

Published online: 10 October 2017

GeLC-MS: A Sample Preparation Method for Proteomics Analysis of Minimal Amount of Tissue

Manousos Makridakis and Antonia Vlahou

Abstract

Application of various proteomics methodologies have been implemented for the global and targeted proteome analysis of many different types of biological samples such as tissue, urine, plasma, serum, blood, and cell lines. Among the aforementioned biological samples, tissue has an exceptional role into clinical research and practice. Disease initiation and progression is usually located at the tissue level of different organs, making the analysis of this material very important for the understanding of the disease pathophysiology. Despite the significant advances in the mass spectrometry instrumentation, tissue proteomics still faces several challenges mainly due to increased sample complexity and heterogeneity. However, the most prominent challenge is attributed to the invasive procedure of tissue sampling which restricts the availability of fresh frozen tissue to minimal amounts and limited number of samples. Application of GeLC-MS sample preparation protocol for tissue proteomics analysis can greatly facilitate making up for these difficulties. In this chapter, a step by step guide for the proteomics analysis of minute amounts of tissue samples using the GeLC-MS sample preparation protocol, as applied by our group in the analysis of multiple different types of tissues (vessels, kidney, bladder, prostate, heart) is provided.

Keywords GeLC-MS, Proteomics analysis, Sample preparation, Tissue proteomics

1 Introduction

Advances in LC-MS/MS based approaches have vastly contributed to the comprehensive analysis of tissue proteome [1–3]. Shotgun approaches are the leading methodologies in the field of proteomics at the moment. Sensitive, hybrid mass spec instruments with high mass resolution and accuracy (e.g., orbitrap) coupled to nano flow liquid chromatography systems provide massive amount of data. The aforementioned systems are dominant in the proteomics methodology nowadays [4].

LC-MS/MS analysis begins with the separation of tryptic peptides by liquid chromatography to be followed by their electrospray ionization (ESI) and injection to the mass spectrometer [4]. Tryptic peptides are produced by trypsin digestion of the solubilized tissue samples (usually overnight in-solution incubation with trypsin). Many variations for this type of protocols have been described in the literature. However, these can be practically split to two main approaches [5, 6]: Classical in-solution digestion protocols and FASP (Filter Aided Sample Preparation).

Classical in-solution digestion protocols utilize a relatively mild buffer for tissue homogenization and protein solubilization. Most frequently, 50 mM ammonium bicarbonate pH 8.5 supplemented with 0.05% SDS is being used for this purpose. Higher concentrations of detergents and/or reducing agents (DTE, TCEP) are usually not preferable as they can interfere/dectivate recombinant trypsin utilized for protein digestion. A minimal amount of detergent (usually SDS) can be combined with sodium deoxycholate to perform better protein solubilization. However, the final concentration of this compound cannot exceed 0.1% for a successful outcome of the trypsinization process. The major pitfall of this approach is the relatively mild buffer that is being utilized for the protein solubilization purposes compromising the protein extraction efficiency. In addition, the buffer has to be compatible with the subsequent trypsinization process plus a minimum final volume of buffer (e.g., 50–100 μL) defined by an optimum enzyme: protein ratio (ratios vary from 1:10 up to 1:100) have to be applied. In cases where high concentrations of detergents (up to 4% SDS) or chaotropic agents (up to 8 M urea) are utilized for the homogenization purposes, a subsequent precipitation step is required to remove detergents and/or chaotropic agents prior to trypsinization. However, precipitation increases sample variability and leads to protein loss which should be avoided in the analysis of minute amount of tissue samples.

On the other hand, FASP is a variation of in solution digestion developed by Mann's group [6]. Sample preparation is being conducted in special filters of regenerated cellulose (Amicon-Meck-Millipore) or polyethersulfone membrane (Sartorius) with specific molecular weight cut off (usually 30 kDa). This protocol utilizes in solution digestion but the trypsinization is being performed inside the filter unit. Filters allow utilization of detergents (SDS, Urea) and reducing agents (DTE, TCEP) at higher concentrations (up to 4% SDS, 7 M Urea, 0.1 M DTE, 0.1 M TCEP) compared to the classical in solution digestion protocol, making protein solubilization much more efficient. Detergents and other interfering compounds are being removed from the sample after centrifugations of the filter (buffer exchange). This protocol is superior to the classical one but its major drawback is that it requires an initial amount of total protein equal to 200 μg to be loaded into the filter. In the protocol developed by Mann's group, it is reported that in a single filter device 0.2–200 μg can be processed. However, due to the many centrifugation steps required for the buffer exchange, alkylation etc. until the final step of trypsinization, this protocol is associated with protein loss. In our hands this protocol did not perform very well in terms of the LC-MS/MS analysis when starting material was below 100–200 μg.

Considering that the most prominent challenge in tissue proteomics analysis is the minimal amount of tissue obtained usually by biopsies, another sample preparation method which is actually gel based (SDS PAGE) can be alternatively performed. A slightly modified protocol of the GeLC-MS method described by Lawlor et al. [7] has been extensively used by our group with high quality and reproducible results obtained for many different types of tissue samples (kidney, bladder, vessels, heart, prostate). The major advantage of this protocol is that it can perform very well with minimal amount of starting material (5–10 μg total protein). This is to some extent linked to the utilization of detergents and reducing agents in the solubilization buffer in concentrations such as those regularly used in FASP, allowing more efficient protein extraction. This GeLC-MS protocol will be described in detail in the following pages.

Table 1 provides a summary of the advantages and drawbacks of the three methodologies that were briefly described in this section.

Table 1
Advantages and drawbacks of the reported sample preparation methodologies for LC-MS/MS proteomics analysis

Sample preparation method for LC-MS/MS analysis	Advantages	Disadvantages
In-solution digestion	1. No severe protein loss 2. Compatible with label free and isobaric-isotopic labeling methods 3. Relative fast protocol	1. Not optimum protein extraction efficiency
FASP	1. Efficient protein extraction 2. Sample purification 3. Compatible with label free and isobaric-isotopic labeling methods	1. Increased amount of starting material required for good results (100–200 μg) 2. Protein loss in the filter units
GeLC-MS	1. No severe protein loss 2. Efficient protein extraction 3. Minimal amount of starting material required (e.g., 5–10 μg) 4. Sample purification	1. Time-consuming and laborious procedure

2 Materials

All solutions should be prepared with ultrapure water (18 MΩ cm at 25 °C) and analytical grade reagents (≥99% purity). Prepare and store all buffers at room temperature unless specified otherwise. Please use the appropriate precautions as stated in the material safety data sheet when handling the chemical reagents and solvents described in this section.

2.1 Lysis Buffer

Two different buffers containing high concentration of detergent (SDS) or chaotropic agent (urea) can be used for the homogenization process (see Note 1). Buffer 1: 0.1 M Tris–HCl, pH 7.6, 4% v/v SDS, 0.1 M DTE, 3.6% v/v protease inhibitors (upon lysis, see Note 2). Buffer 2: 7 M Urea, 2 M Thiourea, 4% w/v CHAPS, 1% w/v DTE, 2% v/v IPG buffer (ampholytes), 3.6% v/v protease inhibitors (upon lysis, see Note 1).

2.2 Homogenization

After selecting the suitable buffer for the best solubilization performance, the homogenization method plays a key role in the sample preparation process. There are five different types of homogenization methods that can be utilized in tissue proteomics experiments: (1) Powderizing tissue in liquid nitrogen with pestle and mortar (see Note 3), (2) Teflon potter homogenizers (see Note 4), (3) ultrasonics produced by water bath sonication or tip sonication (see Note 5), (4) mechanical blade homogenizers (see Note 6), (5) bead based homgenizers (see Note 7).

2.3 Bradford Assay

Sample protein concentration is often estimated by colorimetric approaches such as Bradford Assay. Bradford reagent is commercially available and the concentration is estimated based on the absorbance reading at 595 nm in a photometer. A standard curve (usually with BSA standards of different concentrations) is used for calibration purposes.

2.4 SDS PAGE

1. Separating gel buffer: 12% v/v Acrylamide–Bis solution (37.5:1), 0.375 M Tris–HCl pH 8.8, 0.1% v/v SDS, 0.05% v/v APS, 0.05% v/v TEMED.
2. Stacking gel buffer: 5% v/v Acrylamide–Bis solution (37.5:1), 0.125 M Tris–HCl pH 6.8, 0.1% v/v SDS, 0.05% v/v APS, 0.05% v/v TEMED.
3. 4× Laemmli loading buffer: 0.2 M Tris–HCl pH 6.8, 8% w/v SDS, 40% v/v Glycerol, 20% v/v β-mercaptoethanol, 0.4% w/v bromophenol blue.
4. Molecular weight markers: Different types of markers covering a wide range of molecular weights are commercially available.
5. TGS running buffer: 0.025 M Tris–HCl pH 8.3, 0.192 M Glycine, 0.1% v/v SDS.

2.5 Gel Fixation and Stain

1. Fixation buffer: 30% v/v methanol, 10% v/v acetic acid.
2. Coomassie colloidal blue stain: 10% v/v orthophosphoric acid, 10% w/v ammonium sulfate, 0.12% w/v Coomassie G250, 20% v/v methanol.

2.6 Buffers for the Processing of Excised Gel Pieces

1. Destain solution: 40% v/v acetonitrile, 0.05 M ammonium bicarbonate.
2. Reduction solution: 0.01 M DTE, 0.1 M ammonium bicarbonate.
3. Alkylation solution: 10 mg/mL iodoacetamide, 0.1 M ammonium bicarbonate.
4. Wash solution: 0.1 M ammonium bicarbonate.
5. Ultrapure water (18 MΩ cm at 25 °C).

2.7 Trypsinization and Peptide Extraction

1. Trypsin solution: 10 ng/μL trypsin (proteomics grade), 0.01 M ammonium bicarbonate, pH 8.5.
2. Peptide extraction solution 1: 0.05 M ammonium bicarbonate.
3. Peptide extraction solution 2: 50% v/v acetonitrile, 5% formic acid.
4. Peptide loading buffer (for the LC-MS/MS analysis): 0.1% v/v formic acid, pH 3.5.

3 Methods

1. Tissue homogenization and protein solubilization.
   - Homogenize tissue in a bead based homogenizer (see Note 8) with the following settings: step 1: Speed 12, Time: 5 min, step 2: Speed 10, Time: 3 min.
   - Use stainless steel homogenization beads (0.9–2.0 mm diameter).
   - After the homogenization process is finished, centrifuge samples at 16,000 × *g* for 10 min at room temperature (see Note 9).
   - Keep the supernatant in a clean tube and add protease inhibitors.
   - Determine protein concentration with a colorimetric assay (e.g., Bradford or Lowry).
   - At this point, samples are quite stable and can be stored at −80 °C until use.
2. Sample loading and SDS-PAGE.
   - Load 10 μg of total protein content from each sample into the polyacrylamide gel (5% stacking, 12% separating) and perform electrophoresis at 40 V for 15 min followed by 120 V for 10 min at room temperature.

- This time is sufficient for the samples to migrate from the stacking to the beginning of the separating gel when using mini electrophoresis system (Fig. 1).
- If different type of electrophoresis systems is utilized (e.g., maxi system) then the time should be adjusted. The main idea is that the samples should not be left to separate in the resolving gel (see Note 10).

3. Fixation and Coomassie colloidal blue stain.
   - Fix the gel with fixation buffer for 30 min at room temperature in a shaking platform.
   - Discard fixation buffer and wash the gel with ultrapure water for 10 min at room temperature in a shaking platform.
   - Repeat the washing step twice.
   - Discard water and add Coomassie colloidal blue stain [8].
   - Incubate the gel with the stain for 1 h at room temperature in a shaking platform (see Note 11).
   - Discard colloidal stain and wash the gel with ultrapure water as previously described to remove the excess of the stain.
   - A volume of 50 mL is sufficient for all the above buffers for a mini gel. If the gel is midi or maxi the volume should be adjusted (75 mL for midi gel, 100 mL for maxi gel). It is important for the gel to be submerged in the different incubation buffers.
4. Band excision and processing of the gel pieces.
   - Place the gel in a clean glass surface, excise the band of each sample (1 band per sample) and cut the band in small pieces (1–2 mm size) by utilizing a scalpel (see Note 12).
   - Keep the gel pieces in a clean tube.
   - Destain the gel pieces with destain solution (RT, in a shaking platform) until they become completely transparent (see Note 13).
   - Discard destain solution.

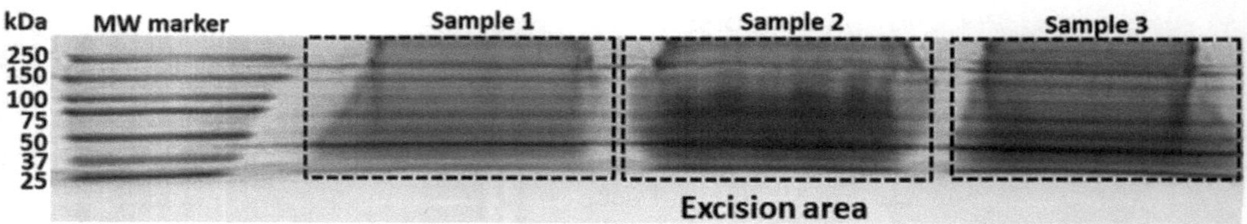

Fig. 1 Coomassie colloidal blue stain of samples that were run with SDS PAGE until they reached the separating gel. Minimum resolution of the samples in the separating gel was allowed. Condensed bands shown for each sample include the total protein content. Band excision area is also depicted

- Reduce with reducing solution for 20 min at room temperature.
- Discard reduction solution.
- Alkylate with alkylation solution for 20 min, at room temperature, in the dark.
- Discard alkylation solution.
- Wash with 0.1 M ammonium bicarbonate for 20 min (RT, in a shaking platform).
- Discard solution.
- Wash with destain solution for 20 min, (RT, in a shaking platform).
- Discard destain solution.
- Wash with ultrapure water for 20 min (RT, in a shaking platform).
- Discard water.
- Dry the samples in a vacuum centrifuge. In all incubation steps mentioned above, the recommended volume of buffers is 200 μL. Volume can be adjusted if necessary so that the gel pieces to be submerged in the different incubation buffers.

5. Trypsin digestion and extraction of the tryptic peptides.
   - Trypsinize samples by adding 600 ng of trypsin (proteomics grade).
   - Use 10 ng/μL tryspin stock solution in 0.01 M ammonium bicarbonate pH 8.5 (60 μL trypsin stock solution are required, see Note 14).
   - Incubate gel pieces with trypsin overnight in a humidified container at room temperature in the dark.
   - The next day proceed with peptide extraction from the gel pieces.
   - Trypsin solution will be absorbed by the gel pieces which will now be swollen.
   - Extract the peptides as follows: Incubate with peptide extraction solution 1 (0.2 mL) for 15 min (RT, in a shaking platform). Keep the supernatant in a clean tube (see Note 15). Incubate with peptide extraction solution 2 for 15 min (RT, in a shaking platform). Mix this supernatant with the previous one. Repeat the previous step. Add the supernatant to the same tube where the previous ones were kept. The final volume of extracted peptide solution will be 0.6 mL. Clean each peptide solution with PVDF filters (0.22 μm) (see Note 16).
   - Dry the clean peptide solution in a vacuum centrifuge (see Note 17) and resuspend the dried peptides in the peptide loading buffer (see Note 18).

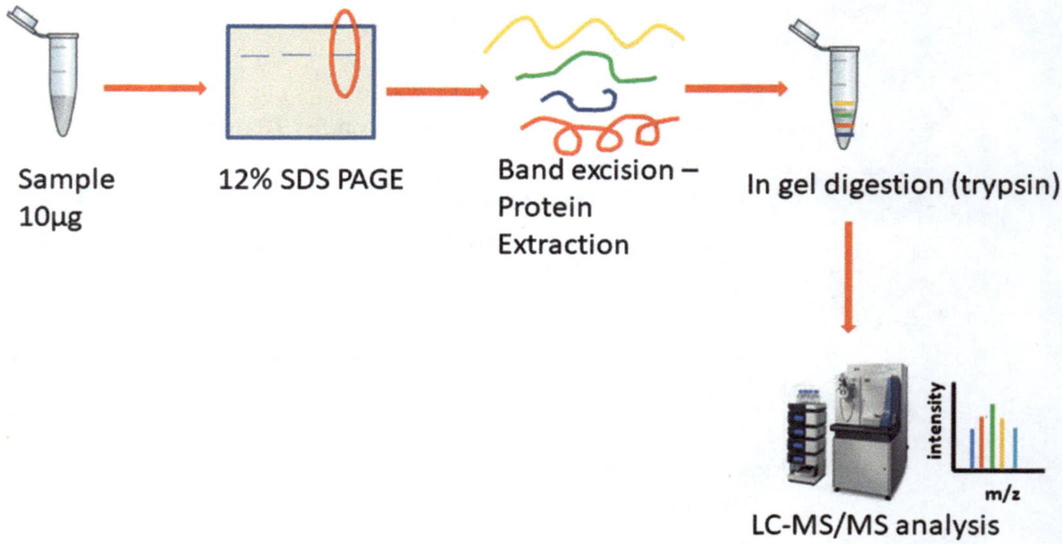

Fig. 2 Workflow of the GeLC-MS sample preparation method

- Resuspension of the dried peptides should be performed just prior the LC-MS/MS analysis. If the samples are not analyzed immediately it is preferably to store them as pellet at −20 °C rather than solubilizing them in the peptide loading buffer.

The general workflow of the GeLC-MS protocol is shown in Fig. 2.

4 Notes

1. Lysis buffers described in Sect. 2 are very efficient in solubilizing tissue samples and extracting proteins mainly due to their high concentration in detergent and reducing agent. It should be noted that if lysis buffer 1 is selected, samples should be diluted at least 20 times (1:20 dilution) prior to Bradford Assay for the determination of protein concentration. High SDS concentration interacts with the Coomassie included in Bradford reagent leading to overestimation of protein content in samples. Existence of high SDS concentration in the sample produces a vivid green color when sample is mixed with Bradford reagent resulting in overestimation of protein concentration.
2. Lysis buffer is being prepared by adding all components besides protease inhibitors at room temperature and stored in aliquots

at −20 °C. Protease inhibitors are added fresh upon tissue sample lysis.

3. This is a difficult and time consuming method leading to major losses in protein yield obtain. Moreover, it is not optimal for minimal amount of tissue. Contamination among different samples can also be a problem if no extensive wash of pestle and mortar is performed after each homogenization procedure.
4. This homogenization method is working properly with soft tissue (e.g., brain). However, if the tissue is not soft enough (e.g., bladder, vessels) it does not work optimal. Finally, in most cases of this homogenization process, the sample needs to be homogenized in a big volume of lysis buffer thus resulting to sample dilution.
5. Ultrasonics provide good results in homogenization procedure. Water bath sonication is milder than tip sonication which is more aggressive. Usually a combination of water bath sonication and tip sonication is being performed for better results. During sonication process (especially tip sonication) samples tend to get overheated and thus working on ice and introducing pauses among tip pulses are mandatory.
6. Mechanical blade (blender) homogenization produces relatively good results but it is time consuming, requires thorough wash of the mechanical blender parts among the analysis of different samples and its reproducibility is not always optimal. Overheating problems may occur in this method as well.
7. Bead-based homogenization relies on collision of the beads with tissue. Beads of different material (stainless steel, zirconium oxide, glass etc.) are commercially available and can be selected depending on the tissue type. Soft tissues can be solubilized with glass beads, zirconium oxide works better with intermediate types of tissue and stainless steel is recommended for the more tough ones. Beads are usually inserted in the tube containing the sample and their acceleration is being produced by vortexing and/or shaking in the homogenizer. This type of homogenizers can analyze simultaneously up to 24 samples with very good reproducibility and results obtained in our hands. Air cool system is protecting the samples from overheating. Contamination effect is prevented since the beads are disposable and are used only one time.
8. Tissue homogenization can be performed with any of the methods described in Sect. 2.2. However, we have noticed that bead based homogenization works better and is more reproducible compared to the other methods. It is also highly recommended when the tissue sample is of minimal amount. Last but not least, this method is much faster (up to 24 samples can be processed at the same time under the same conditions in

about 10 min). Selection of the appropriate beads depends on the type of tissue. Stainless steel beads have a wide range of application and are suitable for soft (brain tissue) and tough type of tissues (vessels, bladder). Specific type of tubes (such as safe lock Eppendorf tubes) should be used with the bead based homogenizers in order to ensure that the caps will remain closed after the vigorous shaking and vortexing that is performed in the bead homogenizer.

9. Centrifugation should not be performed at low temperature (e.g., 4 °C) because detergents (SDS) and/or chaotropic agents (urea) included in the lysis buffer tend to form crystals and precipitate at low temperature. This effect can modify the concentration of the compounds in lysis buffer. It is also possible for some proteins to coprecipitate with urea, SDS crystals thus reducing the amount of solubilized protein obtained.
10. In this approach every sample is represented by one condensed band which practically includes all of its protein content. In this way a lot of time is being saved in the LC-MS/MS analysis. Every sample which is represented by one band is analyzed in one run. If the sample is allowed to be separated in the resolving gel then various bands (at least 10) from each sample will have to be processed and analyzed which makes the procedure more laborious and time consuming especially for the LC analysis.
11. Gel can be stained with colloidal stain from 1 h up to overnight without affecting the quality of the stain.
12. It is highly recommended to cut each band in small pieces (1–2 mm size) since this will speed up the destain process that follows and will lead to better trypsinization and peptide extraction.
13. It is very important to completely destain the gel pieces. Remaining of Coomassie stain in the gel pieces will inhibit trypsinization procedure resulting in a reduced number of tryptic peptides.
14. If this volume of trypsin stock solution (60 μL) is not sufficient to submerge the gel pieces, add some volume of 0.01 M ammonium bicarbonate. By addition of trypsin solution the dried gel pieces get rehydrated and absorb trypsin solution. This is the basic principle of the in-gel digestion which is utilized in the protocol described herein.
15. When aspirating the supernatant into a clean tube precaution should be taken so that gel pieces are not also transferred with the supernatant in the clean tube. Acetonitrile which is included in peptide extraction solution 2 dehydrates gel pieces resulting in their shrinkage for more efficient peptide extraction. At the same time, shrinkage of the gel pieces makes it

possible for some of them to be transferred upon aspiration of the supernatant. This should be avoided as it may result to clogging effects in the capillaries of the LC-systems.

16. This type of filters is suitable for any particle removal from aqueous and mildly organic solutions.
17. When the peptide solution is speed vac to dry a tiny pellet would be visible in the bottom of the tube.
18. For proper resuspension of the dried peptides, add the desired amount of volume from the peptide loading buffer (usually specific volume is added in order to have a final concentration of 1 μg/μL in the tryptic peptides; considering that 10 μg of total protein per sample were loaded into the gel, 10 μL of peptide loading buffer should be added). Mix by pipetting and perform water bath sonication for 30 s. Light-vortex, spin down, and perform LC-MS/MS analysis.

Acknowledgments

Supported in part by FP7 SysVasc (grant number: 603288).

References

1. Uhlen M, Fagerberg L, Hallstrom BM et al (2015) Proteomics. Tissue-based map of the human proteome. Science 347:1260419
2. Kim MS, Pinto SM, Getnet D et al (2014) A draft map of the human proteome. Nature 509:575–581
3. Wilhelm M, Schlegl J, Hahne H et al (2014) Mass-spectrometry-based draft of the human proteome. Nature 509:582–587
4. Aebersold R, Mann M (2003) Mass spectrometry-based proteomics. Nature 422:198–207
5. Leon IR, Schwammle V, Jensen ON et al (2013) Quantitative assessment of in-solution digestion efficiency identifies optimal protocols for unbiased protein analysis. Mol Cell Proteomics 12:2992–3005
6. Wisniewski JR, Zougman A, Nagaraj N et al (2009) Universal sample preparation method for proteome analysis. Nat Methods 6:359–362
7. Lawlor K, Nazarian A, Lacomis L et al (2009) Pathway-based biomarker search by high-throughput proteomics profiling of secretomes. J Proteome Res 8:1489–1503
8. Candiano G, Bruschi M, Musante L et al (2004) Blue silver: a very sensitive colloidal Coomassie G-250 staining for proteome analysis. Electrophoresis 25:1327–1333

Methods in Molecular Biology (2018) 1788: 177–184
DOI 10.1007/7651_2017_111

Published online: 02 December 2017

Targeted Proteomics Driven Verification of Biomarker Candidates Associated with Breast Cancer Aggressiveness

Iva Procházková, Juraj Lenčo, and Pavel Bouchal

Abstract

Breast cancer is the most common and molecularly well-characterized malignant cancer in women; however, its progression to metastatic cancer remains lethal for 78% of patients within 5 years of diagnosis. Identifying novel markers in high risk patients using quantitative methods is essential to overcome genetic, inter-tumor, and intra-tumor variability, and to translate novel findings into cancer diagnosis and treatment. Using untargeted proteomics, we recently identified 13 proteins associated with some key factors of breast cancer aggressiveness: estrogen receptors, tumor grade, and lymph node status. Here we verified these findings in a set of 96 tumors using targeted proteomics based on selected reaction monitoring with mTRAQ labeling (mTRAQ-SRM). This study highlights a panel of gene products that could contribute to breast cancer aggressiveness and metastasis, and can help develop more precise breast cancer treatments.

Keywords Breast cancer, mTRAQ, Selected reaction monitoring

Abbreviations

DDA	Data-dependent acquisition
ER	Estrogen receptor
G1	Tumor grade 1
G3	Tumor grade 3
iTRAQ	Isobaric tags for relative and absolute quantitation
MIDAS™	MRM initiated detection and sequencing
mTRAQ	Mass differential tags for relative and absolute quantification
mTRAQ-SRM	Selected reaction monitoring with mTRAQ labeling
PR	Progesterone receptor
SRM	Selected reaction monitoring
TEAB	Triethylammonium bicarbonate

1 Introduction

Breast cancer is the most serious oncological cause of death among women worldwide [1]. Currently, several clinicopathological parameters are being used in clinical practice as prognostic and predictive factors of breast cancer: age at diagnosis; tumor size; tumor grade; presence of lymph node metastases; distant metastases; and status of estrogen receptors (ER), progesterone receptors (PR), and Her2/Neu (HER2) receptors [2]. Currently used classification systems are effective in treatment predictions; however, they fail in some clinical scenarios. It is thus essential to recognize additional markers and molecular targets associated with various aspects of breast cancer aggressiveness to improve clinical decisions. In the last decade, involvement of shotgun proteomics with Isobaric Tags for Relative and Absolute Quantitation (iTRAQ) labeling to discover such novel targets has led to a high number of proteins presented as potential biomarkers [3]. Due to the limited reproducibility of data-dependent acquisition (DDA) [4, 5] as well as co-fragmentation of co-eluting peptides (which may affect protein quantification), it is appropriate to verify quantitative findings of DDA-based proteomics with an independent method to highlight the most suitable candidates for validation in sample cohorts. Targeted mass spectrometry based on selected reaction monitoring (SRM), a method by Nature Methods in 2012 [6], is a sensitive, quantitative, and high throughput method for both verification and validation steps. The Mass-differential Tags for Relative and Absolute Quantification (mTRAQ) labels have been designed for relative quantification in SRM and are commercially available in triplex format (mTRAQ-Δ0, mTRAQ-Δ4, and mTRAQ-Δ8). In a typical biomarker verification experiment, a global internal standard (pooled from aliquots of all samples) is prepared by labeling it with an mTRAQ-Δ8 label, whereas individual samples are labeled with any of the two remaining variants (mTRAQ-Δ0, mTRAQ-Δ4). The relative quantification is then based on the ratio of signal peak areas of the product ions originating from differently tagged peptides. In our recent breast cancer discovery study [7], we employed iTRAQ-based shotgun proteomics to identify proteins connected with lymph node status, specifically of luminal A tumors. In the current study, we further verify our previous screening findings using the mTRAQ-SRM driven targeted proteomics approach. The validated proteins may be connected to mechanisms of tumor aggressiveness and metastasis, and can be considered as molecular targets of more personalized treatment.

2 Materials

2.1 Tissue Processing and Patient Characteristics

Note: Tissues were frozen in liquid nitrogen within 20 min after surgical removal and stored at −180 °C in tissue bank at MMCI.

A set of 96 preoperatively untreated breast carcinomas of 11–20 mm maximum diameter (pT1c) was selected for the study.

Note: The sample set included 48 grade 1 tumors both ER and PR positive, 24 of them having lymph node metastases at the time of surgery. Another 48 cases were grade 3 carcinomas, 24 of them lymph node positive and 24 lymph node negative.

2.2 Sample Preparation

Frozen breast cancer tissue (approx. 20 mm^3).

Lysis buffer (6 M guanidine hydrochloride; 0.1 M Na-phosphate buffer, pH 6.6; 1% Triton X-100).

MM301 mechanic homogenizer (Retsch, Germany) using a metal ball for 2 × 2 min at 20 s^{-1}.

Bandelin 2200 Ultrasonic homogenizer, Bandelin, Germany; 30 × 0.1 s pulses at 50 W.

RC-DC protein assay (Bio-Rad, USA).

Note: a modified Lowry method that includes protein precipitation to avoid interferences of reducing agents and detergents [8].

Modified filter aided sample preparation protocol.

8 M urea in 0.5 M triethylammonium bicarbonate (TEAB), pH 8.5.

Vivacon 500 filter device, cut-off 10 K (Sartorius Stedim Biotech GmbH, Germany).

5 mM tris(2-carboxyethyl)phosphine in 8 M urea in 0.5 M TEAB, pH 8.5.

10 mM S-methyl methanethiosulfonate in 8 M urea in 0.5 M TEAB, pH 8.5.

0.5 M TEAB, pH 8.5 containing L-1-tosylamide-2-phenylethyl chloromethyl ketone (TPCK) treated trypsin (AB Sciex, USA) (trypsin to protein weight ratio 1:30).

Vacuum concentrator (SpeedVac, Thermo Scientific).

2.3 Labeling

0.5 M TEAB, pH 8.5

2-Propanol

Water

Vacuum concentrator

2.4 Cation-Exchange Chromatography

Cation-exchange cartridge supplied as a part of ICAT kit (POROS® 50 HS, 50-μm particle size, 4.0 mm × 15 mm; Applied Biosystems, USA).

Mobile phase A (10 mM KH_2PO_4 in 25% acetonitrile, pH 3).

Ultrasonic bath.
Infinity 1260 LC system (Agilent, USA).
0.1% formic acid.
C-18 MicroSpin columns (300 Å, The Nest Group, Inc., USA).

2.5 LC-SRM Assay Development

Skyline software, version 1.4 or higher (MacCoss Lab, University of Washington, Seattle, USA) [9].

UniProt database.

2.5.1 Detectability Screening

Tempo nano MDLC capLC system coupled to a 4000 QTRAP mass spectrometer (both AB Sciex, USA).

100 μm × 25 mm trapping column filled with 3 μm/100 Å pore size Atlantis dC18 particles (Waters, USA).

2% ACN, 0.1% TFA.

75 μm PicoTip Emmiter (New Objective, Woburn, USA).

100 mm column of a 3 μm/100 Å pore size Atlantis dC18 stationary phase.

MASCOT search engine (version 2.1 or higher, UniProt database).

2.5.2 Selection of Final SRM Transitions

nanoHPLC and 4000 QTRAP mass spectrometer.

2.5.3 Final LC-SRM Analysis

The most intense global internal standard peptide EYQDLLNVK from desmin.

3 Methods

3.1 Sample Preparation

1. Homogenize frozen breast cancer tissue in 150 μl of lysis buffer in MM301 mechanic homogenizer.
2. Keep on ice for 1 h.
3. Centrifuge the sample, and measure protein concentration from the supernatant using the RC-DC protein assay.
4. Use modified filter aided sample preparation protocol to digest the lysate as below:
5. Mix aliquots of the lysate corresponding to 60 μg of the protein with 200 μl of 8 M urea in 0.5 M triethylammonium bicarbonate (TEAB), pH 8.5 on Vivacon 500 filter device.
6. Centrifuge the device at 14,000 × *g* at 20 °C for 20 min.
7. Add 100 μl of 5 mM tris(2-carboxyethyl)phosphine in 8 M urea in 0.5 M TEAB, pH 8.5 to filter.
8. Incubate at 37 °C for 60 min at 600 rpm, and then centrifuge (see point 6 for conditions).

9. Add 100 μl of 10 mM S-methyl methanethiosulfonate in 8 M urea in 0.5 M TEAB, pH 8.5 to the filter.
10. Incubate at 20 °C for 10 min, and then centrifuge.
11. Dilute the resulting concentrate with 100 μl of 8 M urea in 0.5 M TEAB, pH 8.5. Centrifuge the sample again.
12. Repeat this washing step twice.
13. To subject the concentrate to proteolytic digestion, add 100 μl of 0.5 M TEAB, pH 8.5 containing L-1-tosylamide-2-phenylethyl chloromethyl ketone (TPCK) treated trypsin (AB Sciex, USA) (trypsin to protein weight ratio 1:30).
14. Incubate 37 °C for 16 h.
15. Collect digests by centrifuging into clean tubes.
16. Dry in vacuum concentrator.

3.2 Labeling

1. Reconstitute evaporated digests in 24 μl of 0.5 M TEAB, pH 8.5.
2. Process two aliquots from each sample corresponding to 10 μg of digested protein.
3. Label one sample group with mTRAQ-Δ0, and label the other sample group with mTRAQ-Δ8.
4. Prepare stock solution of each label by adding 50 μl of 2-propanol per one unit of the label.
5. For each labeling reaction, add 0.15 units to 10 μg of digested protein (pH of the mixture has to be ~8.3) [10].
6. Incubate at room temperature for 1 h.
7. Stop the labeling by adding 15 μl of water.

   Note: Faster quenching of non-reacted mTRAQ labels may be achieved by adding TFA or hydroxylamine.
8. Pool samples labeled with mTRAQ-Δ8 to create global internal standard, and then add to each individual mTRAQ-Δ0 labeled sample.
9. Dry samples in vacuum concentrator.

3.3 Cation-Exchange Chromatography

1. Clean samples from excess label and other possible contaminants using cation-exchange cartridge from ICAT kit as follows:
2. Reconstitute samples in 650 μl of mobile phase A (10 mM KH_2PO_4 in 25% acetonitrile, pH 3).
3. Sonicate samples in ultrasonic bath for 2 min.
4. Insert cartridge into Infinity 1260 LC instrument.
5. Set up the gradient as follows (flow rate 1 ml/min if not otherwise stated):

- 3 min 0% B (1 M KCl in mobile phase A, pH 3) at 0.5 ml/min (loading)
- 2 min 0% B (washing)
- 2 min 35% B (elution)
- 2 min 100% B (cleaning)
- 2 min 0% B
- 1 min 100% B (cleaning)
- 3 min 0% B (equilibration)

6. Collect single eluents in one fraction at 6.6–8.1 min from elution phase, and then vacuum-dry.
7. Reconstitute the eluents in 200 μl of 0.1% formic acid, and desalt on C-18 columns using water-acetonitrile solvents system [11].

3.4 LC-SRM Assay Development

The LC-SRM assay development, method refinement, and final data analysis were all performed in Skyline software, version 1.4 and higher [9].

3.4.1 Selection of Surrogate Peptides

1. To screen potential surrogate peptides for detectability in mTRAQ-Δ0 labeled sample (1 μg of digested total protein), pool aliquots of all samples to select the best detectable candidates.
2. Use Tempo nano MDLC capLC system coupled to a 4000 QTRAP mass spectrometer for screening.
3. Pre-concentrate peptides in a 100 μm × 25 mm trapping column filled with 3 μm/100 Å pore size Atlantis dC18 particles at a flow rate 2 μl/min for 10 min with 2% ACN, 0.1% TFA.
4. Use gradient ranging from 5 to 40% of 98% ACN with 0.1% FA to resolve peptides over 60 min at a flow rate 360 nl/min in a 75 μm PicoTip Emmiter filled in-house with 100 mm column of a 3 μm/100 Å pore size Atlantis dC18 stationary phase.
5. Apply spray voltage of 2400 V with source temperature at 175 °C.
6. Operate mass spectrometer in unscheduled SRM mode with both Q1 and Q3 set to unit mass resolution.
7. For initial SRM analyses, use 12 ms dwell time for each transition with 5 ms interscan delay between ranges.
8. Record the enhanced product ion spectrum if the transition of a precursor reached a signal intensity threshold of 5.000 cps.
9. Isolate precursors in Q1 at low resolution, and then record enhanced product ion spectra at a scan speed of 4000 amu/s between 250 and 1400 *m/z* with enabled dynamic fill time.
10. Evaluate MIDAS™ triggered MS/MS spectra using the MASCOT search engine (version 2.1 and above) with the precursor

tolerance at 5 ppm*, fragment ion tolerance at 0.5 *m/z*, and additional settings reflecting the aforementioned criteria on surrogate peptides selection.

*Note: This tight *tolerance can be used because theoretical m/z of precursors is defined in MS method and kept in the resulting MS file.*

11. Create linear equation for retention time prediction from retention time of peptides confirmed by the MIDAS™ workflow (precursor mass error below 1 ppm and MASCOT Expectation value ≤ 0.05).

3.4.2 Detectability Screening

1. Select a maximum of three peptides per protein based on the signal intensity observed in the screening runs.
2. Use 4–5 product ions ($m/z >$ precursor)-1 per precursor showing the most intense transitions confirmed by MIDAS™ experiments.
3. Do not select transitions with potential matrix interference (e.g., those with excessive mTRAQ-Δ0/mTRAQ-Δ8 ratio).

   Note: For assessing potential matrix interference use peak areas of individual transitions normalized to their sum. Contribution of each transition must be the same for mTRAQ-Δ0 and mTRAQ-Δ8 peptides across all injections.
4. Import MS/MS library of mTRAQ peptides into Skyline software in .dat format for MASCOT output.

   Note: Other formats from various search engines can be used to create libraries in newer versions of Skyline software.
5. Define transitions based on observed *m/z* of precursor and fragments stored in the MS/MS library.
6. Add 4–5 of the most intense fragments ($m/z >$ precursor)-1 per precursor for SRM transitions of successfully detected library peptides into the final LC-SRM assay.
7. Use predicated retention times to detect SRM peaks in the LC-SRM chromatograms.

   Note: Retention time calibration kits have been recently developed to predict more precisely retention time of peptides to be quantified.

3.4.3 Selection of Final SRM Transitions

1. Inspect all transitions from three replicates (termed A, B, and C), and exclude injections with unsatisfactory results from further analysis as follows:
   - Select most intense global internal standard peptide EYQDLLNVK from desmin as first quality control measure. A peak for this peptide must be detected in all injections. Injections without this peak being well detected were not further considered.
   - Only consider peptides unambiguously detected in prediction retention time windows.

2. Use only peptides with an average peak area higher than 8.000 for quantification.
3. For each peptide, select single mTRAQ-Δ8/mTRAQ-Δ0 transition pair providing the signal as high as possible and CV as low as possible from 4 to 5 measured transition pairs for relative quantification. Manually check that the selected transition pair is free of interferences [12].
4. Calculate change in protein level showing as fold change as an average mTRAQ-Δ8/mTRAQ-Δ0 ratio of each analyzed group.
5. Determine statistical significance of observed fold-change ratios ($\log_2$) from Mann-Whitney *U*-Test.

Acknowledgments

We would like to thank Rudolf Nenutil for his pathological guidance. We also thank Parhom Towfighi (UCSF Medical Centre) for his work on editing for the book. This work was supported by Czech Science Foundation (Project No. 17-05957S).
No conflict of interests.

References

1. Ferlay J et al (2013) Cancer incidence and mortality patterns in Europe: estimates for 40 countries in 2012. Eur J Cancer 49(6):1374–1403
2. Bundred NJ (2001) Prognostic and predictive factors in breast cancer. Cancer Treat Rev 27 (3):137–142
3. Parker CE, Borchers CH (2014) Mass spectrometry based biomarker discovery, verification, and validation – quality assurance and control of protein biomarker assays. Mol Oncol 8(4):840–858
4. Tabb DL et al (2010) Repeatability and reproducibility in proteomic identifications by liquid chromatography-tandem mass spectrometry. J Proteome Res 9(2):761–776
5. Leitner A, Aebersold R (2013) SnapShot: mass spectrometry for protein and proteome analyses. Cell 154(1):252–252.e1
6. Marx V (2013) Targeted proteomics. Nature Methods 10:19–22
7. Bouchal P et al (2015) Combined proteomics and transcriptomics identifies carboxypeptidase B1 and nuclear factor kappaB (NF-kappaB) associated proteins as putative biomarkers of metastasis in low grade breast cancer. Mol Cell Proteomics 14(7):1814–1830
8. Lowry OH et al (1951) Protein measurement with the Folin phenol reagent. J Biol Chem 193(1):265–275
9. MacLean B et al (2010) Skyline: an open source document editor for creating and analyzing targeted proteomics experiments. Bioinformatics 26(7):966–968
10. Holzmann J et al (2009) Stoichiometry determination of the MP1-p14 complex using a novel and cost-efficient method to produce an equimolar mixture of standard peptides. Anal Chem 81(24):10254–10261
11. Bouchal P et al (2009) Biomarker discovery in low-grade breast cancer using isobaric stable isotope tags and two-dimensional liquid chromatography-tandem mass spectrometry (iTRAQ-2DLC-MS/MS) based quantitative proteomic analysis. J Proteome Res 8 (1):362–373
12. Domanski D et al (2012) MRM-based multiplexed quantitation of 67 putative cardiovascular disease biomarkers in human plasma. Proteomics 12(8):1222–1243

Methods in Molecular Biology (2018) 1788: 185–192
DOI 10.1007/7651_2017_88

Published online: 15 November 2017

Combination Strategy of Quantitative Proteomics Uncovers the Related Proteins of Colorectal Cancer in the Interstitial Fluid of Colonic Tissue from the AOM-DSS Mouse Model

Guixue Hou, Yang Wang, Xiaomin Lou, and Siqi Liu

Abstract

Quantitative proteome analysis using iTRAQ is an important technique to find disease-related proteins. As an important component of tumor microenvironment, tissue interstitial fluid (TIF) has drawn a great attention for its potential as a source for exploration of the solid tumor biomarkers. On the basis of a mouse model of colorectal cancer (CRC) that was induced by the carcinogenetic reagents, we adopted a quantitative proteome analysis with iTRAQ to discover the CRC-related proteins in the TIFs and with MRM to evaluate the corresponding abundance changes in the individual mouse TIF and serum samples.

Keywords AOM-DSS mouse model, Colorectal cancer, iTRAQ, MRM, Serum biomarker, Tissue interstitial fluid

1 Introduction

The development and progression of solid tumors depend on the malignancy of tumor cells, and also on the tumor microenvironment as well, which is mainly comprised of tumor-infiltrating cells, extracellular matrix, matrix-associated molecules, vasculature, and tissue interstitial fluid (TIF) [1, 2]. Of these components of the tumor microenvironment, TIF has drawn great attention for its potential value in clinical application because it contains many secreted proteins that are likely associated with tumor cells and tissues and are relatively easy in protein analysis due to the absence of high-abundance proteins. TIF proteomics therefore provides a new and promising avenue for discovering tumor biomarkers in serum [3–5] an alternative source of tumor TIFs.

Animal models possess many advanced characteristics in cancer study, such as identical genetic background for the selected groups, well-controlled conditions to induce pathologic changes, and relatively plentiful tissue samples. The azoxymethane-dextran sodium sulfate (AOM-DSS)-induced mouse model is a typical inflammation-related CRC model, in which adenoma and adenocarcinoma are induced using an intraperitoneal injection of low dose of AOM followed by oral administration of DSS [6, 7]. -AOM-DSS treatment in mice dramatically shortens the latency time for CRC induction and rapidly recapitulates the adenoma-carcinoma consequence that occurs in human CRC. This model is therefore widely accepted for the study of colon carcinogenesis, especially for inflammation-related CRC. In this study, we proposed that an analysis of TIF-based proteomics toward the colon tissues of AOM-DSS mouse model might benefit to discover new candidates of the CRC biomarker.

The traditional verification methods used for protein biomarkers are antibody-based assays, such as immunohistochemistry, Western blot analysis, and enzyme-linked immunosorbent assays. The efficiency of the traditional verification methods is dependent on the antibody quality, either commercial or laboratory sources. As a result of recent developments in mass spectrometry-based technologies, multiple reaction monitoring (MRM) has emerged as a powerful tool for targeted protein quantification [8]. In contrast to antibody-based techniques, MRM can easily expand protein targets for verification purposes, potentially to over one hundred candidate proteins. Coupling discovery proteomics with MRM-based verification proteomics is a promising technique that can be used for protein biomarker discovery and verification. In this chapter, we described how to screen and discover the CRC-related proteins in the TIF of AOM-DSS mouse model using iTRAQ-labeling quantification and to verify the CRC-related candidates in TIFs and sera using targeted MRM quantification.

2 Materials

2.1 Establishment of an AOM-DSS Mouse Model and Preparation of TIF Proteins

1. Preparation of AOM stock solution: Dissolve AOM in distilled water at 10 mg/mL. Aliquot and store at −20 °C.
2. Preparation of AOM working solution: Thaw aliquots directly before use and dilute the stock solution 1:10 in sterile isotonic saline (i.e., to obtain a final concentration of 1 mg/mL).
3. Preparation of DSS solution: Dissolve fine-grained DSS powder at the concentration of choice. To obtain a 2.5% (wt/vol) solution, dissolve 2.5 g of DSS per 100 mL autoclaved water. DSS solution can be stored at 4 °C for up to 1 week. Calculate the amount needed, allowing 5 mL DSS solution per mouse per day (i.e., 350 mL for ten mice per week).
4. PBS buffer: Prepare 1 L PBS stock (10×) by adding 80 g NaCl, 2 g KCl, 14.4 g Na_2HPO_4, and 2.4 g KH_2PO_4, dissolve in 800 mL dH_2O and adjust the pH to 7.4, add dH_2O until a total volume of 1 L and autoclave, and prepare 1 L 1× PBS buffer from 10× PBS stock as follows—dilute 100 mL of the 10× PBS stock buffer in a total volume of 1 L dH_2O.

2.2 LC-MS Analysis

1. Buffer A (5% ACN with 0.1% FA): Add 50 mL acetonitrile (Optima™ LC/MS Grade, Fisher), 950 mL dH_2O, and 1 mL formic acid, mix, and sonicate for 15 min.
2. Buffer B (95% ACN with 0.1% FA): Add 950 mL acetonitrile (Fisher), 50 mL dH_2O, and 1 mL formic acid, mix, and sonicate for 15 min.

3 Methods

3.1 Establishment of an AOM-DSS Mouse Model and Preparation of TIF Proteins

1. C57BL/6 male mice (7 weeks old) were intraperitoneally injected with AOM (Sigma-Aldrich, St. Louis, USA) at 12.5 mg/kg body weight.
2. One week after AOM injection, the mice were divided into three groups. All three groups were treated with the so-called DSS cycle [9], which comprised two steps in which the mice were fed with 2.5% (w/v) DSS (molecular weight 36,000–50,000, MP Biomedicals, Santa Ana, USA) for 5 days followed by 16 days of normal water feeding, but they were processed for different numbers of cycles. In the Cycle I group, the mice were treated with only one DSS cycle, while mice in the Cycle II and Cycle III groups were treated with two and three DSS cycles, respectively.
3. After treatment, the mice were sacrificed to collect colon tissues and blood. In the pathological evaluation and preparation of TIFs and mouse sera to evaluate pathology, the middle and distal colons were dissected out, opened longitudinally, rinsed with PBS, and fixed using 10% neutral buffered formalin overnight before H&E staining.
4. Dissected the colon tissues from each mouse into 0.3 cm two pieces and carefully rinsed with PBS three times.
5. Incubated the dissected tissues with PBS containing protease inhibitor (Calbiochem/Merck Millipore, Darmstadt, Germany) in a humidified CO_2 incubator for 1 h at 37 °C.
6. The samples were then centrifuged at 1000 × *g* for 5 min. The supernatant was centrifuged twice, once at 2000 × *g* for 5 min and once at 20,000 × *g* for 20 min, to remove cell debris. The resulting supernatant was referred to as TIF, and its protein concentration was measured using a BCA assay.
7. Mouse blood samples were individually collected without anti-coagulant, and the serum was isolated after incubation at room temperature for 1 h and centrifugation at 5000 × *g* for 20 min.

3.2 iTRAQ Labeling and Fractionation Using IPG-IEF

1. Seventy-five micrograms of individual TIF proteins from each group (four mice per group) was pooled, reduced using 10 mM dithiothreitol, and alkylated using 55 mM iodoacetamide.

2. The treated proteins were precipitated in 80% acetone at −20 °C for 2 h and after the solutions were centrifuged at 20,000 × *g*.
3. The pellets were resuspended in 500 mM triethylammonium bicarbonate containing 0.1% SDS, and protein concentrations were then measured.
4. One hundred micrograms of treated TIF proteins from each different group were digested using trypsin (1:30 w/w, Promega, Madison, USA) at 37 °C for 16 h.
5. The tryptic peptides in each group were individually labeled using the iTRAQ Reagents for 8 plex (Applied Biosystems, Foster City, USA), reporters 114, 116, 118, and 121 for control, Cycle I, Cycle II, and Cycle III, respectively.
6. Allow iTRAQ® Reagents for 8 plex to reach room temperature. Spin to bring the solution to the bottom of the tube.
7. To each iTRAQ Reagent for 8 plex, add 50 μL of isopropanol. Vortex to mix and then spin.
8. Transfer the contents of one iTRAQ Reagent for 8 plex vial to one sample tube. Vortex to mix and then spin. If pH is <7.5, add up to 5 μL dissolution buffer. Incubate at room temperature for 2 h.
9. Combine the contents of each iTRAQ Reagent—8 plex-labeled sample tube in a fresh tube. Vortex to mix and then spin.
10. After the labeling reactions were performed, the four labeled peptides were mixed, lyophilized, and resuspended in 2% H_3PO_4.
11. The peptides were further purified using Strata-X-C (Phenomenex, Torrance, USA) and then lyophilized.
12. Fractionation of labeled peptides using immobilized pH gradient-isoelectric focusing (IPG-IEF) A 24 cm linear IPG strip (pH 3–6) (Bio-Rad, Hercules, USA) was rehydrated for 8 h at 20 °C with labeled tryptic peptides in 0.5% IPG buffer. Focusing was then performed for a total of 100 kVh using an Ettan IPGphor 3 IEF System (GE Healthcare, Little Chalfont, UK).
13. After isoelectric focusing, the strip was sliced into 24 equal pieces, and the peptides in each piece were eluted by vortexing them for 2 h in MilliQ water, for 1 h in 50% ACN, and for 1 h in 100% ACN. All eluates from the same gel piece were pooled, lyophilized, and redissolved in MilliQ water.
14. They were then desalted using Strata-X (Phenomenex, Torrance, USA).

3.3 Depleted High-Abundance Proteins in Serum by ProteoMiner

1. A ProteoMiner protein enrichment kit (Bio-Rad, Hercules, USA) was used to enrich the serum proteins present at lower levels.
2. Before trypsin digestion, bovine serum albumin (BSA, Thermo Scientific Pierce, Waltham, USA, 20 fmol per 1 μg sample) was added to individual TIFs and serum samples as an internal standard for use in quality control.
3. These enriched serum proteins were digested using the filter-aided sample preparation (FASP) method [10] in Vivacon 500 ultrafiltration spin columns with a 10 kDa membrane cutoff (Sartorius, Goettingen, Germany).

3.4 Identification of Peptides Using LC-MS/MS

1. The peptides extracted from gel pieces were delivered in duplicate onto a nano RP column mounted in a nano HPLC system (SHIMADZU, Kyoto, Japan) and then eluted using an ACN gradient from 5 to 77% containing 0.1% formic acid (FA) for 95 min at 400 nL/min.
2. The eluates were directly entered in a microTOF-Q II mass spectrometer (Bruker Daltonics, Billerica, USA). Electrospray was performed by setting the spray voltage at 4.5 kV. The time-of-flight pressure was maintained at less than 3×10^{-7} Torr. Scanning was set in positive ion mode with full MS scan from 50 to 2000 m/z and a collision energy of 10 eV. The MS/MS spectra were acquired in a data-dependent mode on the three most abundant ions exceeding 1000 cts. Parent ions that recurred within 0.3 min were excluded from repeated CID MS/MS analysis.

3.5 Verification of Differentially Expressed Proteins Related to CRC Using LC-MRM-MS

1. A TripleTOF 5600 MS (AB SCIEX, Concord, USA) equipped with a nano LC system (SHIMADZU, Kyoto, Japan) was used to identify the peptides with significant MS/MS signals which could be used as target peptides in following MRM experiment.
2. ProteinPilot (AB SCIEX, Framingham, USA) was used to search against the IPI_mouse v3.87 (59,534 entries) database, and Skyline v2.1 (MacCoss Lab, University of Washington, Seattle, USA) [11] was used to establish the MRM transition list.
3. The selected transitions were adopted to survey the protein digests obtained from the individual mouse colonic TIFs and serum samples. TIF MRM assays were performed using a QTRAP 5500 MS (AB SCIEX, Framingham, USA) equipped with a nanoAquity UPLC system (Waters, Milford, USA). The mobile phases consisted of solvent A (2% ACN with 0.1% FA) and solvent B (98% ACN with 0.1% FA). The peptides were separated in a nanocapillary column (75 μm × 20 cm, 1.7 μm particles, 200 Å aperture) (Waters, Milford, USA) at 300 nL/

min using a gradient of 8–30% solvent B for 15 min, 30–60% solvent B for 5 min, and 60–90% solvent B for 5 min.

4. The MS parameters were set as follows: Ion spray voltage at 2500 V, curtain gas at 35.0, ion source gas 1 at 20.00, ion source gas 2 at 0.0, collision gas at high, interface heater temperature at 150.0 °C, entrance potential at 10.00, and Q1 and Q3 at unit resolution. Serum MRM assays were performed using a QTRAP 6500 MS (AB SCIEX, Framingham, USA) equipped with an ekspert nanoLC 425 system (AB SCIEX, Framingham, USA). The peptides were separated using an Eksigent column (75 μm × 15 cm, 3 μm particles, 120 Å aperture) (AB SCIEX, Framingham, USA) at 300 nL/min with a gradient of 5–40% solvent B for 30 min and 40–80% solvent B for 15 min. The MS parameters were set as follows: ion spray voltage at 2300 V, curtain gas at 35.0, ion source gas 1 at 25.0, ion source gas 2 at 0.0, collision gas at high, interface heater temperature at 150.0 °C, entrance potential at 10.0, and Q1 and Q3 at unit resolution.

3.6 Bioinformatics for Analysis of the Mass Spectrum Signals

1. For iTRAQ analysis, tandem mass spectra were extracted using DataAnalysis 4.0 SP1 (BrukerDaltonics, Billerica, USA).
2. The generated .MGF files were searched using Mascot v2.3.01 (Matrix Science, London, UK) [12] against the IPI mouse v3.87 database (59,534 entries). Several parameters were set in Mascot for peptide searching, including a tolerance of two missed trypsin cleavage sites, 0.1 Da for peptide tolerance as well as MS/MS, carbamidomethyl of cysteine and iTRAQ 8-plex of lysine and the n-terminus were set as fixed modifications, while Gln->pyro-Glu of the n-terminus and oxidation of methionine and iTRAQ 8-plex of tyrosine were set as variable modifications.
3. After searching the database, the .DAT and .OUT files derived from Mascot were loaded into Scaffold Q+ v4.0.3 (Proteome Software Inc., Portland, USA) [13] for quantitative evaluation of peptides and proteins. Protein probabilities were assigned using the Protein Prophet algorithm [14]. In this study, peptide identifications with a probability of >95.0% and protein identifications with a probability of >99.0% and at least two unique peptides were qualified for further quantitative analysis. The fold changes in protein abundance were defined as the median ratios of all spectra significantly matched to the protein with reporter signals between two samples.
4. To estimate the secretion potential, SignalP 4.1 [15] and SecretomeP 2.0 [16] were used to analyze whether the identified proteins were classical (i.e., proteins with signal peptides) or nonclassical secretory proteins. Meanwhile, an in-house "mouse serum protein database" generated from mouse serum proteins

identified in the literature [17, 18] was used to compare the published data with the TIF proteins identified in this study.

5. For MRM analysis, all raw data were imported and processed using Skyline v2.1. To ensure correct peak detection and integration, manual inspection was performed for further peptide filtration. Chemically synthetic peptides (GL Biochem, Shanghai, China) were used to validate the retention times and overlays of detected peaks. The relative abundances of target proteins in individual TIF/serum samples were calculated using the following two steps: (1) the median values for the MRM peak areas of these peptides in control mice were set as normalization references (for human serum MRM experiments, the values obtained for the peak areas of these peptides in one control individual were used), and (2) the relative protein abundance of each target protein was adjusted according to the abundance of spiked BSA in the corresponding sample. Mann-Whitney U tests were used to determine statistical significance.

4 Notes

1. Avoid repeated freeze and thaw cycles. AOM is more stable in water or saline than in PBS and is more stable in glassware than in plastic tubes.
2. DSS is very toxic to the colon of rodents, resulting in bloody feces for several days even after stop of DSS administration.
3. Improper dose administration of AOM could lead to high mortality in animal mortality. Also, some mouse strains are highly susceptible to AOM, such as age-, body weight-, or DSS concentration-dependent. Generally, reduction of AOM dose could switch to mouse strain to lower susceptibility. Redo experiment could be considered by using older and/or lower body weight mice, decreased DSS concentrations, or reduced period of administration.
4. If there is no or few tumors found in mouse colons, the phenomenon is likely caused by lower dose of AOM and higher susceptibility of AOM of mouse strain. Administration of higher dose of AOM or alternation of mouse strain may assist in solving such problem.
5. Be sure to check and adjust pH to achieve optimal efficiency in iTRAQ labeling. The pH in the labeling buffer must be greater than 7.5. If the pH is lower than 7.5, labeling efficiency would be significantly reduced.

6. Before LC/MS/MS analysis, clean up the sample mixture using cation exchange chromatography. If the sample mixture is complex, fractionation is required using high-resolution cation exchange column.

Acknowledgment

This work was supported by the National Key Basic Research Program of China (2011CB910704) and the National Natural Science Foundation of China (81372601).

References

1. Peddareddigari VG, Wang D, Dubois RN (2010) The tumor microenvironment in colorectal carcinogenesis. Cancer Microenviron 3:149–166
2. Taketo MM (2012) Roles of stromal microenvironment in colon cancer progression. J Biochem 151:477–481
3. Wiig H, Tenstad O, Iversen PO, Kalluri R, Bjerkvig R (2010) Interstitial fluid: the overlooked component of the tumor microenvironment? Fibrogenesis Tissue Repair 3:12
4. Gromov P, Gromova I, Olsen CJ, Timmermans-Wielenga V, Talman ML, Serizawa RR, Moreira JM (2013) Tumor interstitial fluid – a treasure trove of cancer biomarkers. Biochim Biophys Acta 1834:2259–2270
5. Gromov P, Gromova I, Bunkenborg J, Cabezon T, Moreira JM, Timmermans-Wielenga V, Roepstorff P, Rank F, Celis JE (2010) Up-regulated proteins in the fluid bathing the tumour cell microenvironment as potential serological markers for early detection of cancer of the breast. Mol Oncol 4:65–89
6. Okayasu I, Ohkusa T, Kajiura K, Kanno J, Sakamoto S (1996) Promotion of colorectal neoplasia in experimental murine ulcerative colitis. Gut 39:87–92
7. Rosenberg DW, Giardina C, Tanaka T (2009) Mouse models for the study of colon carcinogenesis. Carcinogenesis 30:183–196
8. Lange V, Picotti P, Domon B, Aebersold R (2008) Selected reaction monitoring for quantitative proteomics: a tutorial. Mol Syst Biol 4:222
9. Neufert C, Becker C, Neurath MF (2007) An inducible mouse model of colon carcinogenesis for the analysis of sporadic and inflammation-driven tumor progression. Nat Protoc 2(8)
10. Wisniewski JR, Zougman A, Nagaraj N, Mann M (2009) Universal sample preparation method for proteome analysis. Nat Methods 6:359–362
11. MacLean B, Tomazela DM, Shulman N, Chambers M, Finney GL, Frewen B, Kern R, Tabb DL, Liebler DC, MacCoss MJ (2010) Skyline: an open source document editor for creating and analyzing targeted proteomics experiments. Bioinformatics 26:966–968
12. Sadeh NM, Hildum DW, Kjenstad D et al (1999) Mascot: an agent-based architecture for coordinated mixed-initiative supply chain planning and scheduling[C]. In: Workshop on agent-based decision support in managing the internet-enabled supply-chain, at agents' 99
13. Searle BC (2010) Scaffold: a bioinformatic tool for validating MS/MS based proteomic studies. Proteomics 10(6):1265–1269
14. Nesvizhskii AI, Keller A, Kolker E, Aebersold R (2003) A statistical model for identifying proteins by tandem mass spectrometry. Anal Chem 75:4646–4658
15. Petersen TN, Brunak S, von Heijne G, Nielsen H (2011) SignalP 4.0: discriminating signal peptides from transmembrane regions. Nat Methods 8:785–786
16. Bendtsen JD, Jensen LJ, Blom N, Von Heijne G, Brunak S (2004) Feature-based prediction of non-classical and leaderless protein secretion. Protein Eng Des Sel 17:349–356
17. Hood BL, Zhou M, Chan KC, Lucas DA, Kim GJ, Issaq HJ, Veenstra TD, Conrads TP (2005) Investigation of the mouse serum proteome. J Proteome Res 4:1561–1568
18. Lai KK, Kolippakkam D, Beretta L (2008) Comprehensive and quantitative proteome profiling of the mouse liver and plasma. Hepatology 47:1043–1051

Methods in Molecular Biology (2018) 1788: 193–214
DOI 10.1007/7651_2017_112

Published online: 19 December 2017

Multiple Reaction Monitoring Using Double Isotopologue Peptide Standards for Protein Quantification

Azad Eshghi and Christoph H. Borchers

Abstract

Multiple reaction monitoring (MRM) is a technique used in tandem mass spectrometry where the first mass analyzer preselects parent ions for fragmentation and the second mass analyzer transmits selected product ions to the detector. This targeted technique has found widespread application in bottom-up proteomics for monitoring target peptides in a complex enzymatic digest. Quantitative MRM can be performed on enzymatically digested samples using spiked-in synthetic peptide standards, providing unsurpassed quantitative accuracy and a dynamic range of four orders of magnitude, often eliminating the need for prior depletion of high-abundance proteins. The development of MRM assays requires technical rigor, and this chapter details a methodology for sample preparation, data acquisition, and analyses to successfully perform quantitative MRM assays using two distinct isotopologue peptide standards to quantify proteins in mouse plasma and heart tissue.

Keywords Heart, Isotopologue, Monitoring, Mouse, MRM, Multiple, Multiplex, Peptide, Plasma, Protein, Proteomics, Quantification, Reaction

1 Introduction

Multiple reaction monitoring (MRM) is a technique used in tandem mass spectrometry to target a precursor ion and a corresponding product ion(s) [1]. While MRM can be performed on any instrument capable of tandem mass spectrometry, triple quadrupole mass spectrometers are typically used [2]. To increase the number of ions monitored (hundreds), upstream and online LC separation can be performed to separate analytes prior to entry into the mass spectrometer [3]. Upon ionization and entry into quadrupole mass filter 1, ions are gated into the quadrupole collision cell (q2) for fragmentation, and resulting product ions are selected by quadrupole mass filter 3 (Q3) to reach an ion detector. The signal intensity of the transmitted ions is transformed into chromatographic peak heights or, more often, peak areas which can then be integrated to determine the relative abundance of product ions and—by inference—the abundance of the parent ion [4].

The precision of quantitative MRM can be improved by spiking known concentrations of isotopically labeled analogues into the sample, thereby allowing comparison of the integrated peak areas between the endogenous analyte and the spiked-in analyte, to determine the endogenous analyte concentration [1]. To achieve accurate and precise measurement of endogenous analyte abundance, the choice of standard(s) is critical. For MRM of peptides, stable isotope-labeled standard (SIS) peptides serve as ideal standards for quantitation of their endogenous native counterparts (END) [5] due to identical retention times on the LC column, identical chemical, physical properties, and fragmentation patterns, after accounting for the mass shifts due to the isotopic labels. These attributes allow the standards to be spiked in early in sample preparation, controlling for variability in sample processing and any inherent variation in LC-MS/MS, including retention time shifts, ion suppression, and fragmentation efficiency [6].

The precision of MRM assays can be further enhanced by including two distinct isotopologue standards where one standard peptide serves as a normalizer for the peak area measurement and the second is used to generate a dilution series [7]. Using this strategy, calibration curves can be generated in the sample matrix and subsequently used to measure endogenous analyte concentration with improved precision [7].

Quantitative MRM has been extensively used for protein quantification in complex biological matrixes, with workflows utilizing bottom-up proteomics [8–12]. Existing protein/peptide mass spectra libraries may be consulted to select proteotypic peptides and corresponding transitions [13]. Alternatively, the parent protein may be used to select MRM compatible peptides using peptide picking software [14], and the corresponding synthetic peptides can subsequently be optimized for quantitative MRM [15]. A double isotopologue peptide standard strategy is utilized by incorporating heavy stable isotope-labeled ($^{13}C^{15}N$) amino acids into peptides during synthesis [7]. One peptide standard is synthesized which carries a $^{13}C^{15}N$-labeled amino acid on the C-terminus, and a corresponding peptide standard is synthesized which carries an additional $^{13}C^{15}N$-labeled amino acid adjacent to the C-terminal $^{13}C^{15}N$-labeled amino acid [7].

For protein quantification, the double stable isotope-labeled standard (SIS2) is maintained at a constant concentration across all samples and serves as a normalizer, while a dilution series of the single stable isotope-labeled standard (SIS1) is spiked into pooled samples, which serve as background matrix. The peak area ratios of SIS1 to SIS2 measured in a pooled-sample background matrix are used to create calibration curves which are utilized to measure the endogenous peptide concentration in individual samples (peak area normalization to SIS2). This strategy was recently used to precisely measure the concentrations of 31 human plasma proteins in plasma [7].

The following procedure describes a methodology modified from that described in LeBlanc et al. [7], for performing double isotopologue peptide standard (SIS1/SIS2) MRM quantification of proteins in mouse plasma and heart tissue.

2 Materials

All solvents and reagents must be LC-MS grade. Solution preparation and sample manipulation must be compatible with proteomic workflows, with special care being taken to avoid use of chemicals during sample processing that might interfere with subsequent LC-MS/MS. Use new polypropylene tubes for solution preparation and sample manipulation when possible. If using common laboratory glassware for solvents other than water, rinse with LC-MS grade water and allow to air dry.

2.1 Tissue and Plasma Sample Preparation Components

1. Tris 1 M, pH 8.0 (100 mL): To a 250 mL beaker, add 12.12 g Tris base and 50 mL of H_2O, and stir using a magnetic stirrer until fully dissolved. Reduce the pH to 8.0 using 12 M HCl while maintaining stirring. In a graduated cylinder, adjust the volume to 100 mL with H_2O. Store the solution in a 250 mL screw-top bottle. Store at ambient laboratory temperature for daily usage and at 4 °C for long-term storage (up to 6 months).
2. Tris 100 mM, pH 8.0 (100 mL): In a 250-mL screw-top bottle, dissolve 10.0 mL of 1 M Tris pH 8.0 in 90.0 mL H_2O.

2.2 Trypsin Digestion Components

1. Urea 9 M, Tris 300 mM, pH 8.0 (10 mL)—Solution A (*see* **Note 1**): Weigh 5.41 g of urea and transfer to a 15-mL polypropylene conical tube. Add 3 mL 1 M Tris (pH 8.0) and adjust the volume to 10 mL with H_2O. Urea dissolution is endothermic and will require significant mixing using a vortex mixer for 5–10 min or 10–15 min on a tube shaker/rotor.
2. Dithiothreitol (DTT) 0.5 M (605 μL) (*see* **Notes 1** and **2**): Weigh 46.7 mg of DTT and dissolve in 605 μL of Solution A in a 1.5-mL conical tube.
3. Urea 9 M, Tris 300 mM, pH 8.0, DTT 20 mM (1 mL)—Solution B (*see* **Notes 1** and **2**): Add 40 μL 0.5 M DTT to 960 μL of Solution A and mix by using a vortex mixer.
4. Iodoacetamide (IAA) 0.5 M (605 μL) (*see* **Notes 1** and **3**): Weigh 56.0 mg of IAA and dissolve in 605 μL of Solution A in a 1.5-mL conical tube.
5. L-1-Tosylamide-2-phenylethyl chloromethyl ketone (TPCK)-treated trypsin 1 μg/μL (~1 mL) (*see* **Note 4**): Pre-weigh a 1.5-mL-conical tube and add ~1 mg TPCK-treated trypsin (by estimating). After adding the trypsin, close the conical

tube lid, and briefly centrifuge the tube to collect the trypsin to the bottom of the tube. Record the weight of trypsin by weighing the 1.5-mL conical tube containing trypsin and subtracting from the pre-trypsin addition weight. Add a volume of 100 mM Tris pH 8.0 to achieve 1 μg/μL TPCK-treated trypsin.

6. Formic acid (FA) 10% (2 mL) (*see* **Note 5**): Add 0.2 mL FA to 1.8 mL H_2O in a 2-mL microcentrifuge tube and mix using a vortex mixer.

2.3 Stable Isotope-Labeled Peptide Standard and Solid Phase Extraction Components (See Note 7)

1. Formic acid (FA) 0.1% (50 mL) (make two) (*see* **Note 5**): Add 50 μL FA to 49.95 mL H_2O in a 50 mL polypropylene conical tube and mix using a vortex mixer.
2. FA 0.1% acetonitrile (ACN) 60% (50 mL) (*see* **Notes 5** and **6**): Transfer 30 mL of ACN to a 50-mL polypropylene conical tube, add 19.55 mL H_2O, and mix by gentle manual shaking. Add 50 μL FA and mix as above.
3. Methanol 100% (100 mL) (*See* **Note 6**).
4. FA 0.1% ACN 30% (FA-ACN) (50 mL) (*See* **Note 5**): Transfer 15 mL of ACN to a 50-mL polypropylene conical tube, add 34.55 mL H_2O, and mix by manual and gentle shaking. Add 50 μL FA and mix as before.
5. Stock solution of single stable heavy isotope-labeled peptide standard (SIS1), 1 nmol/μL in 0.1% FA 30% ACN, at least 6 μL (*see* **Note 8**): Dissolve lyophilized peptide standard (synthesized or purchased) to a concentration of 1 nmol/μL by adding the required volume of 0.1% FA 30% ACN. For example, add 300 μL 0.1% FA 30% ACN to 300 nmol of lyophilized peptide and mix with a vortex until peptide is fully dissolved.
6. Stock solution of double stable heavy isotope-labeled peptide standard (SIS2)—50 pmol/μL in 0.1% FA 30% ACN—at least 6 μL (*see* **Note 8**): Dissolve lyophilized peptide standard (synthesized or purchased) to a concentration of 50 pmol/μL by adding the required volume of 0.1% FA 30% ACN. For example, add 300 μL 0.1% FA 30% ACN to 15 nmol of lyophilized peptide and mix by vortex until peptide is fully dissolved.
7. Single stable heavy isotope-labeled peptide standard (SIS1) working solution 2000 fmol/μL in 0.1% FA consisting of 250 unique SIS1 peptides: Prepare a 1.5-mL low-binding microcentrifuge tube by creating a hole in the center of the lid. This can be done using an 18 gauge needle. Inspect the hole and remove any particulates with high-pressure nitrogen gas. Add 1 μL of each SIS1 peptide standard from the stock solution to the modified microcentrifuge tube. For a SIS1 peptide mix containing 250 peptides, the final volume will be 250 μL. Close the lid of the conical tube minimizing movement that

would cause loss of sample through the hole in the lid. Seal the hole in the lid using Parafilm, and briefly centrifuge to collect sample to the lower portion of the conical tube. Place in a rack that will provide insulation (rack must fit into a Labconco Fast-Freeze Flask™), and freeze at −80 °C for 1 h. Afterward, remove the Parafilm, and place the frozen sample (still in the rack) in a Labconco Fast-Freeze Flask™ that has been previously cooled to −20 °C, and lyophilize overnight. The following day, add 500 μL 0.1% FA to the dried 250 SIS1 peptide standards to obtain a 2000 fmol/μL working solution. Seal the hole in the conical tube lid with Parafilm, gently vortex to ensure peptides are fully dissolved, and aliquot 5 × 100 μL in low-binding microcentrifuge tubes. Storage conditions will vary based on the stabilities of the individual peptides; for long-term storage (weeks to months), store at −80 °C.

8. Double stable isotope-labeled heavy peptide standard (SIS2) working solution, 100 fmol/μL in 0.1% FA, consisting of 250 unique SIS2 peptides: Prepare a microcentrifuge tube as described in step 7. Add 2 μL of each SIS2 peptide standard (from the 50 pmol/μL stock solution) to the modified microcentrifuge tube. For a 250 SIS2 peptide mix, the final volume will be 500 μL. Lyophilize the SIS2 peptide mixture as described in step 7. Add 1000 μL 0.1% FA to the dried 250 SIS2 peptide standard to obtain a 100 fmol/μL working solution. Seal the hole using Parafilm and gently vortex, ensuring that the peptides are fully dissolved, and divide into 10 × 100 μL aliquots in low-binding microcentrifuge tubes. Storage conditions will vary based on individual peptide stabilities; for long-term storage (weeks to months), store at −80 °C.

3 Methods

Perform all experiments at ambient laboratory temperature (20–26 °C), unless otherwise indicated.

3.1 Heart Sample Preparation for Protein Digestion

This section describes in detail the preparation of whole mouse heart obtained for subsequent protein digestion using TPCK-treated trypsin. The main considerations for sample processing are disruption of macrostructures within tissues that maintain cell-cell contact and the lysis of cells to release total protein, followed by protein denaturation, reduction, and alkylation for efficient digestion by TPCK-treated trypsin. In this methodology, tissue macrostructure disruption and cell lysis are achieved by homogenization. Subsequent denaturation of proteins is achieved through the use of chemicals which have only minimal interference with ensuing LC-MS/MS.

Ideally, tissue samples should be perfused to minimize "contamination" with blood proteins. The number of biological replicates will be unique to the specific research project, and reagent and solution volumes should be scaled accordingly. The method detailed here is modified from Percy et al. [16] and is for the use of six mouse hearts (in this case, obtained from three male mice and three female mice).

1. Place each mouse heart separately in a small wax weighing boat, and rinse with 300 μL of 100 mM Tris pH 8.0. Using tweezers, transfer each whole heart to a ~2-cm^2 absorbent gauze pad and dab dry. Weigh the heart tissue in a new dry wax weighing boat, and record the weight.
2. Transfer the tissue to a 2-mL conical tube and—for each 100 mg of heart tissue—add 300 μL of 100 mM Tris pH 8.0.
3. Add a 3-mm stainless steel ball to each heart sample, close the lid of the tube, and place on ice for 5 min.
4. Perform homogenization using a Retsch MM400 CryoMill. Transfer samples from step 3 to a milling cup, and place the milling cup in the center of the milling clamp device and clamp firmly.
5. Operate the Retsch CryoMill at 23 Hz for 1 min. Remove the samples, and incubate them on ice for 5 min. Repeat this sequence a total of three times. Proceed immediately to tryptic digestion to minimize protein degradation due to endogenous protease activity (Sect. 3.3).

3.2 Blood Plasma Sample Preparation for Protein Digestion

Appropriately prepared plasma is free of blood cells, so the homogenization step is not necessary. Hence, plasma can be processed directly for subsequent digestion with TPCK-treated trypsin digestion. Ideally, plasma from multiple biological replicates should be used, and this protocol describes processing of mouse plasma obtained from three female mice and three male mice ($N = 6$).

3.3 Tryptic Digestion

1. Use 22.5 μL of each plasma or tissue homogenate, performing 12 digestions in total. Transfer the 22.5 μL from each sample to individual 2.0-mL microcentrifuge tubes.
2. To each sample, add 67.5 μL Solution B, and incubate at 23 °C for 30 min on a shaking heating block (500 rpm).
3. Add 7.2 μL of 0.5 M IAA to each sample, and vortex to mix. Incubate at 23 °C for 30 min, protected from light.
4. Add 7.3 μL of 0.5 M DTT to each sample, and vortex to mix. Incubate at 23 °C for 5 min.
5. Add 840.3 μL of 100 mM Tris pH 8.0 to each sample, and vortex to mix.

6. Add 56.3 μL of trypsin solution (*See* **Note 9**) to each sample, and incubate on a shaking heating block at 37 °C and 500 rpm for 18 h.
7. Samples should be clear and transparent after step 6 is completed. If debris is observed in samples, centrifuge at 16,000 × *g* for 10 min, and transfer the supernatant to a new 2-mL microcentrifuge tube.
8. Add 112.5 μL of 10% aqueous FA to each replicate, and vortex to mix. There should be no visible solid particles in the samples after the addition of 10% FA. If solid particles are observed, centrifuge the samples as in step 7, and transfer the supernatants to new 2-mL microcentrifuge tubes.

3.4 Synthetic Peptide Standards and Determination of the Lower Limit of Quantification (LLOQ)

To quantify endogenous peptides in a sample, a double isotopologue peptide standard strategy provides superior quantitative precision [7]. Isotopologue peptides are synthesized so that one peptide standard (SIS1) contains a single $^{13}C^{15}N$ heavy isotope amino acid on the C-terminus and the second peptide standard (SIS2) contains the same C-terminus heavy amino acid plus an additional upstream $^{13}C^{15}N$-labeled amino acid. The SIS2 peptides are spiked at a constant concentration across all samples, while the SIS1 peptides are spiked in the background matrix to form a dilution series. The peak area ratio, SIS1/SIS2, is plotted versus SIS1 peptide concentration (in the dilution series) to generate a unique standard curve for each peptide pair. The equation of the line calculated from the linear curve will subsequently be used to determine the concentration of the endogenous peptide (END) in the sample by using the ratio of the endogenous peptide's peak area normalized to that obtained for SIS2 (i.e., END/SIS2). The constant SIS2 concentration across all samples controls for experimental variation including, but not limited to, solid phase extraction efficiency, ion suppression, and other variables that reduce precision in quantification.

The following methodology describes the use of 500 peptides (250 SIS1 peptides and 250 SIS2 peptides) for the quantitation of 250 unique END peptides. The method is adapted from LeBlanc et al. [7], and the number of peptides quantified can be modified.

1. Pool 350 μL from each plasma digest (Sect. 3.3, step 8) to obtain 2.1 mL of pooled digest. Do this separately for heart digests as well. These pooled samples will be used as background matrix for determination of the lower limit of quantification (LLOQ) and for generating calibration curves used to measure the END peptide concentration in the individual samples.
2. Aliquot 200 μL from pooled digests into 6 × 1.5 mL microcentrifuge tubes. Do this in parallel for the individual plasma and heart samples. Store the remaining 0.9 mL of the pooled digest and the 763.6 μL of the individual tryptic digests at −80 °C.

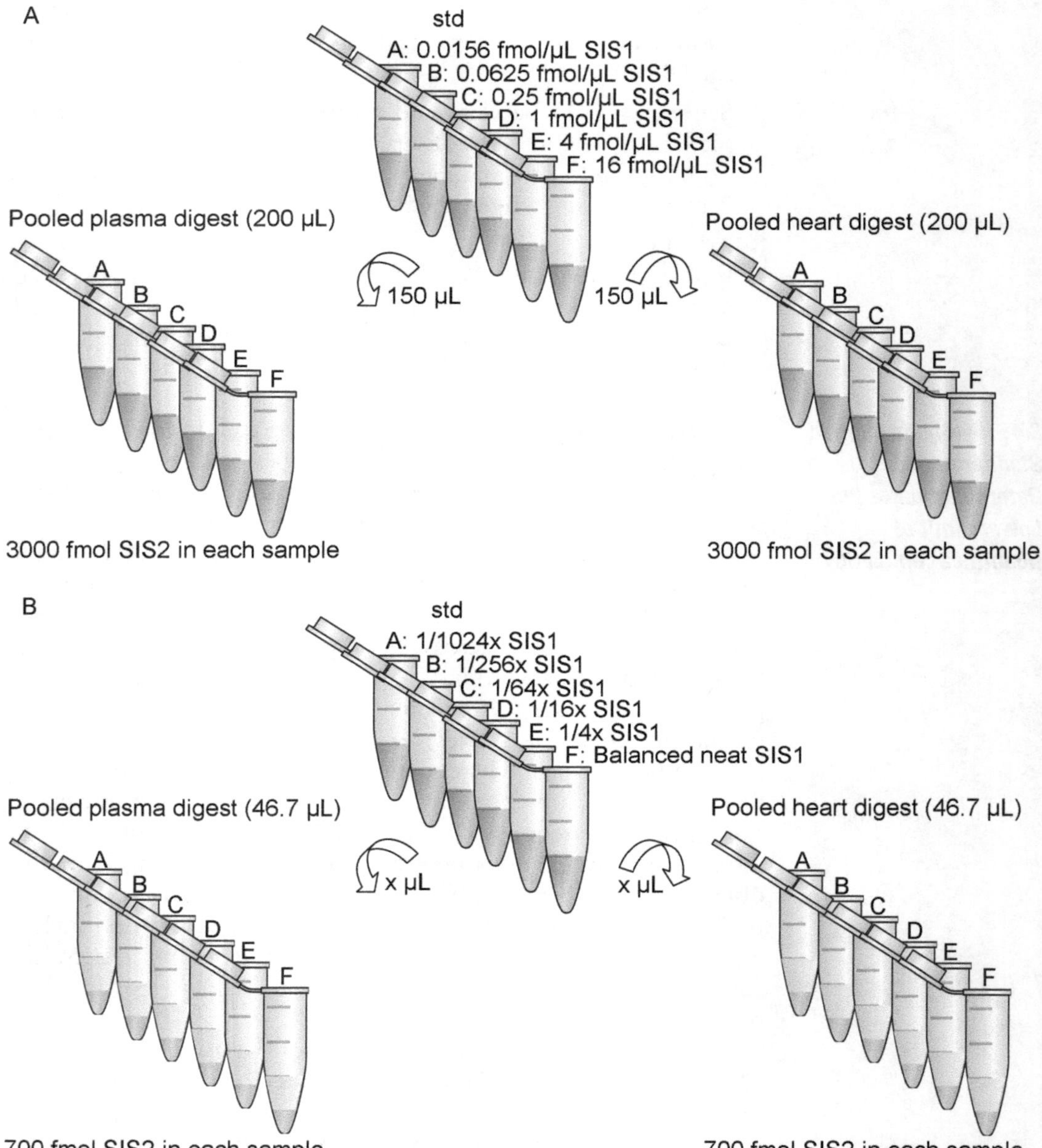

Fig. 1 Schematic for equimolar and balanced dilution series. (**a**) Dilution series used for determination of LLOQ and interfernce testing. (**b**) Dilution series used for quantification

3. In parallel, perform the following dilution series (Fig. 1a) from the 2000 fmol/µL SIS1 working stock. Label six 1.5-mL conical tubes A through to F.
   (a) To tube F, add 436.48 µL 0.1% FA and 3.52 µL of 2000 fmol/µL SIS1 working solution.

(b) To tubes A through E, add 330 μL 0.1% FA.

(c) Perform a dilution series using fourfold dilutions by adding 110 μL from tube F to tube E and so on through tube A, vortexing after each additional step.

(d) After addition of the final 110 μL, aliquot to tube A and vortex, remove 110 μL from tube A, and discard (or save for future use).

4. Add 150 μL of standard A to a single aliquot of pooled plasma and heart tissue homogenate from step 2. Repeat with the remaining plasma and heart pooled digests where one aliquot from each pooled sample receives 150 μL from a single dilution standard. Label tubes to distinguish between spiked peptide standards.
5. Add 30 μL of 100 fmol/μL SIS2 working solution to each aliquot from step 4, and mix with a vortex mixer. This completes the spiking of the samples with the peptide standards. These will subsequently be used to determine the LLOQ.

3.5 Solid Phase Extraction

Solid phase extraction (SPE) can be performed using columns packed with C_{18} or a similar packing material. The following procedure describes SPE using Waters OASIS HLB 10 mg × 1 cc columns. If operating a negative pressure vacuum manifold, maintain a pressure of −20 kPA, and if using a positive pressure vacuum manifold, maintain a pressure of 5 kPA.

1. Label 12 SPE columns to differentiate between samples spiked with peptide standards A through F.
2. Place the SPE columns on either a negative or positive pressure vacuum manifold.
3. Add 1 mL of 100% methanol to each SPE column, and elute using the vacuum manifold (*see* **Note 10**). Flow through is waste.
4. Add 1 mL of 0.1% FA to each SPE column and elute using the vacuum manifold (See Note 10). Flow through is waste.
5. Add the entire volume of sample to the correspondingly labeled SPE column, and incubate for 10 min on the manifold at atmospheric pressure allowing the sample to pass through the column packing by gravity (*see* **Note 11**).
6. After 10 min of flow by gravity, elute any remaining sample using the vacuum manifold (*see* **Note 10**). The flow through is waste.
7. Add 1 mL of H_2O to each SPE column and draw/push using the vacuum manifold (*see* **Note 10**). Repeat a total of two times. The flow through is waste.
8. Label 12 1.5-mL microcentrifuge tubes to be subsequently used for collecting samples to be eluted from step 7.

9. Place the SPE columns directly into the microcentrifuge tubes (*see* **Note 12**).
10. Add 300 μL 60% ACN 0.1% FA to the SPE column, and incubate for 10 min to allow elution by gravity (*see* **Note 13**). Retain the eluates.
11. After a 10-min incubation, dispense any remaining elution volume in the column into the 1.5-mL conical tubes using the technique described in Note 13. This completes the solid phase extraction of the peptides.
12. Place the samples into an 80-well polypropylene microcentrifuge rack, and freeze at −80 °C.

3.6 Determination of the Retention Times of the Standard Peptides

The methodology described here is compatible with the Agilent 6490 series triple quadrupole mass spectrometer coupled to an Agilent 1290 Infinity high-performance liquid chromatography system equipped with a Zorbax Eclipse Plus C18 rapid resolution HD 2.1 × 150 mm 1.8 μm column. The optimal number of transitions that can be concurrently monitored by an Agilent 6490 MS is in the range of 70–90 transitions. Hence, a sample containing 500 peptides where three transitions are monitored per peptide (1500 transitions) cannot be analyzed concurrently. Prior online separation of peptides on an LC system makes monitoring of 1500 transitions possible by splitting the number of concurrent transitions monitored by the 6490 MS into retention time windows where a targeted peptide is only monitored as it elutes from the online LC system. This approach, called "scheduled MRM," requires prior knowledge of the retention times of each peptide being monitored. The instructions given below are for determining the retention times of the isotopically labeled peptide standards, which will be identical to the retention time of their endogenous peptide counterparts. These steps assume an operational LC-MS system and knowledge of the data acquisition and data analysis software packages, MassHunter and Skyline Daily, respectively. Skyline Daily is used to select the peptide transitions by taking advantage of mouse tandem MS libraries such as that provided by the National Institute of Standards and Technology (NIST). As a starting point for selecting three transitions per peptide, the three product ions displaying the highest relative intensities in the NIST library are selected for each peptide. If transitions are not available in the tandem MS library for a particular peptide, transitions will require experimental verification by monitoring all product y and b ions in preliminary LC-MS analysis. The following steps assume that three transitions have been preselected for each endogenous, SIS1, and SIS2 peptide.

1. Add 125 μL of SIS2 working solution (100 fmol/μL) to a 300 μL PP screw vial (*see* **Note 14**).

2. Add 6.25 μL of SIS1 working solution (2000 fmol/μL) to the same vial used in step 1.
3. Add 118.75 μL of 0.1% FA to obtain a 250 μL solution of 50 fmol/μL SIS1 and SIS2.
4. Screw on the cap, and place the vial in the "Vial 1" position of the autosampler rack on the LC system.
5. Load an unscheduled MRM method using the MassHunter data acquisition software.
6. In the method editor tab under the HiP Sampler tab, set the injection volume to 10 μL.
7. Under the Binary Pump tab, set the flow rate to 0.4 mL/min. Mobile phase A should be 0.1% FA in H_2O, and mobile phase B is 0.1% FA in ACN.
8. Under the same tab, modify the gradient to create a 90-min gradient as follows: 3–20% B in 50 min, 20–30% B (65 min), 30–50% B (85 min), 50–70% B (87 min), 70–100% B (89 min), and 100–2% B (90 min).
9. Under the same tab, set the post run time to 4 min and the stop time to 90 min.
10. In the QQQ tab, paste in a maximum of 70 transitions, and adjust the cycle time to provide a minimum dwell time of 10 ms, ensuring MRM is selected. Save the method.
11. Repeat step 10 for the remaining 1430 transitions, saving a unique method after pasting in a new set of 70 transitions. This will require 22 methods, each with ≤70 transitions.
12. Create a 22-step worklist in MassHunter using the methods created in step 11, set the injection position to "Vial 1," and commence acquisition. Include blanks at the end of the worklist as required by the specific laboratory SOP.
13. After data acquisition, import the data files into Skyline Daily, and verify that each SIS1 and SIS2 transition was detected.
14. After verification, create a scheduled method within Skyline Daily using the retention times identified for each of the SIS1 and SIS2 peptides. Concurrent transitions should not exceed 120 when creating this method, with 70 concurrent transitions being optimal. Set the retention time window to 1 min. Adjust the cycle time to obtain a minimum dwell time of 10 ms.
15. Load the method created in step 14 in MassHunter, and perform dynamic MRM acquisition, injecting 10 μL of sample from Vial 1.
16. After data acquisition is complete, import the results into Skyline Daily, and confirm retention times and transitions for all 250 SIS1 and SIS2 peptide pairs.

17. Using the newly acquired retention times, create a new method as in step 14, but increase the retention time window from 1 to 5 min. This will require multiple method files that will be used to check the retention times in plasma.

3.7 Response Curve

Samples prepared in Sect. 3.5, step 12 will initially be used to perform a retention time check in plasma using the data acquisition method created in Sect. 3.6, step 17. Subsequently, a single method will be created for data acquisition which will be used to generate a response curve for each SIS1 peptide in order to determine the LLOQ for each of the 250 assays.

1. Using scissors, precut Parafilm into 12 × ~1 cm^2. Remove the samples from the −80 °C freezer (*see* Sect. 3.5, step 12), open the lids of the microcentrifuge tubes, seal with the precut Parafilm, and poke 3–5 holes using an 18 gauge needle. Work quickly, and immediately place the samples back into the −80 °C freezer.
2. Place a Fast-Freeze lyophilizer flask (large enough to accommodate the sample rack used in step 1) in a −20 °C freezer for 30 min.
3. Place the samples in the freeze flask, seal the freeze flask top, and lyophilize for a minimum of 4 h.
4. Add 300 μL of 0.1% FA to each sample and vortex to solubilize and centrifuge to drive the samples to the bottom of the microcentrifuge tubes. Using a 1-mL pipette, transfer the samples to correspondingly labeled PP screw vials (*see* **Note 14**). Place the samples in the autosampler rack of the HPLC system.
5. Modify the scheduled methods created in Sect. 3.6, step 17 by changing the sample injection volume to 20 μL. Perform dynamic MRM as detailed in Sect. 3.6 using sample F from plasma and from the heart.
6. Import the data into Skyline Daily, and—using the newly acquired retention times—generate a single scheduled method for data acquisition in the heart and another for plasma. If the retention times across all 250 peptide pairs are nearly identical between the two matrixes, then a single method can be used to perform dynamic MRM in both matrixes.
7. Using the scheduled method created in step 7, perform dynamic MRM on the heart and plasma samples A through F in triplicate (use a 20-μL injection volume).
8. After data acquisition is complete, import the data into Skyline Daily. In the peptide setting quantification tab, ensure normalization is performed using SIS2.
9. Data analysis includes confirming that all three transitions are present for the SIS1 and SIS2 peptides, that peaks are assigned

correctly, and that transitions are free of interference from endogenous non-specific transitions. Transitions displaying equal fragment-ion ratios for all of the peptides (END, SIS1, and SIS2) and with identical retention times are considered to be interference-free.

10. Having manually verified each transition and removed transitions showing interference, open the document grid pane in Skyline Daily, and view the transition results. Skyline Daily calculates the area for each defined transitions, and these values will be used to calculate a coefficient of variation for each transition using the three replicate injections per sample.
11. A coefficient of variation >20% indicates that the LLOQ is at a higher concentration—i.e., if standard A is >20%, move to standard B, and so on, until the CV is ≤20%.
12. Save the Skyline Daily file using a different name, and replace SIS1 transitions with the m/z values of the native ($^{12}C^{14}N$) isotope counterparts.
13. Create a new method file which will be used to perform dynamic MRM to monitor endogenous and SIS2 transitions.
14. Use the method created in step 13 to perform data acquisition using three injections, one each from plasma and heart samples A, C, and F (Fig. 2).
15. Analyze the data for the transitions of the endogenous and SIS2 peptides, as in step 9. If an interference is detected for a transition of an endogenous peptide, the corresponding transitions will require modification (i.e., removal) in the method for the SIS1 and SIS2 peptide. Transitions of the endogenous peptide take priority over the transitions of the corresponding SIS1 and SIS2 peptides.
16. After data validation, use the ratio of the endogenous peptide transition's peak area to that of the SIS2 peptide to determine the concentration of the endogenous peptide. Keep in mind that 20 μL of sample was injected on column, which corresponds to 0.27 μL of the original digest (Fig. 2).

3.8 Calibration Curve and Endogenous Peptide Quantification

The LLOQ determined in Sect. 3.7 will be used as the lowest concentration that can be used to quantify the endogenous peptide in a biological matrix. The LLOQ will be used as a reference to create a six-point dilution series of SIS1 peptide, covering a dynamic range of at least three orders of magnitude. The dynamic range will be centered at the concentration of endogenous peptide which was determined in Sect. 3.7, in order to obtain equal numbers of data points above and below the concentration of endogenous peptide. The pooled plasma and heart samples will be spiked with six different concentrations of SIS1 peptides from a dilution series and a single concentration of SIS2 peptide. Standard curves

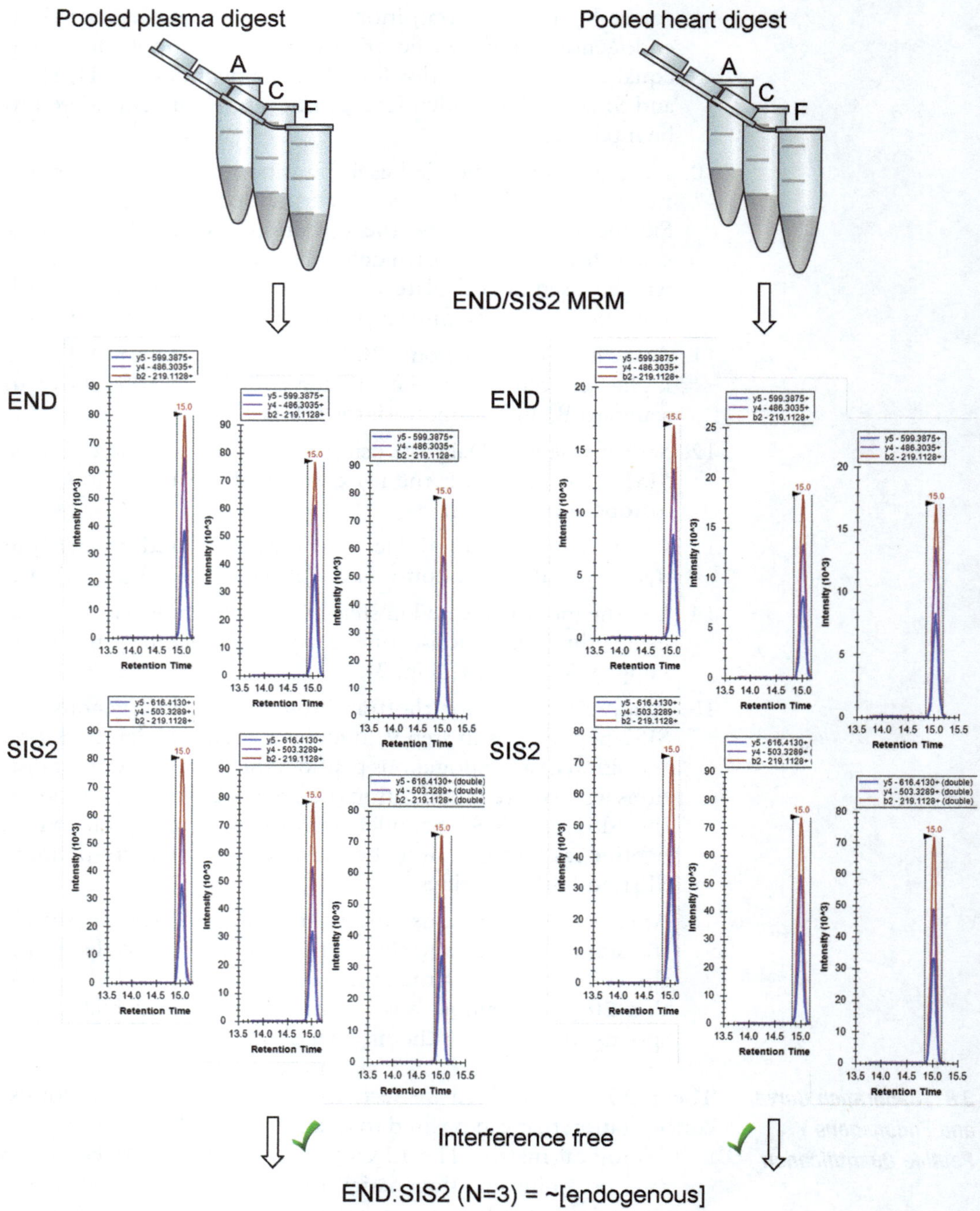

Fig. 2 Schematic for interference screening and intitial estimation of endogenous peptide (END) concentration in mouse plasma and heart digests. Samples used to determine LLOQ are subsequently used for interference screening and for single-point quantitation of the endogenous peptide. Consistent product-ion ratios and identical END and SIS2 retention times indicate intereference-free transitions. Triplicate injections of samples A, C, and F and subsequent peak area ratio comparisons of END and SIS2 allow single-point measurement of the END concentration in each matrix

are generated by plotting SIS1-SIS2 peak area ratios versus SIS1 concentration. Individual plasma and heart samples will be spiked with three concentrations of SIS1 peptide and a single concentration of SIS2. The three concentrations of SIS1 spikes in each sample will serve as quality controls to measure the precision and coefficient of variation of the calibration curve for each sample. In the same samples, the endogenous SIS2 peak area ratio will be used to calculate the concentration of the endogenous peptide in each plasma and heart sample using the calibration curves generated in pooled samples.

1. Use the endogenous peptide concentrations which were determined in Sect. 3.7, step 16 to calculate the SIS1 peptide concentration required to generate standard curves for each of the 250 peptides. For a six-point calibration curve, the concentration of the endogenous peptide should ideally lie between concentration points C and D (*see* **Note 15**).
2. Use the 2000 fmol/μL (or 1 nmol/μL if needed) SIS1 peptide mixture to make a new working solution of SIS1 peptides (a concentration-balanced mixture). This working stock solution or concentration-balanced mixture will contain the high concentration (standard F) of SIS1 peptide (as determined for each peptide in step 1); the different SIS1 peptides in the mixture will have different concentrations (Fig. 1b).
3. Make the balanced SIS1 mixture as described for the equimolar 2000 fmol/μL SIS1 working solution—by adding the required volume from the SIS1 working or stock solution. Enough of the balanced SIS1 mixture should be prepared to formulate the following:
   (a) Six spikes that will be spiked into pooled samples and used to generate data for the upper point (standard F) on the curve (performed in triplicate for pooled plasma and heart).
   (b) Standard F will also be used to perform a fourfold dilution to prepare standard E. Standard E will be used to make a fourfold dilution to prepare standard D and so on through to standard A (Fig. 3).
   (c) The balanced SIS1 mixture will also be used to prepare samples that will be used to obtain low-, medium-, and high-quality control (QC) data points for each (individual/non-pooled) plasma and heart digests. On a six-point calibration curve, the low, medium, and high QC data points should fall between the two lowest, two middle, and two highest concentration points on the curve, respectively.

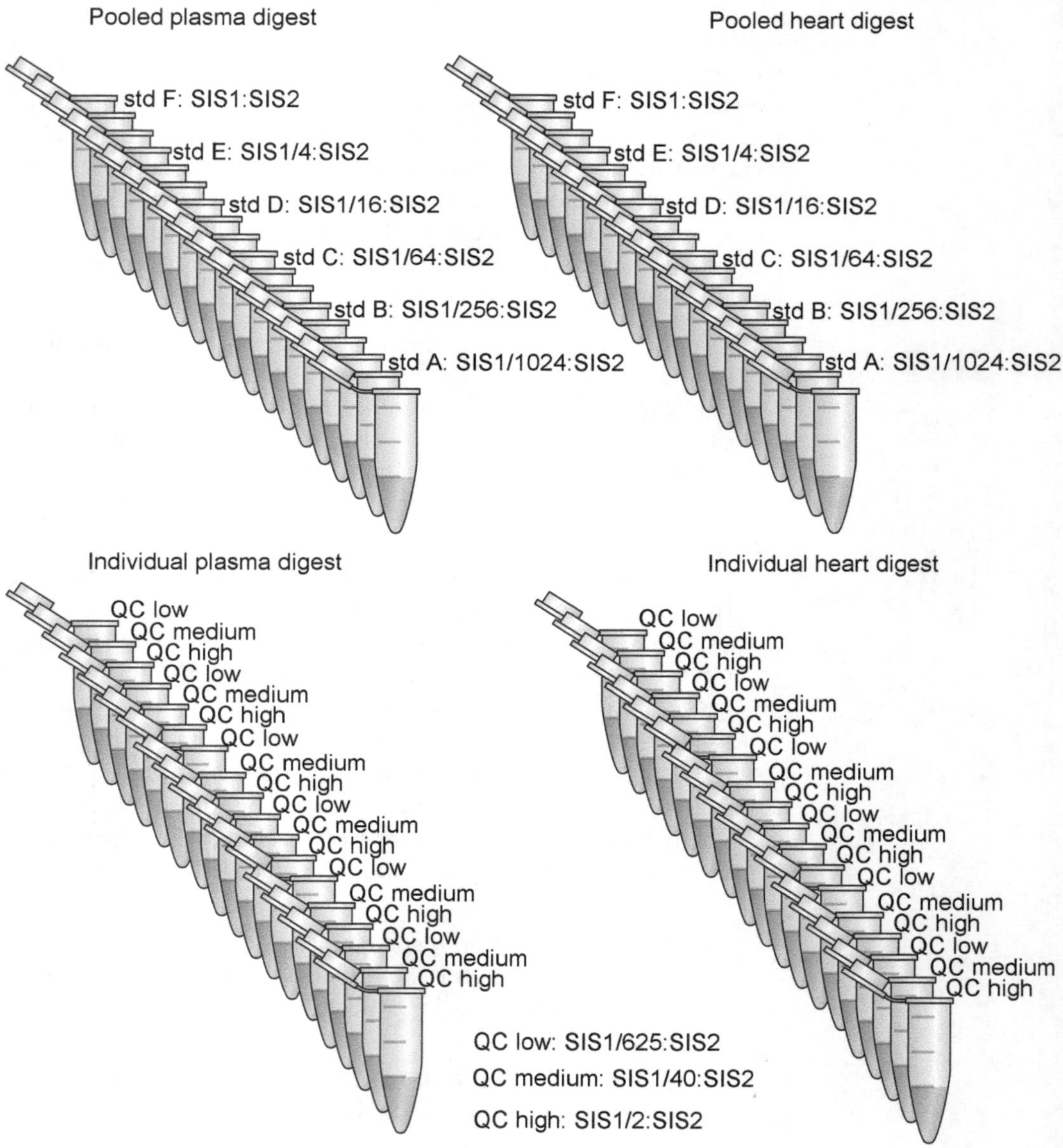

Fig. 3 Schematic for preparation of standard and quality control (QC) samples. Balanced SIS1 mix is used to make up standard (std) sample F, which will subsequently be used in a fourfold dilution series to make up standards E through A. These samples are spiked in triplicate into pooled digests of plasma and heart for obtaining data which will be used to generate linear regression curves. Similarly, the balanced SIS1 mixture is used to prepare QC samples containing low (625-fold dilution), medium (40-fold dilution), and high (twofold dilution) concentrations of SIS1 peptide. These (QC) samples are spiked into the individual plasma and heart digests. Individual and pooled digests are also spiked with constant [SIS2], which serves as a normalizer

4. After adding the respective volumes of SIS1 peptides, lyophilize and subsequently dissolve in 0.1% FA as detailed for the 2000 fmol/μL working solution of SIS1 peptides.

5. Prepare 72 labeled microcentrifuge tubes as follows: individual plasma and heart samples will be spiked with low, medium, and high SIS1 peptides (corresponding to low, medium, and high QC data points). For six plasma and six heart digest samples, this equals 36 samples. The remaining 36 microcentrifuge tubes will be used for the six-point calibration curve prepared by using each plasma and heart pooled digest, in triplicate.
6. Remove the pooled and individual plasma and heart digest samples from the −80 °C freezer (*see* Sect. 3.4, step 2). Aliquot 46.7 μL of sample into triplicate vials for each individual digest and in 18 replicates for each pooled plasma and heart sample (Fig. 3), using the pre-labeled microcentrifuge tubes from step 5.
7. The total amount of standard peptide spiked in will depend on the sample volume being injected onto the LC column. For this method, the injection volume will be 20 μL. Spike each sample with 7 μL of SIS2 100 fmol/μL working solution (will equate to 200 fmol of SIS2 injected on column).
8. In parallel, make a six-point dilution series covering three orders of magnitude using the concentration-balanced SIS1 mixture prepared in steps 1 through 4. A dilution series of a factor of 4 between data points on the calibration curve will suffice. Since samples will be lyophilized after SPE, when making this dilution series, the final volume that the sample will be dissolved in and the volume injected onto the LC column must be known. The suggested volumes are 70 and 20 μL in this procedure, respectively (*see* **Notes 16** and **17**).
9. Similar to step 8, use the concentration-balanced mixture of SIS1 peptides diluted by 625, 40, and 2, which will be used to spike samples to prepare the low, medium, and high QC samples, respectively.
10. Spike each replicate from the individual digests with only one of three (QC) SIS1 concentrations, where one replicate is spiked with low (625-fold dilution), the second with medium (40-fold dilution), and the third with high (2-fold dilution) SIS1 concentrations. Do this for all replicates so that a total of six replicates at each of the three QC concentrations are spiked into each matrix background (Fig. 3).
11. Spike the aliquots of the pooled matrix with the dilution series made in step 8. Each standard should be prepared in triplicate (Fig. 3).
12. Perform SPE on all samples and lyophilize as described in Sects. 3.5 and 3.7, respectively (*see* **Note 18**).
13. After the lyophilization step, dissolve samples in 70 μL 0.1% FA and vortex to mix, followed by brief centrifugation to collect the sample at the bottom of microcentrifuge tubes.

14. Using a P200 pipette, transfer the samples to corresponding pre-labeled 300 μL PP screw vials or a 96-well plate compatible with the LC autosampler. Place samples in the autosampler.
15. Open MassHunter Data Acquisition and navigate to the work-list tab. Create a work list using the scheduled dynamic MRM methods for monitoring the SIS1 and SIS2 transitions in plasma and heart and the corresponding methods for monitoring the endogenous and SIS2 transitions. Data acquisition will be performed twice on each QC sample, once using dynamic MRM to monitor the SIS1 and SIS2 transitions and a second time (in a subsequent injection) to monitor the endogenous and SIS2 transitions. Dynamic MRM data acquisition will be performed on standards, with monitoring of transitions for SIS1 and SIS2 only. Within the work list, add steps for blank injections of 0.1% FA using the data acquisition method for the endogenous and SIS2 transitions. Include these blanks after every three injections of sample to monitor carryover of endogenous peptides. Similarly, include these steps using data acquisition for monitoring the SIS1 and SIS2 transitions after every three injection of standards. This will measure carryover of SIS peptides between sample injections used to generate calibration curves.
16. After data acquisition is complete, import the data into Skyline Daily sequentially, creating two separate files, one for analyzing the acquisition of the endogenous and SIS2 peptides and a second file for analyzing the data acquisition of the SIS1 and SIS2 peptides. For both files, analyze the data as follows: Confirm that the peaks are correctly assigned by the software, that peak integration is being performed at equal distances on each side of the peak across all replicates, and that all three transitions are present.
17. Save the Skyline Daily files by renaming the files.
18. Quantitation should be performed by using the product ion displaying the highest intensity across all three peptides (endogenous, SIS1 and SIS2). To do so, reexamine all transitions and delete the two product ions displaying relatively lower intensities across all three isotopologue peptides.
19. Navigate to View in Skyline Daily and select Document Grid.
20. In the new Document Grid window select View and then Replicates.
21. Assign replicates with the respective sample type and concentration, and for the endogenous peptides, leave analyte concentration empty. Exit the Document Grid window.
22. For both files, navigate to Skyline Daily Settings and select Peptide Settings. In the modifications tab ensure the SIS2 peptide is selected as the internal standard.

23. In the Skyline Daily file, used for calculating the endogenous to SIS2 peak area ratio, navigate to the View pane and select Document Grid. In the new window, select View, then Peptide Ratio Results, and export the data and save. The ratio to standard values in this table will be used to calculate the concentration of the endogenous peptide from the regression line of the calibration curves generated in the Skyline Daily file used to analyze the standards and QC samples.
24. In the Skyline Daily file used to analyze the standards and QC samples, navigate to Settings and then Peptide Settings. Under the Quantification tab, select the following: Regression Fit, Linear; Normalization Method, Ratio to SIS2; and Regression Weighting, $1/(x \times x)$, MS Level 2.
25. View the calibration curve for each peptide and analyze for outliers; R^2 values should be ≥ 0.95. Navigate to Document Grid, and in the new window select Peptides with Accuracy and Stats. Ideally, the measured CVs for all six QC concentrations should be $\leq 20\%$. Calibration curves with $R^2 \geq 0.95$, CV $\leq$ 20%, can be used to calculate endogenous peptide concentrations.
26. In the same Document Grid window, navigate to View and then to Peptide Quantification. Using the criteria in step 25, export the equations of the regression lines in CSV format, and save the file.
27. Use the equation of the regression line from step 26 and the ratio of endogenous to SIS2 from step 23 to calculate the endogenous peptide concentration in the plasma and heart tissue, taking into account the equivalent volume of plasma or mass of heart tissue injected on the LC column.

4 Notes

1. This solution should be prepared immediately prior to use. Solubilized urea should be stored and used at ambient laboratory temperatures or lower. Do not heat urea as it may increase protein carbamylation which affects downstream LC-MS/MS analysis.
2. All manipulation of DTT should be performed in a fume hood.
3. This solution should be prepared immediately prior to use. Protect IAA solution from light by wrapping container with aluminum foil.
4. Prepare this solution immediately prior to use. Lyophilized trypsin aerosolizes easily; wear a respiratory mask in addition to standard level 1–2 laboratory personal protective equipment. Weigh TPCK-treated trypsin in a room with minimum airflow. It is imperative to use TPCK-treated trypsin to minimize chymotryptic activity.

5. Work with formic acid in a fume hood.
6. Work with acetonitrile and/or methanol in a fume hood.
7. Synthesis of peptide standards is not described in this chapter. If not synthesized in-house, custom peptide standards can be purchased from various vendors. Ideally, vendors who validate their peptides using MRM should be chosen.
8. Peptides must be fully solubilized; otherwise quantitation will be inaccurate due to loss of solid peptide particulates during the elution step of solid phase extraction. If a peptide standard is not soluble in 0.1% FA 30% ACN, the concentration of ACN can be increased to 60%, or an alternative solvent (such as DMSO) can be used, and/or the pH can be increased by substituting 0.1% ammonium hydroxide for FA. For difficult peptides, solvent combinations described above all require empirical testing. If all solvent combinations fail to solubilize a peptide standard, an alternate peptide standard should be designated for synthesis.
9. The amount of trypsin required to obtain efficient digestion has been empirically determined for experiments in mouse plasma and heart. Protein concentration in other samples should be determined prior to tryptic digestion, and trypsin should subsequently be added at a trypsin-protein ratio of 1:20 (w:w). For example, a sample containing a total of 500 μg of protein would require 25 μg of trypsin.
10. When drawing or pushing liquid through an SPE column, try to avoid completely removing the liquid out of the SPE column to minimize drying the column packing.
11. To facilitate flow by gravity, ensure that there are no air gaps within the column. Flow can be initiated by a brief exposure to vacuum and subsequent equilibration to atmospheric pressure by allowing air back into the vacuum chamber.
12. Before placing the column in the 1.5-mL microcentrifuge tube, ensure that any residual liquid remaining in the tip of the column is removed.
13. If elution is not observed after 1 min, initiate flow by using a 1-mL pipette fitted with a 1-mL pipette tip. Insert the pipette tip into the column far enough to create a seal between the outer surface of the pipette tip and the open rim of the SPE column. Slowly depress the pipette to initiate flow, allowing the remaining liquid to flow by gravity.
14. Any vial compatible with the Infinity 1290 autosampler can be used.
15. Endogenous peptides with concentrations determined to be lower than the LLOQ will not be quantitated.

16. These volumes have been calculated to achieve a matrix concentration similar to that used for creating scheduled dynamic MRM acquisition methods with a 1-min retention time window. Altering these volumes may result in retention time shifts of peptides and should therefore be avoided.
17. Perform all dilutions of SIS peptides in 0.1% FA.
18. To increase throughput, individual SPE columns can be substituted for a 96-well SPE plate. Proceed according to manufacturer instructions.

Acknowledgments

We are grateful to Genome Canada and Genome British Columbia for financial support (project codes 181MRM GAPP, 234DMP-DIG, 204PRO for operations and 214PRO for technology development). CHB is grateful for support from the Leading Edge Endowment Fund (University of Victoria) and for support from the Segal McGill Chair in Molecular Oncology at McGill University (Montreal, QC, Canada). CHB is also grateful for support from the Warren Y. Soper Charitable Trust and the Alvin Segal Family Foundation to the Jewish General Hospital (Montreal, QC, Canada). We would like to thank Helena Petrosova for technical editing and Sarah A. Michaud for providing sample preparation protocols for mouse tissues. Helena Petrosova and Sarah A. Michaud are affiliated with the University of Victoria Genome BC Proteomics Centre.

References

1. McReynolds JH, Anbar M (1977) Isotopic assay of nanomole amounts of nitrogen-15 labeled amino acids by collision-induced dissociation mass spectrometry. Anal Chem 49 (12):1832–1836
2. Shi T, Su D, Liu T, Tang K, Camp DG, Qian WJ, Smith RD (2012) Advancing the sensitivity of selected reaction monitoring-based targeted quantitative proteomics. Proteomics 12 (8):1074–1092
3. Vogg G, Achatz S, Kettrup A, Sandermann H (1999) Fast, sensitive and selective liquid chromatographic-tandem mass spectrometric determination of tumor-promoting diterpene esters. J Chromatogr A 855(2):563–573
4. Lange V, Picotti P, Domon B, Aebersold R (2008) Selected reaction monitoring for quantitative proteomics: a tutorial. Mol Syst Biol 4:222
5. Kondrat RW, McClusky GA, Cooks RG (1978) Multiple reaction monitoring in mass spectrometry/mass spectrometry for direct analysis of complex mixtures. Anal Chem 50 (14):2017–2021
6. Barr JR, Maggio VL, Patterson DG, Cooper GR, Henderson LO, Turner WE, Smith SJ, Hannon WH, Needham LL, Sampson EJ (1996) Isotope dilution—mass spectrometric quantification of specific proteins: model application with apolipoprotein A-I. Clin Chem 42 (10):1676
7. LeBlanc A, Michaud SA, Percy AJ, Hardie DB, Yang J, Sinclair NJ, Proudfoot JI, Pistawka A, Smith DS, Borchers CH (2017) Multiplexed MRM-based protein quantitation using two different stable isotope-labeled peptide Isotopologues for calibration. J Proteome Res 16 (7):2527–2536
8. Kuzyk MA, Smith D, Yang J, Cross TJ, Jackson AM, Hardie DB, Anderson NL, Borchers CH (2009) Multiple reaction monitoring-based, multiplexed, absolute quantitation of 45 proteins in human plasma. Mol Cell Proteomics 8 (8):1860–1877

9. Domanski D, Percy AJ, Yang J, Chambers AG, Hill JS, Freue GV, Borchers CH (2012) MRM-based multiplexed quantitation of 67 putative cardiovascular disease biomarkers in human plasma. Proteomics 12 (8):1222–1243
10. Percy AJ, Chambers AG, Parker CE, Borchers CH (2013) Absolute quantitation of proteins in human blood by multiplexed multiple reaction monitoring mass spectrometry. Methods Mol Biol 1000:167–189
11. Chambers AG, Percy AJ, Yang J, Camenzind AG, Borchers CH (2013) Multiplexed quantitation of endogenous proteins in dried blood spots by multiple reaction monitoring-mass spectrometry. Mol Cell Proteomics 12 (3):781–791
12. Percy AJ, Yang J, Chambers AG, Simon R, Hardie DB, Borchers CH (2014) Multiplexed MRM with internal standards for cerebrospinal fluid candidate protein biomarker quantitation. J Proteome Res 13(8):3733–3747
13. Kusebauch U, Campbell DS, Deutsch EW, Chu CS, Spicer DA, Brusniak MY, Slagel J, Sun Z, Stevens J, Grimes B, Shteynberg D, Hoopmann MR, Blattmann P, Ratushny AV, Rinner O, Picotti P, Carapito C, Huang CY, Kapousouz M, Lam H, Tran T, Demir E, Aitchison JD, Sander C, Hood L, Aebersold R, Moritz RL (2016) Human SRMAtlas: a resource of targeted assays to quantify the complete human proteome. Cell 166(3):766–778
14. Mohammed Y, Domanski D, Jackson AM, Smith DS, Deelder AM, Palmblad M, Borchers CH (2014) PeptidePicker: a scientific workflow with web interface for selecting appropriate peptides for targeted proteomics experiments. J Proteome 106:151–161
15. Wu C, Shi T, Brown JN, He J, Gao Y, Fillmore TL, Shukla AK, Moore RJ, Camp DG 2nd, Rodland KD, Qian WJ, Liu T, Smith RD (2014) Expediting SRM assay development for large-scale targeted proteomics experiments. J Proteome Res 13(10):4479–4487
16. Percy AJ, Michaud SA, Jardim A, Sinclair NJ, Zhang S, Mohammed Y, Palmer AL, Hardie DB, Yang J, LeBlanc AM, Borchers CH (2017) Multiplexed MRM-based assays for the quantitation of proteins in mouse plasma and heart tissue. Proteomics 17(7)

Methods in Molecular Biology (2018) 1788: 215–224
DOI 10.1007/7651_2017_77

Published online: 07 October 2017

Isobaric Labeling-Based LC-MS/MS Strategy for Comprehensive Profiling of Human Pancreatic Tissue Proteome

Chih-Wei Liu and Qibin Zhang

Abstract

The pancreas is an organ with both endocrine and exocrine functions, and various pathologies, such as pancreatic cancer and diabetes are associated with this organ. Owing to the limited pancreatic biopsy samples available for research, it is critical to make the best use of cadaveric pancreatic tissue for biomarker studies and mechanistic understanding of pancreas-related pathologies. Discovery-phase quantitative proteomics has attracted a lot of attention for its capabilities in large-scale protein identification and accurate protein quantification. Here, we describe a workflow using isobaric labeling (tandem mass tag or TMT) based quantitative proteomics to confidently identify and quantify human pancreatic tissue proteome, including sample preparation, isobaric tag labeling, peptide level fractionation, LC-MS/MS, database search, and statistical analysis.

Keywords Fractionation, Human pancreatic tissue, Isobaric labeling, LC-MS/MS, MaxQuant, Perseus, Tandem mass tag, Tissue proteome

1 Introduction

Large-scale proteomic analysis of clinical tissue samples has been widely applied to uncover the molecular level changes of pathological conditions [1, 2]. Various quantitative methods such as label free, metabolic labeling, isotopic tags, and isobaric tags have been developed for protein quantification [3]. Among those methods, isobaric tag labeling provides high precision compared to the label free method. In addition, the multiplexing capability of isobaric labeling enables an increased analysis throughput, for instance, ten samples can be analyzed simultaneously using tandem mass tag (TMT)-10plex-based peptide level labeling strategy [1, 4–6]. Furthermore, as missing value issues are common in label free proteomics, TMT10-plex strategy has the advantage of much less missing values due to the coisolation and fragmentation of the isobaric precursor ions, which provides high quality data for statistical analysis and in turn better statistical power to identify the differentially expressed proteins from the limited number of clinical samples.

In this guideline, the TMT10-plex based isobaric labeling LC-MS/MS strategy is described for comprehensive identification of human pancreatic tissue proteins. This guideline covers the entire workflow of isobaric labeling-based bottom-up proteomics strategy, which include tissue homogenization and protein digestion, peptide level TMT isobaric labeling, offline fractionation to reduce the complexity of individual proteome sample, LC-MS/MS data acquisition and database search for protein identification and quantification. Finally, the proteins with differential expressions between different pathological conditions are identified using statistical analysis.

2 Materials

All reagents and chemicals are available from Sigma-Aldrich (St. Louis, MO, USA), unless otherwise specified. Protease inhibitor cOmplete® is from Roche Life Sciences (Indianapolis, IN, USA). The BCA protein assay and sequencing-grade trypsin are respectively obtained from ThermoFisher Scientific (Rockford, IL, USA) and Promega (Madison WI, USA). All solvents used are HPLC-grade. Details of reagents and materials used in each step are listed below.

2.1 For Protein Digestion

1. Lysis buffer: 8 M urea, 75 mM NaCl in 100 mM ammonium bicarbonate pH 7.8 with 10 mM NaF, phosphatase inhibitors cocktail 2 and 3, and protease inhibitor cOmplete® at manufacturer suggested concentrations of 1% and 4% of final volume, respectively.
2. 375 mM dithiothreitol (DTT): weigh 11.6 mg of DTT in a microcentrifuge tube, and add 200 μL DI water (*see* **Note 1**).
3. 1 M iodoacetamide (IAA): weigh 18.5 mg of IAA in a microcentrifuge tube, and add 100 μL DI water (*see* **Note 2**).
4. 500 mM $CaCl_2$: weigh 5.55 g $CaCl_2$ into a container with 100 mL DI water, store at room temperature.
5. Trypsin stock solution: prepare in 1 μg μL^{-1} in 50 mM acetic acid, and store at −20 °C before use.

2.2 For TMT-10plex Isobaric Labeling

The amounts listed below are sufficient for labeling ten samples/channels.

1. 1 M TEAB: prepare stock solution of TEAB (triethyl ammonium bicarbonate) by weighing 1.63 g in 15 mL plastic centrifuge tube and add DI water to a final volume of 10 mL, store at room temperature.
2. 100 mM TEAB: dilute 1 M TEAB stock solution to 100 mM TEAB before use. 150 μL of 1 M TEAB adds into a microcentrifuge tube with 1350 μL of DI water.

3. TMT labeling reagent: for the 0.8 mg vials, add 41 μL of anhydrous acetonitrile to each tube. For the 5 mg vials, add 256 μL of anhydrous acetonitrile to each tube (*see* **Note 3**).
4. 5% hydroxylamine: the stock solution of hydroxylamine is 50%, add 10 μL of 50% hydroxylamine into a microcentrifuge tube with 90 μL of 100 mM TEAB, freshly prepared.

2.3 For High-pH RPLC Fractionation

1. Xbridge C18 analytical and guard columns are from Waters (Milford, MA, USA) or other C18 columns.
2. Buffer A: 10 mM TEAB in water (dilute from the 1 M TEAB stock solution).
3. Buffer B: 10 mM TEAB in 90% CH_3CN (*see* **Note 4**).

3 Methods

Seven major procedures (Fig. 1) are involved in analysis of pancreatic tissue proteome. As a demonstration, tissue samples from five healthy subject and five type 1 diabetic patients are included in one TMT experiment for quantitative comparison. In the case of more than ten samples to be compared, multiple TMT experiments will need to be performed, typically with a pooled sample serve as the common reference [6].

3.1 Protein Digestion

1. Approximately 30 mg of cadaveric pancreatic tissue from each subject is pulverized and homogenized in 500 μL of Lysis buffer.
2. The tissue sample mixture is sonicated for 3 min in an ice water bath.
3. Incubate the sample mixture with 13.7 μL 375 mM DTT (final concentration at 10 mM) for reduction of disulfide bonds at 37 °C for 1 h.
4. Add 10.5 μL 1 M IAA (final concentration at 20 mM) and incubate at room temperature for 1 h in the dark.

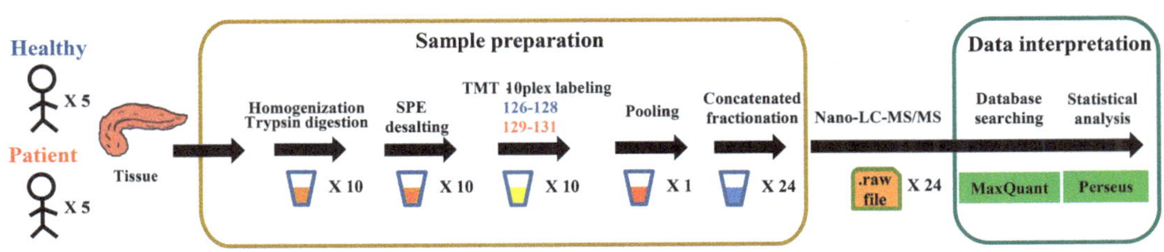

Fig. 1 Schematic representation of TMT10-plex based isobaric labeling proteomics workflow for tissue proteome study. Samples from five healthy subject and five patients are included in one TMT labeling experiment for quantitative comparison. Each procedure is described in Sect. 3

5. The sample mixture is further diluted eightfold with 50 mM NH_4HCO_3 to reduce the urea concentration <1 M (*see* **Note 5**).
6. Trypsin (0.3 mg, 1:100 ratio of enzyme to substrate) and 8.5 μL 500 mM $CaCl_2$ (final concentration at 1 mM) are added for digestion carried out at 37 °C for 4 h.
7. A second step of trypsin digestion (add 0.3 mg, 1:100 ratio of enzyme to substrate) is performed overnight at 37 °C.
8. The digestion mixture is acidified with 225 μL 20% TFA (final concentration at 1%) to stop digestion before C18 SPE desalting.

3.2 C18 SPE Desalting

The SPE desalting methods may vary depending on SPE materials and manufacturer's instructions. As an example, a 1 mg/1 mL C18 SPE cartridge (SUPELCO) can be used to desalt protein digests up to 5 mg (*see* **Note 6**).

1. Place glass tubes at the lower section of the manual SUPELCO SPE vacuum station below "needles" for collecting solvent used for conditioning and rinsing the column (*see* **Note 7**).
2. Add 1 mL methanol (MeOH) to condition the column and control the flow of MeOH solvent as fast as 15–20 drops per minute by adjusting vacuum level. Repeat twice (In total 3 mL MeOH).
3. Rinse the column with 1 mL acidified water (0.1% TFA) and control the flow of 0.1% TFA solvent as fast as 10–15 drops per minute. Repeat once (In total 2 mL 0.1% TFA).
4. After rinsing, the digestion mixture from Sect. 3.1 is loaded on to the column and flow should be controlled at a speed at least twofold lower than that used in the rinsing step.
5. Add 4 mL (1 mL for 4 times) of H_2O–CH_3CN (95:5) with 0.1% TFA into the column to wash away the salts. The flow should be at a speed similar to that for loading digestion mixture.
6. Take out the glass tube with solvent waste and put a new 2 mL microcentrifuge tube in the vacuum chamber for collecting desalted samples.
7. Add 1.2 mL of CH_3CN–H_2O (80:20) with 0.1% TFA into column to elute the peptides from column at a speed as slow as 8–10 drops per minute (*see* **Note 8**).
8. After collecting the sample, turn off vacuum and take out the sample collection tubes, and concentrate the peptide samples in vacuum concentrator before peptide concentration measurement.

3.3 TMT-10plex Isobaric Labeling

TMT-10plex labeling on peptides is performed according to the manufacturer's instructions, brief instructions are shown below.

1. 50 μg of peptides from each sample dissolved in 100 μL 100 mM TEAB is used for labeling, and 20 μL of the TMT reagent is used for each 50 μg peptide sample by adding the reagent into peptide solution (*see* **Note 9**).
2. Incubate the reaction at room temperature for 1 h.
3. Add 8 μL of 5% hydroxylamine to the sample and then vortex and incubate at room temperature for 15 min to quench the reaction.
4. Combine all samples (up to 10 samples/channels) in a new microcentrifuge tube and store at −80 °C before high-pH RPLC fractionation.

3.4 High pH RPLC Fractionation of Peptide Mixture

Many fractionation methods have been reported for peptide fractionation in order to reduce sample complexity before LC-MS/MS analysis. The high pH RPLC fractionation using C18 column is demonstrated here.

1. Xbridge C18 analytical (5 μm particles, 250 mm × 4.6 mm) and guard (20 mm × 4.6 mm) columns are used for peptide separation.
2. Prepare buffer A and B (Sect. 2.3), and degas for those two buffers before use.
3. Aliquot the labeled samples from Sect. 3.3 to create the quality control (QC) samples for optimizing peptide separation conditions (*see* **Note 10**).
4. QC samples are used to evaluate the separation performance, and further optimize the effective LC separation gradient if needed.
5. A typical starting LC gradient condition at a flow rate of 0.5 mL min^{-1} can be 0 to 5% buffer B in 10 min, from 5 to 35% B in 60 min, from 35 to 70% B in 15 min, and hold at 70% B for an additional 10 min for washing column before using 0% B for 10 min to reequilibrate column.
6. Collect 96 fractions of the peptides to cover the entire peptide elution time window (*see* **Note 11**).
7. Concatenate the 96 fractions into 24 fractions (*see* **Note 12**).
8. Dry the final 24 fractions in vacuum concentrator and reconstitute in 0.1% formic acid (FA) at a final concentration of 0.2 μg $μL^{-1}$ before LC-MS/MS analysis (*see* **Note 13**).

3.5 LC-MS/MS Analysis

Protocols for LC-MS/MS analysis can vary because of diversity of LC systems (manufacturer, column, solvent composition, gradient, flow rate, etc.) and MS instruments (manufacturer, electrospray condition, fragmentation, MS parameters, analyzer, etc.). The following is the practice routinely used in our laboratory.

1. LC-MS/MS platform consists of an UltiMate 3000 RSLCnano system and a Q Exactive HF mass spectrometer coupled with an EASY-Spray ion source (ThermoFisher Scientific).
2. Peptide separation is performed on a PepMap C18 analytical column (2 μm particle, 50 cm × 75 μm, ThermoFisher Scientific). Injection volume is 2.5 μL (0.5 μg peptide amount loaded into column) per fraction (*see* **Note 14**).
3. A binary solvent system consisting of 0.1% FA in water (solvent A) and 0.1% FA in CH_3CN (solvent B) is used at a flow rate of 250 nL min^{-1} (*see* **Note 15**).
4. LC separation is performed using the following gradient setting: hold at 4% B for 3 min (for desalting), from 4 to 8% B in 0.1 min, 8 to 40% B in 90 min (effective gradient), 40 to 90% B in 0.1% min, hold at 90% B for 10 min (for washing column), 90 to 4% B in 0.1 min, and hold at 4% B for 17 min for reequilibrating column (*see* **Note 16**).
5. MS data are acquired in profile mode and resolution for full scan (400–2000 m/z) is set to 120,000 (at m/z 200) with maximum ion injection time of 50 ms, and automatic gain control (AGC) target of 1e6.
6. MS/MS data are acquired with data-dependent top 15 method. An isolation window of 1.4 m/z is used to isolate precursor ions for fragmentation by higher-energy collisional dissociation (HCD) at normalized collision energy of 32 (*see* **Note 17**). Resolution for MS/MS spectrum is set to 60,000 (at m/z 200) with maximum ion injection time of 100 ms (*see* **Note 18**). AGC target for MS/MS scans is 1e5.
7. Precursor ions with single, seven and higher charge states are excluded from fragmentation, and dynamic exclusion time is set to 20 s.

3.6 Database Search for Protein Identification and Quantification

Many database search software packages are available for this purpose. MaxQuant is demonstrated here [7].

1. The acquired datasets (.raw files) are analyzed using MaxQuant and the built-in Andromeda search engine against a UniProt human database (*see* **Note 19**).
2. Variable modifications include protein N-terminal acetylation and methionine oxidation.
3. Fixed modifications contain cysteine carbamidomethylation and TMT labeled N-terminus and lysine residue.
4. A maximum of two missed cleavages are allowed for the search.

5. Trypsin/P is selected as the semispecific proteolytic enzyme (*see* **Note 20**).
6. The false discovery rate (FDR) cut off used for both peptides and proteins is 0.01 (1%) using decoy database.
7. The precursor intensity fraction is set as 0.75 to minimize influence of the co-eluting peptides in quantification [8].
8. Only the razor/unique peptides are used for quantitative calculations.
9. The other parameters are the default settings in MaxQuant software for processing orbitrap-type data (*see* **Note 21**).

3.7 Statistical Analysis

The search results in ProteinGroups.txt generated by MaxQuant are directly processed by Perseus software [9]. The differentially expressed proteins are identified by statistical analysis tools built in Perseus.

1. Import the quantitative data from ProteinGroups.txt into Perseus.
2. The potential contaminants, reverse hits and proteins only identified by modification site are excluded.
3. A filtering criterion is set to keep the identified proteins with the quantified values of all ten reporter ions (no missing value) in the final identification list.
4. The protein intensities are log2-transformed and normalized by subtracting the median intensity in each column/sample.
5. Categorize the samples into two groups: healthy subjects and patients.
6. Principal component analysis (PCA) is performed based on protein intensities to differentiate two groups.
7. Two-samples tests coupled with Benjamini–Hochberg (FDR cutoff of 0.05) correction are performed to identify the differentially expressed proteins [10].

4 Notes

1. Stock DTT solution should be freshly prepared before use.
2. Keep the IAA solution in the dark.
3. The unused reagent could be dried and stored at −20 °C for future use.
4. When preparing buffer B (10 mM TEAB in 90% CH_3CN), add CH_3CN slowly to 10 mM TEAB because 100% CH_3CN can precipitate TEAB.

5. High concentration of urea affects the trypsin digestion efficiency, trypsin activity is not affected when reducing the urea concentration to <1 M.
6. The columns should not be dried in the entire experimental processes except for the last eluting step.
7. The vacuum pressure used in the chamber should not exceed 20" Hg for the specific SUPELCO SPE cartridge.
8. When the column is almost dry, increase the vacuum pressure to elute the peptide sample completely to increase the recovery.
9. Equilibrate the TMT label reagents to room temperature before use. For the 0.8 mg vials, add 41 μL of anhydrous acetonitrile to each tube. For the 5 mg vials, add 256 μL of solvent to each tube. Allow the reagent to completely dissolve for 5 min with occasional vortexing, and briefly centrifuge the tube to gather the solution. 41 μL of reagent can label 100 μg peptides, and the reagent volume can be further adjusted according to the amount of peptides used for labeling. Make sure that the same amount of TMT reagents for labeling each sample.
10. The LC gradient condition may need optimization in order to have the peptides elute relative evenly during the entire LC gradient time. In the case of 500 μg of labeled peptide mixture from Sect. 3.3, 300 μg could be used as QC samples (100 μg per injection) for optimization of LC gradient.
11. The fraction collector is set to 4 °C while collecting the peptide fractions.
12. After collection of 96 fractions, 4 out of 96 fractions are pooled into one fraction, referred to as concatenation. For instance, fraction 1st, 25th, 49th, and 73rd are combined into one. Do so for the other 23 pooled fractions.
13. A rough estimation of solvent volume used for reconstitution of peptides in each fraction can be determined by the amount of peptides loaded onto column for fractionation and the number of final fractions. For instance, 100 μg divided by 24 fractions to have 4.2 μg per fraction, and to prepare sample at 0.2 μg μL^{-1} need to add 20.8 μL solvent for reconstitution.
14. The injection volume depends on the sample loop of autosampler, column loading capacity, and MS detector, therefore injection volume need to be adjusted based on the actual setup.
15. A flow rate of 250 nL min^{-1} for C18 50 cm × 75 μm i.d. column results in around 550–600 bar column pressure when heating column at 35 °C.
16. The gradient used for peptide separation can be modified depending on separation performance. However, all samples

must be run under the same condition to limit variations between samples.

17. The normalized collision energy used for TMT-labeled peptides is usually higher than those for label free peptides (without labeling) in order to generate high quality reporter ions in low mass region, the normalized collision energy could be further optimized according to the number of peptides identified and the quantification accuracy.
18. To ensure 6 mDa differences between reporter ions in TMT-10plex, at least a resolution of 35,000 is needed to clearly resolve the reporter ion signals. The resolution of 60,000 is the optimal setting in Q Exactive HF MS.
19. The database information needs to include the type, sequence entry number, and releasing date of database.
20. The selection of enzyme used for search is based on the enzyme that is chosen for protein digestion in Sect. 3.1.
21. For TMT-10plex labeling, different batches of TMT reagents have slightly different isotope impurity which needs to be included for database search to correct the reporter ion ratio. The information of isotope impurity can be found in the reagent kit.

Acknowledgment

This work was supported by the National Institute of Diabetes and Digestive and Kidney Diseases of the National Institutes of Health under Award Number R01DK114345. Cadaveric human pancreata were obtained from the Network for Pancreatic Organ Donors with Diabetes (nPOD), a collaborative type 1 diabetes research project sponsored by the JDRF. Organ Procurement Organizations (OPO) partnering with nPOD to provide research resources are listed at http://www.jdrfnpod.org/for-partners/npod-partners/.

References

1. Liu CW, Atkinson MA, Zhang Q (2016) Type 1 diabetes cadaveric human pancreata exhibit a unique exocrine tissue proteomic profile. Proteomics 16(9):1432–1446
2. Chung L, Baxter RC (2012) Breast cancer biomarkers: proteomic discovery and translation to clinically relevant assays. Expert Rev Proteomics 9(6):599–614
3. Xie F, Liu T, Qian WJ, Petyuk VA, Smith RD (2011) Liquid chromatography-mass spectrometry-based quantitative proteomics. J Biol Chem 286(29):25443–25449
4. Svinkina T, Gu H, Silva JC, Mertins P, Qiao J, Fereshetian S, Jaffe JD, Kuhn E, Udeshi ND, Carr SA (2015) Deep, quantitative coverage of the lysine acetylome using novel anti-acetyl-lysine antibodies and an optimized proteomic workflow. Mol Cell Proteomics 14 (9):2429–2440
5. Murphy JP, Stepanova E, Everley RA, Paulo JA, Gygi SP (2015) Comprehensive temporal

protein dynamics during the diauxic shift in Saccharomyces cerevisiae. Mol Cell Proteomics 14(9):2454–2465

6. Liu CW, Bramer L, Webb-Robertson BJ, Waugh K, Rewers MJ, Zhang Q (2017) Temporal profiles of plasma proteome during childhood development. J Proteome 152:321–328
7. Tyanova S, Temu T, Cox J (2016) The MaxQuant computational platform for mass spectrometry-based shotgun proteomics. Nat Protoc 11(12):2301–2319
8. Michalski A, Cox J, Mann M (2011) More than 100,000 detectable peptide species elute in single shotgun proteomics runs but the majority is inaccessible to data-dependent LC-MS/MS. J Proteome Res 10(4):1785–1793
9. Tyanova S, Temu T, Sinitcyn P, Carlson A, Hein MY, Geiger T, Mann M, Cox J (2016) The Perseus computational platform for comprehensive analysis of (prote)omics data. Nat Methods 13(9):731–740
10. Tusher VG, Tibshirani R, Chu G (2001) Significance analysis of microarrays applied to the ionizing radiation response. Proc Natl Acad Sci U S A 98(9):5116–5121

Methods in Molecular Biology (2018) 1788: 225–241
DOI 10.1007/7651_2017_81

Published online: 21 October 2017

MALDI Imaging Mass Spectrometry of N-glycans and Tryptic Peptides from the Same Formalin-Fixed, Paraffin-Embedded Tissue Section

Peggi M. Angel, Anand Mehta, Kim Norris-Caneda, and Richard R. Drake

Abstract

Matrix-assisted laser desorption/ionization (MALDI) imaging mass spectrometry (IMS) is a unique and well developed tool for probing the protein content of formalin-fixed, paraffin-embedded tissue (FFPE). Integral to this approach is the application of trypsin, and more recently peptide N-glycosidase F, to release tryptic peptides or N-glycans from tissue and report localization of distinct species. This is typically done on serial or adjacent tissue sections, and there is an emerging need to understand the colocalized protein population linked to the exact same regions of N-glycans. Here we describe an approach where N-glycans are first imaged from a tissue section followed by reprocessing of the same tissue section for tryptic peptide MALDI IMS. Strategies for colocalizing peptides to target N-glycans or N-glycan regions are described.

Keywords: Formalin-fixed, Imaging mass spectrometry, MALDI imaging mass spectrometry, Paraffin-embedded tissue imaging, Peptide identification for imaging mass spectrometry, Peptide N-glycosidase F, Proteomics, Tryptic peptide imaging

1 Introduction

MALDI IMS enables high throughput reporting of hundreds to thousands of molecules localized within a single thin tissue section, allowing new insight into many disease processes that are impossible to understand by conventional microscopy [1–3]. Robust workflows using enzymatic approaches are used to access the highly cross-linked proteins of FFPE tissues, allowing the study of stored clinical tissues [4–6]. Spraying a thin molecular layer of trypsin onto tissue sections allows imaging of trypsin produced peptides analogous to solution proteomic workflows, with additional information on localization of the proteomes within the tissue architecture [7]. Robust MALDI IMS workflows using the enzyme PNGase F to release N-glycans from thin tissues sections have been developed [6] and are now routinely used to examine distribution of N-glycosylation related to the pathology of disease [8–10].

There is intense interest in combining the imaging workflows for proteomics and glycomics to understand the protein populations undergoing specific glycosylation changes. Combined workflows working on tissue sections in solution have reported the sequential glycan and proteome composition of single FFPE tissue sections [11]. Here, we describe the process of preparing the same FFPE tissue section for imaging of first N-glycans followed by tryptic peptides from the same tissue section. A similar technique was recently reported to obtain N-glycan and peptide expression from the same tissue section with antigen retrieval performed after application of PNGase F and before trypsin application [12]. The current protocol details a strategy using a single antigen retrieval step prior to PNGase F application, followed by the release and MALDI IMS of N-glycans, specific washing steps to remove all matrix, PNGase F protein, and N-glycans from the tissue section, followed by tryptic digestion and MALDI IMS of tryptic peptides. The imaging data reported here was produced using a MALDI Fourier transform ion cyclotron resonance (FT-ICR) mass spectrometer. In the workflow, the accurate mass for N-glycans are known from databases created by past studies [13–15]. This protocol allows discovery of co-localized N-glycans and tryptic peptides toward a more complete understanding of the tissue microenvironment.

2 Materials

Prepare all solutions in HPLC grade water. Follow all safety and waste disposal regulations.

2.1 MALDI Imaging Mass Spectrometry Solutions

1. Antigen retrieval solution: Pour 50 mL water into a clean 50–100 mL bottle. Add 50 μL of citraconic acid anhydride buffer to the water. Add 4 μL of 12 M HCl and mix. Add water to the bottle for a total of 100 mL and mix. Check that pH is around 3.0 ± 0.5 (*see* **Note 1**). Use the same day (*see* **Note 2**).
2. Preparation of 95% ethanol. Add 950 mL 200 proof ethanol to a clean bottle. Add 50 mL water and mix.
3. Preparation of 70% ethanol: Add 700 mL 200 proof ethanol to a clean bottle. Add 300 mL water and mix.
4. 25% Trifluoroacetic acid (TFA): Add 3 mL water to a clean bottle. Carefully add 1 mL of trifluoroacetic acid to the water and mix.
5. Matrix solvent [50% Acetonitrile (ACN), 0.1% TFA]: Add 25 mL water to a clean 100 mL bottle. Add 400 microliters of 25% TFA (Solution #4) to the water and mix. Add 50 mL of acetonitrile and mix. Add water to a final volume of 100 mL and mix. Store for up to 2 months at room temperature.
6. MALDI matrix for N-glycan or peptide imaging (Alpha-cyano-4-hydroxycinnamic acid (CHCA), 7 mg/mL in 50%

acetonitrile/0.1% TFA): Weigh out 0.0420 ± 0.001 g CHCA. Add the solid CHCA to a clean 50 mL falcon tube. Bring to volume with 6 ml of matrix solvent. Vortex briefly and sonicate 5 min using a benchtop sonicator. Filter CHCA solution using a 13 mm 0.2 μm PTFE hydrophilic syringe filter graded for use with HPLC solvents (*see* **Note 3**).

2.2 Tissue Clearing Solvents

1. High pH Tissue Clearing Solution (10 mM Tris, pH 9). Add 500 mL water to a clean 1000 mL bottle. Add 1.21 ± 0.12 g Tris base and mix. Add water to 900 mL and adjust pH to 9.0 ± 0.03 using 1 M HCl. Add water to a total volume of 1000 mL.
2. Low pH Solution. Pour 50 mL water into a clean 50–100 mL bottle. Add 50 μL of citraconic acid anhydride buffer to the water. Add 4 μL of 12 M HCl and mix. Add water to the bottle for a total of 100 mL and mix. Check that pH is around 3.0 ± 0.5 (*see* **Note 1**). Use the same day (*see* **Note 2**). This is the same solution as the antigen retrieval solution.

2.3 Enzyme Solutions

1. Ammonium bicarbonate solution (25 mM, pH 7.5): Add 50 mL water into a clean 100 mL bottle. Weigh out 0.1976 ± 0.02 g of ammonium bicarbonate and add to water, agitate to mix. Add water to the bottle for a total of 100 mL and mix. The pH should be 7.5 ± 0.5. Store at 4 °C for use up to 2 weeks.
2. Preparation of 1 mL of 0.1 μg/μL PNGase F solution. To 100 μg of PNGase F add 1000 μL of prepared 25 mM ammonium bicarbonate and mix (*see* **Note 4**). Use same day. This prepares enough enzyme to cover four microscope slides.
3. Preparation of 1 mL of 0.1 μg/μL trypsin solution. To 100 μg of trypsin add 1000 μL of prepared 25 mM ammonium bicarbonate and mix (*see* **Note 5**). Use same day. This prepares enough enzyme to cover four microscope slides.

2.4 TM-Sprayer™ solutions

1. Push solvent (50% methanol–water): Add 500 mL methanol to a clean bottle. Add 500 mL water and mix. Solvent may be kept at room temperature during the duration of use (*see* **Note 6**).

3 Methods

Carry out all procedures at room temperature unless otherwise specified.

3.1 Heating and Dewaxing

1. Incubate the slides with tissue face up in a 60 °C oven for 1 h.
2. Prepare Coplin jars of solvent for dewaxing by pouring the following solutions into Coplin jars: Xylenes, two Coplin jars; 200 proof ethanol, USP grade, two Coplin jars; 95% ethanol,

one Coplin jar; 70% ethanol, one Coplin jar; double distilled water, two Coplin jars. Solvent should be added to a level that will allow complete immersion of the tissue sections.

3. Dewax the slides in the following order with the specified times: xylenes, two times at 2 min each; 200 proof ethanol, two times at 1 min each; 95% ethanol solution, one time at 1 min; 70% ethanol solution, one time at 1 min; HPLC grade water, two times at 3 min each. For each step, immerse the slides completely in freshly poured solution for the stated length of time. At the end of the incubation time, agitate the slides briefly as they are removed from the Coplin jars.
4. Dry the slides in a vacuum desiccator for 5 min.

3.2 Slide Scanning

1. This step produces an optical image needed for selecting tissue regions for imaging analysis. Fiducials, or reference points, are needed to accurately "teach" the instrument where the tissue is located on the slide.
2. Use a reflective metallic marker to make a small circle at each corner of the microscope slide. Use a black marker to draw a cross or hash mark on top of each silver circle. The reflective marker provides a contrasting background for clear visualization of the black mark to use as a fiducial.
3. For images that will be acquired by mass spectrometry at $\geq$100 μm spatial resolution, scan the whole slide at a minimum of 1200 ppi resolution. For images that will be acquired with a $\leq$50 μm stepsize scan the slide at a minimum of 2400 ppi resolution. Save the images as JPEG, bitmap (*.bmp) or 8 bit TIFF formats.
4. After scanning and prior to antigen retrieval, slides may be stored overnight in a desiccator. For longer times over 2 days, store the slides at −20 to −80 °C. It is preferable to proceed with the next step immediately.

3.3 Antigen Retrieval

1. This step details antigen retrieval for imaging mass spectrometry using a vegetable steamer. A rice cooker or decloaker may also be used for antigen retrieval.
2. Fill a vegetable steamer to the marked water level and preheat for 5 min prior to antigen retrieval (*see* **Note 7**).
3. Add around ~10 mL of the antigen retrieval solution to a plastic five slide mailer with side opening.
4. Place three slides into each five slide mailer with side opening. Slides in positions 1 and 5 are placed with tissue facing outward to the solution. Position 3 may face either way. This allows good solvent access to the tissue.
5. Fill the slide mailer the rest of the way with the antigen retrieval solution so that all tissue is completely covered.

6. Steam must be released from the slide mailer to prevent rupture during antigen retrieval. A hole may be punched into the mailer lid using an 18-gauge needle. If the mailer has no holes punched in the lid, snap closed only one corner of the mailer. This allows steam to exit. Place the mailer in the center of the vegetable steamer.
7. Heat for 30 min. Temperature should reach 95 °C for a minimum of 20 min (*see* **Note 8**).
8. Place the hot mailer in a container holding cool water (*see* **Note 9**). Cooling water should come midway up the side of the mailer. Cool for 5 min in the water bath.
9. Remove half the buffer from the mailer and replace with distilled water. Cool on the countertop for 5 min. Repeat removal of half the buffer two more times, each with 5 min of cooling. Complete by rinsing in 100% distilled water.
10. Dry the slides for 5 min in a desiccator.

3.4 PNGase F Application by the M3 TM-Sprayer™

1. A syringe pump with 0.05% accuracy in pumping 25 μL/min is used for enzyme application. A glass or plastic 1-mL syringe with Luer lock is used for loading enzyme into the sprayer.
2. Fill a glass or plastic 1-mL syringe with Luer lock syringe with prepared PNGase F solution ensuring that there are no bubbles in syringe (*see* **Note 10**). Fasten the syringe to the TM-Sprayer™ line used for enzyme spraying. Secure the syringe in the pump.
3. Set the pump to a flow rate of 25 μL/min with an inner diameter matching that of the used syringe. Do not turn on the pump at this time.
4. Place the microscope slides with tissue samples on the TM-Sprayer™ sample area, fastening them with lab tape.
5. Turn on the TM-Sprayer™ and then the controlling computer.
6. Open the nitrogen gas tank valve, setting the regulator to 10 psi.
7. In the TM-Sprayer™ software, set the temperature to 45 °C. Temperature will not adjust without the nitrogen gas flowing.
8. Program the TM-Sprayer™ to cover the appropriate number of slides, allowing a 5 mm additional edge distance for sprayhead turn round (*see* **Note 11**).
9. Program the TM-Sprayer™ method for PNGase F to use 15 passes, crisscross pattern, velocity of 1200, 3.0 mm track spacing, and a dry time of zero. The tip of the sprayhead should be 40 mm distance from the surface of the slide.
10. Start the syringe pump. Place a blank microscope slide under the nozzle of the spray head to check the TM-Sprayer™ to

monitor the start of enzyme solution spraying. It generally takes about 1–3 min to start emitting solution.

11. Once moisture is detected on the blank slide, press "Start" in the TM-Sprayer™ software.
12. PNGase F solution will be applied in a thin layer onto target tissues. During application of the enzyme, moisture will be observed on the slide as the sprayhead passes over an area; this should dry within 30 s to maintain localization of released enzymes.

3.5 Incubation for On-Tissue Digestion

1. The same incubation approach is used for both PNGase F and trypsin digestion and is performed immediately after application of PNGase F or trypsin.
2. Prepare an incubation chamber for digestion using a plastic 100 × 15 mm cell culture dish. Fit a single layer paper towel (Wypall 60×) on the bottom of the dish. Fold two 4 × 6 Kimwipes and place at opposite sides of the dish. Using a spray bottle of water, add water to saturate the papertowels and Kimwipes. Stop adding water when excess water accumulates in the dish, observed by tilting the dish to one side. For the specified towels and Kimwipes, this is about 5 mL of water.
3. The incubation chamber should be preheated for 15–30 min in an oven at 37.5 °C and should show a thin layer of condensation on the top of the incubation dish before placing the tissue inside. The incubation dish is heated with lid in place; no lab tape or other sealant is required.
4. Place the slide with the tissue facing upward into the incubation chamber using the Kimwipes as supports. Gently push the slide down slightly so that when the cover is placed on, the tissue does not touch the incubation chamber cover.
5. Incubate 2 h (*see* **Note 12**) in the oven set at 37.5 ± 1.5 °C (*see* **Note 13**).
6. After incubation, remove the slide slowly while holding it parallel with the countertop. Wipe off the condensation to prevent liquid rolling onto the tissue surface and delocalizing N-glycans (or peptides).
7. Store the slide in a five slide mailer to protect the released N-glycans or peptides. If matrix cannot be sprayed the same day, store briefly in a desiccator (6–12 h) or at −20 °C for long term (2–3 days).
8. It is recommended to immediately spray matrix onto the slide.

3.6 MALDI Matrix Application by the M3 TM-Sprayer™

1. MALDI matrix application is performed after PNGAse F digestion and again after trypsin digestion.

2. An isocratic pump with 0.05% accuracy in pumping 100 μL/min is used to for matrix application. A glass 5-mL syringe with Luer lock for loading matrix into the sprayer.
3. Ensure that the isocratic pump is set to pump 100 μL/min (*see* **Note 14**). Solvent may be degassed to limit flow variation.
4. Turn on the TM-Sprayer™ and controlling computer.
5. Open the nitrogen gas tank, setting the regulator to 10 psi.
6. In the TM-Sprayer™ software, set the temperature to 80 °C. Temperature will not adjust without the nitrogen gas flowing.
7. Program the TM-Sprayer™ method for CHCA matrix application to use eight passes, crisscross pattern, velocity of 1300, 2.5 mm track spacing and zero dry time. The tip of the spray-head should be 40 mm distance from the surface of the slide.
8. Place the samples on the TM-Sprayer™ platform, fastening them with lab tape.
9. Program the TM-Sprayer™ to cover the appropriate number of slides, allowing a 5 mm additional edge distance for spray-head turn round (*see* **Note 11**).
10. Fill a glass 5-mL syringe with the filtered MALDI matrix solution, ensuring that there are no bubbles in the syringe.
11. Fasten the syringe to the TM-Sprayer™ line going to the 6-port valve. With the valve switch in "Load" position, inject the MALDI matrix solution into the 5 mL loop.
12. Ensure that the pump is flowing at 100 μL/min and that appropriate pump pressure readouts are stable.
13. Move the 6-port valve switch to "Spray."
14. Use a blank microscope slide to check the TM-Sprayer™ nozzle for spraying of solution. Once matrix is detected as an opaque film on the dummy slide, press "Start" in the TM-Sprayer™ software.
15. CHCA solution will be applied in a thin layer onto target tissues.
16. When finished, matrix coated slides may be imaged immediately by MALDI imaging mass spectrometry or stored in a desiccator until imaging experiments.
17. Figure 1 shows a typical overall average spectrum and example images from the workflow.

3.7 Tissue Clearing of Matrix and Residual N-glycans

1. The purpose of tissue clearing is to remove matrix, PNGase F and N-glycans prior to trypsin digestion.
2. Remove matrix by immerse the slide in 200 proof ethanol for 1 min.

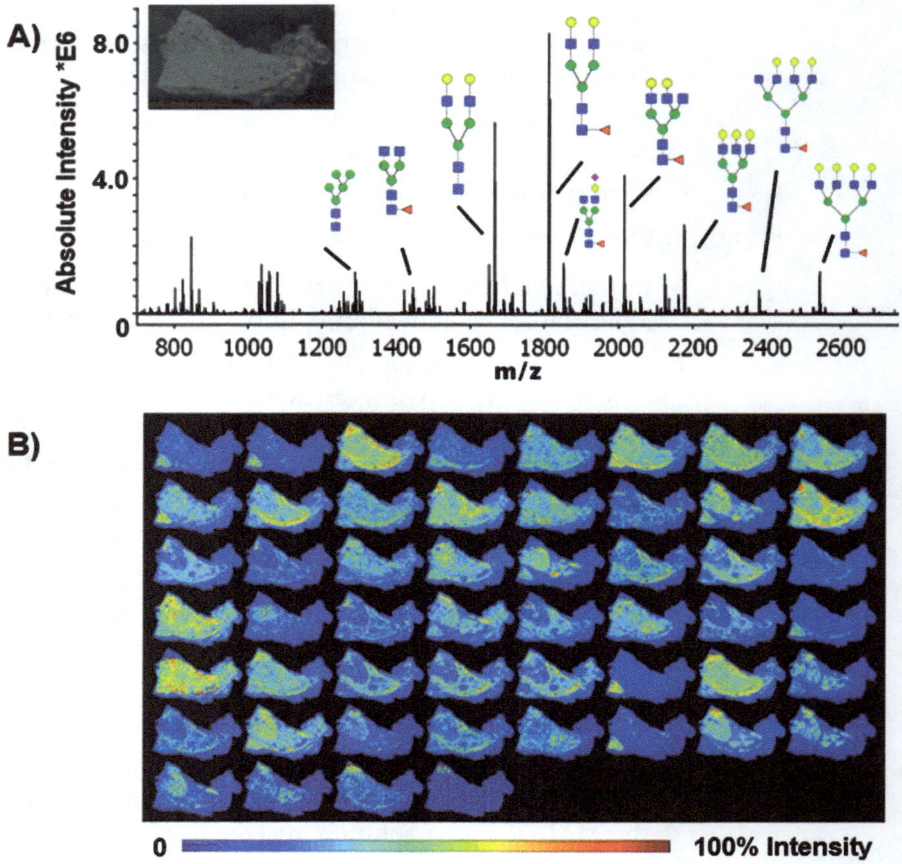

Fig. 1 Example data of N-glycan image data. (**a**) Typical overall average mass spectrum; the N-glycans corresponding to m/z are a known entity in this experiment. Inset is the scanned tissue. (**b**) Examples of N-glycan images created as heat maps of a selected ion's intensity and demonstrating basic pattern motifs on tissue

3. Immerse the slide in 95% ethanol solution for 1 min followed by 70% ethanol for 1 min.
4. Remove hydrophilic N-glycans by incubating the slide in water for 1 min, High pH Tissue Clearing Solution for 1 min, water for 1 min, Low pH Tissue Clearing Solution (citraconic buffer) 1 min, and water 1 min. For each step, agitate the slides 3–5 times at the end of the incubation.
5. Wipe excess water off the back of the slide and dry for 5 min in desiccator.

3.8 Trypsin Application by the M3 TM-Sprayer™

1. A syringe pump with 0.05% accuracy in pumping 25 μL/min is used to for enzyme application. A glass or plastic 1-mL syringe with Luer lock for loading enzyme into the sprayer.

2. Fill a 1 mL glass or plastic 1-mL Luer lock syringe with prepared trypsin solution ensuring that there are no bubbles in the syringe (*see* **Note 15**).
3. Secure the syringe in the pump. Set the pump to a flow rate of 30 μL/min with an inner diameter matching that of the used syringe. Do not turn on the pump at this time.
4. Place the microscope slides with tissue samples on the TM-Sprayer™ platform, fastening them with lab tape.
5. Turn on the TM-Sprayer™ and controlling computer.
6. Open the nitrogen gas tank valve, setting the regulator to 10 psi.
7. In the TM-Sprayer™ software, set the temperature to 45 °C. Temperature will not adjust without the nitrogen gas flowing.
8. Program the TM-Sprayer™ to cover the appropriate number of slides, allowing a 5 mm additional edge distance for spray-head turn round (*see* **Note 11**).
9. Program the TM-Sprayer™ method for trypsin to use 8 passes, crisscross pattern, velocity of 1200, 3.0 mm track spacing, a zero dry time. The sprayhead should be 40 mm from the tip of the sprayhead to the top of the slide surface.
10. Start the syringe pump.
11. Place a blank microscope slide under the nozzle of the spray head to check the TM-Sprayer™ to monitor the start of enzyme solution spraying. It generally takes about 1–3 min to start emitting solution.
12. Once moisture is detected on the blank slide, press "Start" in the TM-Sprayer™ software.
13. Trypsin solution will be applied in a thin layer onto target tissues.
14. Perform trypsin digestion following Sect. 3.5, Incubation for On-Tissue Digestion.
15. After On-Tissue Digestion, complete Sect. 3.6, MALDI Matrix Application by the M3 TM-Sprayer™.
16. Upon completion, matrix coated slides may be imaged immediately by MALDI imaging mass spectrometry or stored in a desiccator until imaging experiments are completed (*see* **Note 16**).
17. Figure 2 shows a typical overall average spectrum and example images from the workflow.
18. Co-localized N-glycans and tryptic peptides may be identified by manual searches comparing the data patterns or using the vendor neutral SCiLS Lab software (Bruker).

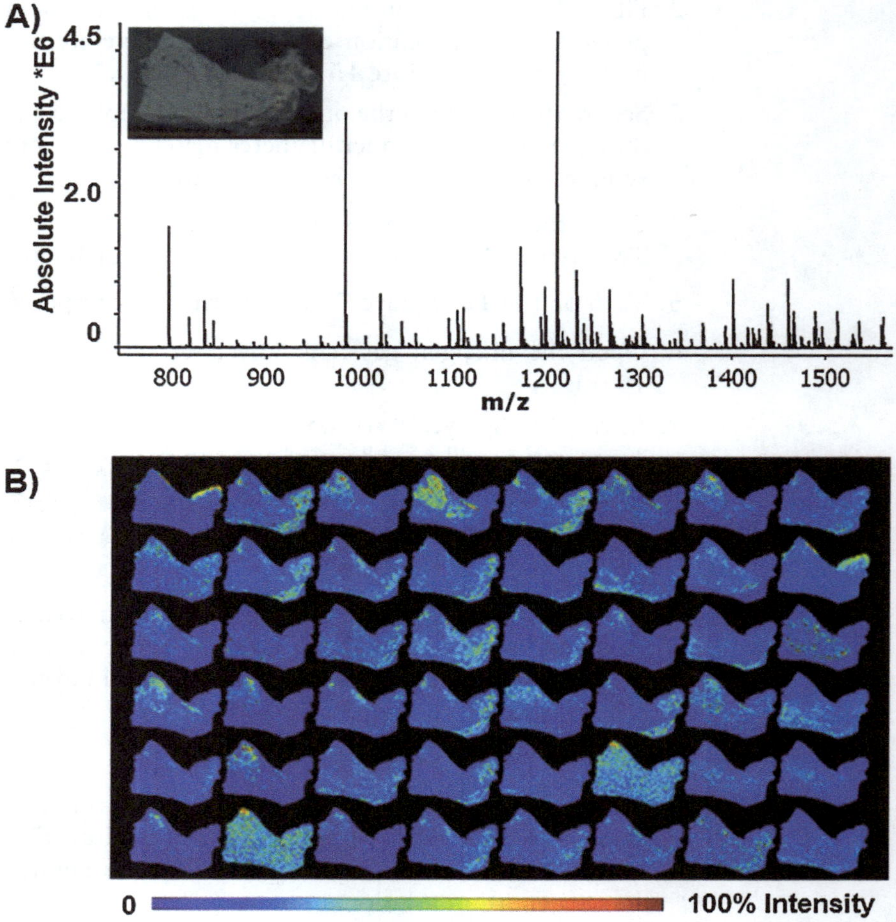

Fig. 2 Example of tryptic peptide image data. (**a**) Typical overall average mass spectrum; the tryptic peptide identifies are an unknown entity in this experiment. Inset is the scanned tissue. (**b**) Examples of tile view of images from tryptic peptides created as heat maps of a selected ion's intensity, showing 48 peptide images out of a total of 4535 detected image patterns

3.9 Identifying Co-localization of Tryptic Peptides with a Target N-glycan

1. Co-localization of a target N-glycan with tryptic peptides may be calculated using the vendor neutral SCiLS software (Bruker).
2. Load the both the N-glycan and trypsin image data into a single instance of SCiLS using the data import wizard; this procedure differs based on instrument type.
3. Load a database of singly charged N-glycan masses using the clipboard function. Go to File → Import → m/z intervals from CVS or clipboard (*see* **Note 17**). After naming the list, click Save.
4. Generate image data for the N-glycans by selecting the N-glycan database list in the "m/z intervals" pulldown menu. Click the

small left arrow beside the menu name. Click the "eye" icon on the far right to open the m/z Image Generator. Click OK to generate images.

5. View images using the "m/z Images" pulldown menu. Click the small arrow to the left of the menu name. A list of m/z will display. Selecting the "eye" icon to the left of the m/z will generate a heat map of the m/z pattern on the tissue. Identify an N-glycan m/z of interest for computing colocalized peptides.
6. On the menu at the top of the software, select Tools → Find Co-Localized m/z Values. Ensure that an appropriate normalization method is selected, and check the box next to Regions. In the Mode menu, select "Correlate with m/z Image." Type in the mass of the N-glycan exactly as found in the imported database list and using the correct interval width. Next to "m/z Intervals", select the regional peak list generated for the image generated by the trypsin workflow. These are all possible peptides that may correlate to the selected N-glycan. Save as an informative name and click "OK."
7. Once the data has calculated, a new peak list will appear under the "m/z Intervals" pulldown menu. Click the small left arrow beside the menu name. Click the "eye" icon to the far right of the menu window. A Threshold bar is displayed with a Correlation Value. Clicking in the bar raises or lowers the correlation value and reports the number of intervals associated with the correlation value. Correlation values of greater than 0.6 should be inspected for co-localization with the target peak. Once a correlation value is selected, Click the "eye" icon on the far right to open the m/z Image Generator. Click OK to generate images.
8. Figure 3 demonstrates an image data result from the procedure.

3.10 Identifying Colocalized Tryptic Peptides and N-glycans Based on Tissue Region

1. Co-localization of N-glycans with tryptic peptides by region on tissue (e.g., tumor) may be calculated using the vendor neutral SCiLS software (Bruker).
2. Load the both the N-glycan and trypsin image data into a single instance of SCiLS using the data import wizard; this procedure differs based on instrument type.
3. Create a Combined Peaklist with defined N-glycans and a SCiLS created Tryptic Peptide Peaklist using a text editor. Add the database of N-glycan mass-to-charges (*see* **Note 17**) to a new instance of a text editor. In SCiLS, collect the Tryptic Peptide Peaklist by going to File → SCiLS Report Table. Use the pulldown tab to select the Tryptic digest Peaklist created by SCiLS. This will have the format of "image data name: Imported Peaks." On the bottom of the report tables window,

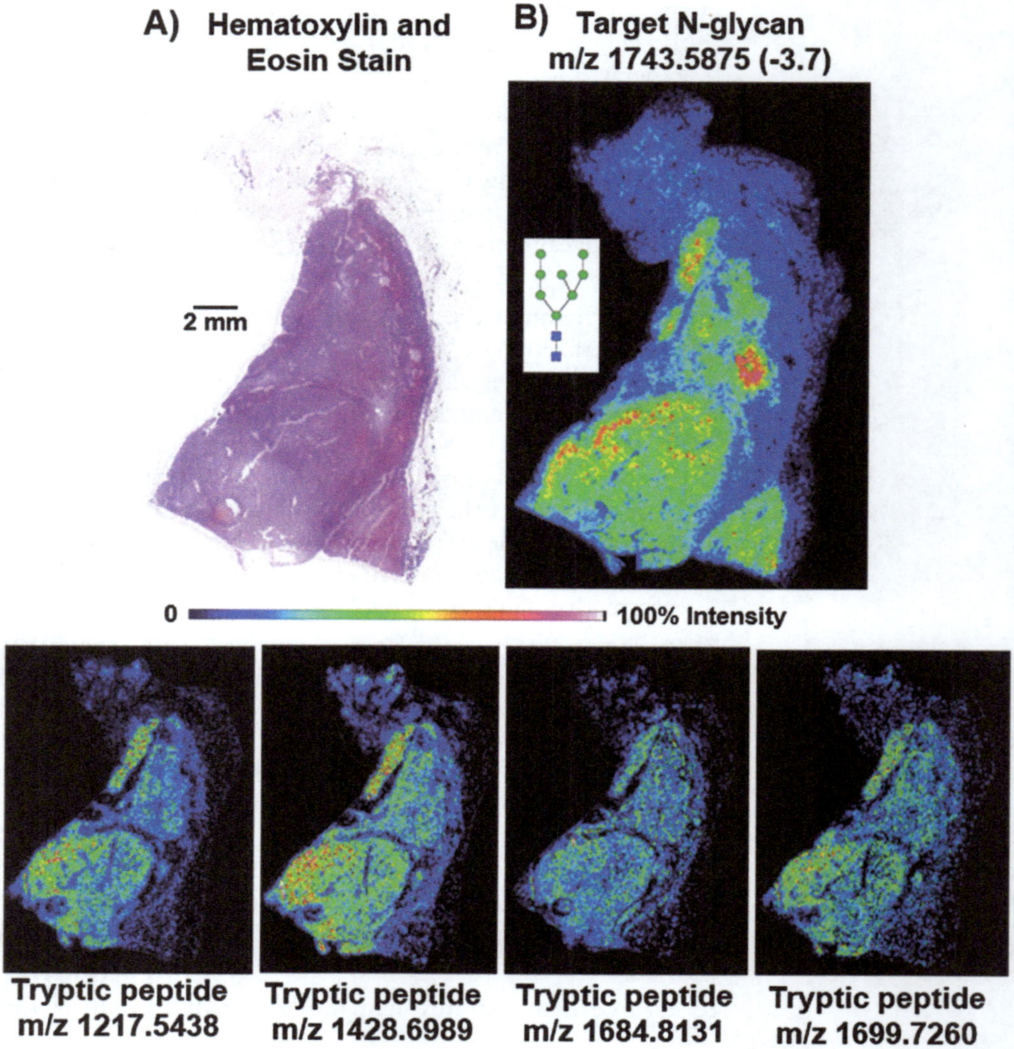

Fig. 3 Example identification of a tryptic peptide corresponding to a single target N-glycan. (**a**) Tissue features shown by hematoxylin and eosin stain of the same tissue section after being treated by PNGase F and trypsin and going through two rounds of imaging experiments. (**b**) Target N-glycan found by MALDI IMS Experiment, identified as a high mannose Man7 structure. *Green* circles represent mannose residues; *blue* squares represent *N*-acetylglucosamine (GlcNAc). Parentheses indicates mass accuracy of N-glycan as found by MALDI FT-ICR. (**c**) Tryptic peptides found using the "Co-localize to m/z Feature" of SCiLS. The unknown tryptic peptide is then a target for identification by tandem mass spectrometry

select the notepad icon. This will copy the data to your clipboard. Paste the data into the text editor appended to the list of N-glycans (*see* **Note 18**). Save the Combined peaklist.

4. Load the combined peaklist into SCiLS. Copy all m/z intervals from the Combined Peaklist. Go to File → Import → m/z intervals from CVS or clipboard. After naming the list, click Save.

5. Define image regions by intensity and expression pattern using image Segmentation. From the top menu, select Tools → Segmentation. Ensure that an appropriate normalization method is selected, and check the box next to Regions. Select the option "Work on all individual spectra." Select the created Combined Peaklist as m/z intervals. For parameters, use the pulldown menu to set denoising as "Weak", keep the method as "Bisecting k-Means" and as metric, use the pulldown menu to select "Manhattan" (*see* **Note 19**). This clusters peaks based on intensity and pattern within the tissue.
6. Segmentation may be evaluated in the Labels menu. Select the calculated segmentation. Double-click on the colored squares to extend the segmentation tree to smaller regions. The number to the right of the box reports how many spectra associate with the regional pattern. Figure 4a shows the results of segmenting out N-glycan image data with trypsin image data based on region.
7. Within the label tree, select a region of interest based by single clicking on the same-colored box. On the bottom of the Label menu, select "Save selected class as Region." Name and save the region. This will generate a mass spectrum for that particular region. Regions are extracted from the segmentation data and appear as highlighted regions in the tissue outline Fig. 4b.
8. On the menu at the top of the software, select Tools → Find Co-Localized m/z Values. Ensure that an appropriate normalization method is selected, and check the box next to Regions. In the Mode menu, select "Correlate with Region." Check the box that corresponds to the region of interest. Next to "m/z Intervals", select the created Combined Peaklist. Save as an informative name and click "OK."
9. Once the data has calculated, a new peak list will appear under the "m/z Intervals" pulldown menu. Click the small left arrow beside the menu name. Click the "eye" icon to the far right of the menu window. A Threshold bar is displayed with a Correlation Value. Clicking in the bar raises or lowers the correlation value and reports the number of m/z intervals associated with the correlation value. Correlation values of greater than 0.6 are good candidates for co-localization with the target region. Once a correlation value is selected, Click the "eye" icon on the far right to open the m/z Image Generator. Click OK to generate images.
10. Figure 4c demonstrates an image data result from the procedure (*see* **Note 12**).
11. Discovery of the tryptic peptides corresponding to N-glycan localization allows targeted LC-MS/MS proteomic experiments for peptide sequencing and protein identification.

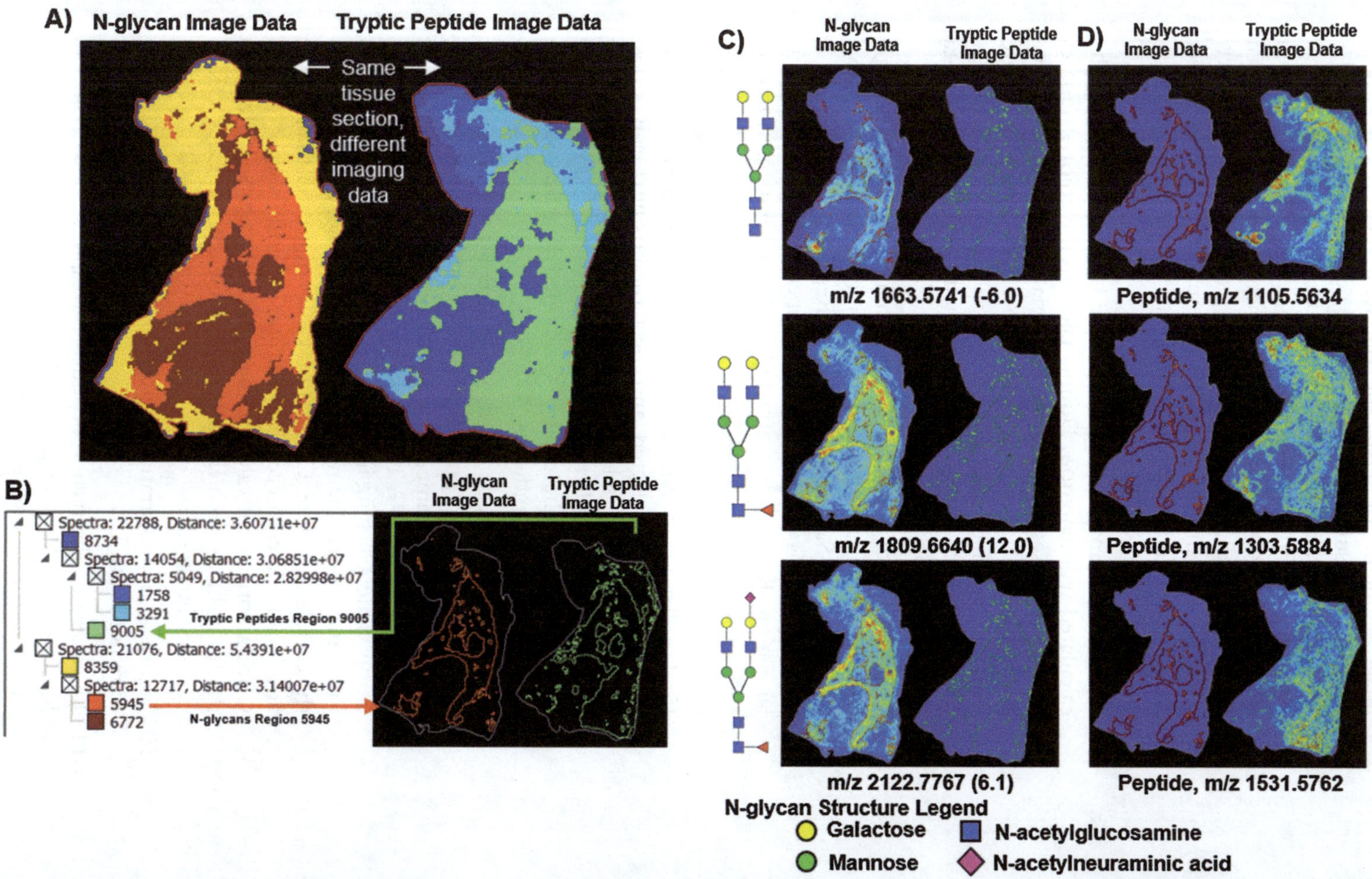

Fig. 4 Example of results from an Image segmentation strategy to identify common regions between N-glycan and tryptic peptides detected by MALDI IMS on the same tissue section. (**a**) Output of image segmentation from the two imaging experiments on the same tissue section using a combined list of known N-glycans and all detected tryptic peptides. Parentheses indicates mass accuracy of N-glycan as found by MALDI FT-ICR. The *orange* region on the N-glycan image data and the light green region on the tryptic peptide image data are very similar. (**b**) Regions are extracted from the segmentation data and appear as highlighted regions in the tissue outline. (**c**) Examples of N-glycan images correlating to the extracted region data. The N-glycan peaks do not appear in the tryptic peptide image data, as expected. (**d**) Examples of tryptic peptide image data correlating to the image regions

4 Notes

1. Ensure that the solution is mixed well. The citraconic acid anhydride solution should not need additional pH adjustment.
2. This volume is enough for filling two slide mailers for antigen retrieval and for tissue clearing of N-glycans.
3. Use of the filter significantly reduces clogging of the solvent lines of the M3 TM-Sprayer™.
4. For some highly active recombinant PNGase F preparations, using water instead of ammonium bicarbonate buffer allows equal efficiency in digestion.
5. One batch allows spraying of four slides, the maximum capacity for one batch of samples on the M3 TM-Sprayer™.
6. When refilling the push solvent, ensure the bottle is thoroughly cleaned out before adding new solvent to prevent bacterial growth.
7. In our laboratory, the vegetable steamer used in antigen retrieval is a Rival Model CKRVSTLM21.
8. Certain tissues with high fat content or those with open areas such as normal skin, breast and lung may detach during lengthy antigen retrieval. To limit loss of tissue from the slide, reduce antigen retrieval time to 20 min.
9. Use hot pads or heat resistant gloves to remove the mailer. It is full of near boiling solution and could be a burn hazard to skin and eyes.
10. To eliminate bubbles within the syringe barrel, and after loading all the PNGase F solution required, pull a small volume of air into the syringe. Gently dispense the syringe until the air bubble is gone.
11. Excess solutions will be deposited at the point of the sprayhead turn around location. Ensuring the turnaround point is located off the tissue prevents delocalization and excess matrix build up.
12. Fibrous tissues such as breast, skin, and fibrotic liver, lung or heart valve may require longer digestion times.
13. Frequently digital readouts on small ovens do not accurately report internal temperature. Ensure that the internal oven temperature is at the correct temperature using a secondary thermometer placed in the oven. This is essential for correct digestion.
14. Solvent should be degassed to limit flow variation.

15. Use sequencing grade modified trypsin to ensure high quality peptide for identification by parallel proteomics workflow after earmarking which peptides are co-localized with N-glycans.
16. Slides may be stored in a desiccator for up to 3 months until peptide fragmentation may be completed.
17. N-glycans are observed in imaging data using the described preparation as the intact precursor plus one or two sodium adducts.
18. The N-glycans are loaded as a specific list to minimize MALDI matrix peaks from being incorporated as peaks of interest during subsequent computation of image patterns.
19. These are general parameters for segmentation that we have found work well with most image data and accurately segment out regions corresponding with tissue histology.

Acknowledgments

This work was supported by the National Institute of Health/National Cancer Institute R21 CA185799 to RRD. PMA appreciates support from the National Institutions of Health through the National Institute of General Medical Sciences P20GM103542.

References

1. Norris JL, Caprioli RM (2013) Analysis of tissue specimens by matrix-assisted laser desorption/ionization imaging mass spectrometry in biological and clinical research. Chem Rev 113 (4):2309–2342
2. Aichler M, Walch A (2015) MALDI imaging mass spectrometry: current frontiers and perspectives in pathology research and practice. Lab Investig 95(4):422–431. 25621874
3. Schwamborn K, Kriegsmann M, Weichert W (2017) MALDI imaging mass spectrometry—from bench to bedside. Biochim Biophys Acta 1865(7):776–783. 27810414
4. Groseclose MR, Andersson M, Hardesty WM, Caprioli RM (2007) Identification of proteins directly from tissue: in situ tryptic digestions coupled with imaging mass spectrometry. J Mass Spectrom 42(2):254–262
5. Heijs B, Carreira RJ, Tolner EA, de Ru AH, van den Maagdenberg AMJM, van Veelen PA, McDonnell LA (2015) Comprehensive analysis of the mouse brain proteome sampled in mass spectrometry imaging. Anal Chem 87 (3):1867–1875
6. Powers TW, Jones EE, Betesh LR, Romano PR, Gao P, Copland JA, Mehta AS, Drake RR (2013) Matrix assisted laser desorption ionization imaging mass spectrometry workflow for spatial profiling analysis of N-linked glycan expression in tissues. Anal Chem 85 (20):9799–9806
7. Casadonte R, Caprioli RM (2011) Proteomic analysis of formalin-fixed paraffin-embedded tissue by MALDI imaging mass spectrometry. Nat Protoc 6(11):1695–1709
8. Briggs MT, Kuliwaba JS, Muratovic D, Everest-Dass AV, Packer NH, Findlay DM, Hoffmann P (2016) MALDI mass spectrometry imaging of N-glycans on tibial cartilage and subchondral bone proteins in knee osteoarthritis. Proteomics 16(11–12):1736–1741
9. Drake RR, Powers TW, Jones EE, Bruner E, Mehta AS, Angel PM (2017) MALDI mass spectrometry imaging of N-linked glycans in cancer tissues. Adv Cancer Res 134:85–116
10. Angel PM, Baldwin HS, Gottlieb D, Su YR, Mayer JE, Bichell D, Drake RR (2017) Advances in MALDI imaging mass spectrometry of proteins in cardiac tissue, including the heart valve. Biochim Biophys Acta 1865 (7):927–935. PMC5527275

11. Turiák L, Shao C, Meng L, Khatri K, Leymarie N, Wang Q, Pantazopoulos H, Leon DR, Zaia J (2014) Workflow for combined proteomics and glycomics profiling from histological tissues. Anal Chem 86 (19):9670–9678. PMC4427244
12. Heijs B, Holst S, Briaire-de Bruijn IH, van Pelt GW, de Ru AH, van Veelen PA, Drake RR, Mehta AS, Mesker WE, Tollenaar RA (2016) Multimodal mass spectrometry imaging of N-glycans and proteins from the same tissue section. Anal Chem 88(15):7745–7753. 27373711
13. Powers TW, Neely BA, Shao Y, Tang H, Troyer DA, Mehta AS, Haab BB, Drake RR (2014) MALDI imaging mass spectrometry profiling of N-glycans in formalin-fixed paraffin embedded clinical tissue blocks and tissue microarrays. PLoS One 9(9):1–11. e106255
14. Powers TW, Holst S, Wuhrer M, Mehta AS, Drake RR (2015) Two-dimensional N-glycan distribution mapping of hepatocellular carcinoma tissues by MALDI-imaging mass spectrometry. Biomol Ther 5(4):2554–2572
15. Holst S, Heijs B, de Haan N, van Zeijl RJM, Briaire-de Bruijn IH, van Pelt GW, Mehta AS, Angel PM, Mesker WE, Tollenaar RAEM (2016) Linkage-specific in-situ sialic acid derivatization for N-glycan mass spectrometry imaging of FFPE tissues. Anal Chem 88 (11):5904–5913

Methods in Molecular Biology (2018) 1788: 243–250
DOI 10.1007/7651_2017_80

Published online: 10 October 2017

Differential Adipose Tissue Proteomics

Kelly J. Shields and Changgong Wu

Abstract

Differential proteomic analysis (comparative quantitative proteomics) is a robust quantitative technique used to detect and identify the proteome of selected tissues. The expression levels (upregulated vs. downregulated) of proteins in tissue samples that differ by experimental design or anatomic location are determined by a series of assays including (1) 2D difference gel electrophoresis (2D-DiGE), (2) protein spot picking based on a priori thresholds, (3) Mass Spectrometry, and (4) follow-up Western Blot for antibody validation (Chen et al., Mol Cell Proteomics 14:2466–2478, 2015). Differential proteomic analysis is a perfect method for analyzing a heterogeneous tissue such as adipose tissue with a composition spectrum consisting of white to brown adipocytes along with a stromal vascular fraction dependent on anatomical location and inflammation. The adipose tissue proteomic protocol outlined here was successful in identifying differentially expressed proteins both significantly upregulated and downregulated between the experimental and control groups (Shields et al., Pulm Circ 6:586–596, 2016).

Keywords Adipose tissue, Mass spectrometry, Proteomics, Western blotting

1 Introduction

Differential proteomic analysis is a robust quantitative technique used to detect and identify the proteome of selected tissues. Overall, the technique relies on a number of steps inherently dependent on each other. Depending on the nature of the experiment, appropriate tissue preparation is critical to ensure that the desired proteins are not removed during this initial process. For instance, salivary gland tissue preparation calls for an acetone precipitation of proteins [1]; however, this would eliminate desired lipid proteins found in adipose tissue [2]. Additionally, differences in sample preparation can affect the ionization efficiency of mass spectrometry lending to the need for a standardization of techniques by tissue [3].

In particular, adipose tissue is no longer considered an inert, energy storage tissue, but is now recognized as an endocrine organ having a role in a myriad of cellular processes from lipid metabolism and glucose homeostasis to participating in localized and systemic inflammatory responses [4, 5]. The composition spectrum of adipose tissue consists of a range of white to brown adipocytes [6, 7] along with a stromal vascular fraction including immune-regulating cells dependent on anatomic location and degree of inflammation [8].

Here we demonstrate that a large number of proteins can be differentially identified from a small volume of heart adipose tissue using differential proteomic analysis between a control and experimental group. The expression levels (upregulated vs. downregulated) of proteins in tissue samples that differ by experimental design or anatomic location are determined by a series of assays including (1) 2D difference gel electrophoresis (2D-DiGE), (2) protein spot picking based on a priori thresholds, (3) Mass Spectrometry, and (4) follow-up Western Blot for antibody validation [9].

2 Materials

Prepare all buffers same day and with ultrapure water and analytical grade reagents [Note 1]. Follow all waste disposal guidelines.

- Phosphate buffered saline.
- 2D lysis buffer.
  - 7 M urea.
  - 2 M thiourea.
  - 2% CHAPS.
  - 50 mM.
  - Tris pH 8.5.
  - 0.5% protease inhibitor.
- Acetone.
- GE healthcare 2D Quant Kit.
- Lumiprobe LLC 3Dye 2D DiGE kit.
  - High grade *N,N* dimethylformamide.
  - 10 mM lysine.
- Rehydration buffer.
  - 7 M urea.
  - 2 M thiourea.
  - 2% CHAPS.
  - 2% dithiothreitol.
  - 0.5% IPG buffer.
  - 0.002% bromophenol blue.
- 24 cm pH 3–10 nL gel strips.
- ETTAN IPGPhor apparatus.

  - 1.0% dithiothreitol.
  - 2.5% iodoacetamide.
- 13.5% Polyacrylamide gels (26 cm × 1 mm × 20 cm) in Ettan DALT six gel apparatus.
  - Sealed with a 0.5% low melting point agarose solution containing bromophenol blue.
  - Running buffer was placed into the upper reservoir.

    50 mM Tris–HCl.

    0.384 M glycine.

    0.2% SDS.
  - Running buffer was placed into the lower reservoir.
    25 mM Tris–HCl.

    0.192 M glycine.

    0.1% SDS.
- Typhoon 9400 scanner.
  - DeCyder 2D 7.0 software.
  - ImageQuant TL 7.0 software.
- Coomassie Blue.
- Dithiothreitol (DTT).
- Iodoacetamide.
- Trypsin.
- C18 ZipTips.
- MALDI-TOF/TOF tandem MS.
- FlexAnalysis and BioTools software.
- Swiss-Prot protein database using a local Mascot search engine.

3 Sample Preparation/2D Cleanup

1. Adipose samples were washed with phosphate-buffered saline.
2. Samples were then cut into smaller, approximately 3 mm × 3 mm × 3 mm volumes using a sterile scalpel.
3. Adipose samples were homogenized under ice-cold conditions.
4. Adipose sample proteins were extracted using a 2D lysis buffer.
   - 7 M urea.
   - 2 M thiourea.
   - 2% CHAPS.
   - 50 mM.

- Tris pH 8.5.
- 0.5% Protease inhibitor.

5. Adipose sample proteins were desalted by ice-cold acetone in a volume ratio of 1:4 and wash the pellet using 80% ice-cold acetone for four times [Note 2].
6. Adipose sample proteins were resolved in 2D lysis buffer and subjected to GE healthcare 2-D Quant kit.
7. Protein concentrations were evaluated using Bradford Method and adjusted to 1 μg/μL.

4 Protein Labeling

1. Lumiprobe LLC 3Dye 2D DiGE kit.
2. Working solution of each CyDye freshly prepared with high grade *N,N* dimethylformamide.
3. 1 μL of Cy2, Cy3 or Cy5 working solution added to 50 μg of sample protein.
   (a) Cy2 = mixed standard.
   (b) Cy3 = control or experimental (for technical reproducibility).
   (c) Cy5 = experimental or control (for technical reproducibility).
4. All samples incubated on ice for 30 min in the dark.
5. 1 μL of 10 mM lysine added to each sample.
   (a) The addition of 10 mM lysine quenches the reaction.
6. Samples incubated on ice for additional 10 min in the dark.

5 Strip Rehydration, Isoelectric Focusing (IEF) and 2D Sodium Dodecyl Sulfate (SDS) PAGE

1. 15 μg of each labeled sample were mixed together and diluted further with 450 μL rehydration buffer.
   - 7 M urea.
   - 2 M thiourea.
   - 2% CHAPS.
   - 2% Dithiothreitol.
   - 0.5% IPG buffer.

- 0.002% bromophenol blue.

2. 24 cm pH 3–10 nL gel strips were rehydrated with the samples.
   (a) Strips were overlaid with mineral oil to prevent drying.
   (b) 8 h at room temperature.
3. Proteins separated using isoelectric focusing (IEF) using ETTAN IPGPhor apparatus.
   (a) Set to 20 °C.
   (b) Focusing voltages/time.
      - 60,000 V h overnight.
4. Strips equilibrated using 1.0% dithiothreitol 15 min.
5. Strips equilibrated a 2nd time with 2.5% iodoacetamide for 15 min.
6. The focused IEF protein strips were separated by molecular weight on 13.5% polyacrylamide gels (26 cm × 1 mm × 20 cm) in Ettan DALT six gel apparatus.
   (a) The focused IEF protein strips were placed on top of the PAGE gel and sealed with a 0.5% low melting point agarose solution containing bromophenol blue.
   (b) Running buffer was placed into the upper reservoir.
      - 50 mM Tris–HCl.
      - 0.384 M glycine.
      - 0.2% SDS.
   (c) Running buffer was placed into the lower reservoir.
      - 25 mM Tris–HCl.
      - 0.192 M glycine.
      - 0.1% SDS.
   (d) Gels were run at 1 W/gel at 15 °C.

6 Image Acquisition and Data Analysis

1. All 2D gels were scanned using a Typhoon 9400 scanner.
   (a) Resolution of 100 μm.
   (b) Photomultiplier tubes set for three color lasers to achieve linear signal.
   (c) Same voltages were used for all gels.
2. Gel images were imported DeCyder 2D 7.0 software.
3. Spot detection was conducted.
   (a) Spot exclusion filters based on peak slope, area and height.

   (b) Manual inspection to eliminate artifacts, spots and unexclude any spots eliminated by the exclusion filters.
4. Spot maps were created and matched to each other automatically.
   (a) Manual inspection to ensure correct matching.
5. Quantitation of spots.
   (a) Ratio of Cy3 and Cy5 to the normalized Cy2.
   (b) Samples from the control and experimental group were randomly assigned to either Cy3 or Cy5 to eliminate preferential labeling with the dyes and avoid dye-specific effects due to glass plates or differences in excitation.
   (c) Multiple gels were run with dye swap.
   (d) All values were expressed in relation to the normalized values for the particular gel (Cy2).
6. The pictures were edited using ImageQuant TL 7.0 software, and
7. Differential in-gel analysis and biological variation analysis of the 2-D DiGE results were performed using DeCyder 2-D 7.0 Software.
   (a) $p < 0.05$ used as a threshold for significance [Note 3].

7 Protein Identification by Mass Spectrometry [10]

1. 200 μg adipose proteins in control or treated samples were used for normal 2D gel and stained with Coomassie Blue for protein identification.
2. Protein spots of interest ($p < 0.05$) from a Coomassie Blue-stained 2D gel were excised corresponding to the 2D DIGE image.
3. The protein plugs were reduced by dithiothreitol (DTT), and
   (a) Alkylated with iodoacetamide, [Note 4].
   (b) Digested with trypsin with the ratio of 1:30 (trypsin–protein) at 37 °C overnight.
   (c) Desalted with C18 ZipTips (Millipore).
4. Both mass spectrometry and tandem mass spectrometry analyses of the digested peptides were performed on a MALDI-TOF/TOF tandem MS (Bruker UltrafleXtreme MALDI TOF/TOF mass spectrometer, Bruker Daltonics, Billerica, MA).
5. The database search and analysis were performed using FlexAnalysis and BioTools software (Bruker Daltonics) against other

Mammalia (other than primate and rodents) in the Swiss-Prot protein database using a local Mascot search engine [10].

8 Western Blotting [Note 5]

1. Proteins were extracted from heart adipose using Western blotting lysis buffer.
   - 50 mm Tris, pH 7.5.
   - 150 mm NaCl.
   - 1% Triton X-100.
   - 1 mm EDTA.
   - 0.5% Protease inhibitor.
2. Protein concentrations were evaluated using Bradford method as above.
   (a) Protein concentrations were adjusted to 1 μg/μL.
   (b) 20 μg of protein from each sample was separated by 15% SDS–polyacrylamide gel electrophoresis.
   (c) The proteins were transferred onto a nitrocellulose membrane.
   (d) The membranes were blocked with 5% milk and probed with anti-C3 antibody (1:2000) and anti-FABP4 antibody (1:3000; Abcam, Cambridge, MA), Actin (Santa Cruz, CA) was used as loading control.
   (e) The membranes were visualized with enhanced chemiluminescent substrate (PerkinElmer, Waltham, MA).

9 Notes

1. All buffers should be prepared using HPLC grade water, to minimize background noise in 2D gel and protein identification.
2. Acetone precipitation is good for clear background and better 2D gel images.
3. Statistical analysis can be performed using a two-tailed unpaired Student's *t*-test with Microsoft Excel. Consider the differences as significant when the *p*-values are less than 0.05 and Cy3/Cy5 ratios of >2 or <0.5 with a 95% confidence interval based on the analytical variations.

4. As the samples were reduced and alkylated after IEF, so the reduction and alkylation can be omitted before trypsin digestion.
5. To get the better Western blotting result, it is better to extract protein using the western lysis buffer in step 1 and not use the 2D lysis buffer extract samples.

References

1. Geguchadze RN, Machen L, Zourelias L, Gallo PH, Passineau MJ (2012) An AAV2/5 vector enhances safety of gene transfer to the mouse salivary gland. J Dent Res 91:382–386
2. Shields KJ, Verdelis K, Passineau MJ, Faight EM, Zourelias L, Wu C et al (2016) Three-dimensional micro computed tomography analysis of the lung vasculature and differential adipose proteomics in the Sugen/hypoxia rat model of pulmonary arterial hypertension. Pulm Circ 6:586–596
3. Calderón-Celis F, Encinar JR, Sanz-Medel A (2017) Standardization approaches in absolute quantitative proteomics with mass spectrometry. Mass Spec Rev 9999:1–23. https://doi.org/10.1002/mas.21542
4. Trayhurn P (2005) Endocrine and signalling role of adipose tissue: new perspectives on fat. Acta Physiol Scand 184:285–293
5. Baker AR, Silva NF, Quinn DW, Harte AL, Pagano D, Bonser RS et al (2006) Human epicardial adipose tissue expresses a pathogenic profile of adipocytokines in patients with cardiovascular disease. Cardiovasc Diabetol 5:1
6. Cypess AM, Lehman S, Williams G, Tal I, Rodman D, Goldfine AB et al (2009) Identification and importance of brown adipose tissue in adult humans. N Engl J Med 360:1509–1517
7. Sharp LZ, Shinoda K, Ohno H, Scheel DW, Tomoda E, Ruiz L et al (2012) Human BAT possesses molecular signatures that resemble beige/brite cells. PLoS One 7:e49452
8. Peinado JR, Pardo M, de la Rosa O, Malagon MM (2012) Proteomic characterization of adipose tissue constituents, a necessary step for understanding adipose tissue complexity. Proteomics 12:607–620
9. Chen CL, Chung T, CC W, Ng KF, JS Y, Tsai CH et al (2015) Comparative tissue proteomics of microdissected specimens reveals novel candidate biomarkers of bladder cancer. Mol Cell Proteomics 14:2466–2478
10. Wu C, Wang Z, Zourelias L, Thakker H, Passineau MJ (2015) IL-17 sequestration via salivary gland gene therapy in a mouse model of Sjogren's syndrome suppresses disease-associated expression of the putative autoantigen Klk1b22. Arthritis Res Ther 17:198

Methods in Molecular Biology (2018) 1788: 251–268
DOI 10.1007/7651_2017_113

Published online: 15 December 2017

Quantification of Breast Cancer Protein Biomarkers at Different Expression Levels in Human Tumors

Yi Chen, David Britton, Elizabeth R. Wood, Stephen Brantley, Michelle Fournier, Marek Wloch, Vonetta L. Williams, Joseph Johnson, Anthony Magliocco, Ian Pike, and John M. Koomen

Abstract

Liquid chromatography-selected reaction monitoring (LC-SRM) mass spectrometry has developed into a versatile tool for quantification of proteins with a wide range of applications in basic science, translational research, and clinical patient assessment. This strategy uniquely complements traditional pathology approaches, like hematoxylin and eosin (H&E) staining and immunohistochemistry (IHC). The multiplexing capabilities offered by mass spectrometry are currently unmatched by other techniques. However, quantification of biomarkers in tissue specimens without the other data obtained from H&E-stained slides or IHC, including tumor cellularity or percentage of positively stained cells inter alia, may not provide as much information that is needed to fully understand tumor biology or properly assess the patient. Therefore, additional characterization of the tissue proteome is needed, which in turn requires the ability to assess protein markers across a wide range of expression levels from a single sample. This protocol provides an example of multiplexed analysis in breast tumor tissue quantifying specific biomarkers, specifically estrogen receptor, progesterone receptor, and the HER2 receptor tyrosine kinase, in combination with other proteins that can report on tissue content and other aspects of tumor biology.

Keywords Biomarkers, Breast cancer, Estrogen receptor, HER2 receptor tyrosine kinase, Liquid chromatography-selected reaction monitoring mass spectrometry, Targeted proteomics

1 Introduction

Proteomics has unique potential to contribute to pathology and patient assessment; specifically, the ability to quantify protein expression, assess the level of posttranslational modification, and compare the amounts of wild-type and mutant protein without the use of specific antibodies for each assay can be implemented in numerous ways ranging from functional mechanistic assessment to detection of clinical biomarkers. As the paradigm for cancer diagnosis and treatment shifts to rely on specific molecular measurements, analytical chemists must provide additional capability for biomarker measurements. One important area for mass spectrometry-based proteomics is to enable a link between a specific clinical biomarker and the rest of the proteome. Because most proteomics approaches utilize bulk tissue specimens, the advantages of antibody-driven, microscopy-visualized techniques cannot be directly replicated. However, multiplexed measurements can be made to link specific biomarkers to the context of the background proteome, evaluating such diverse aspects as tissue content and specific biological interactions.

One major challenge for this type of approach is the ability to implement all of these measurements using a single (often very precious or limited) clinical sample. Proteins are expressed across a very wide range of abundance; often, the most biologically or clinically important proteins are expressed at very low levels. The strategy must be designed to provide the best quality of data for the most important measurements. In this example, we will assess clinical biomarkers for breast cancer: estrogen receptor (ER) [1, 2], progesterone receptor (PR) [3, 4], and human epidermal growth factor receptor 2 (HER2) [5, 6], which are used for patient classification and selection of therapy [7, 8]. Enzyme-linked immunosorbent assays (ELISA) and immunohistochemical (IHC) analysis are typically used in the clinic to evaluate expression of ER, PR, and HER2 in order to classify breast cancers [5, 9–11]. However, these techniques are limited because they require antibodies for each endpoint. As an example, characterization of multiple phosphorylations on a single protein then requires an antibody for each site.

Several novel methods have been developed recently to assess the proteins, including liquid chromatography-selected reaction monitoring (LC-SRM) [12–17] quantification of protein expression for these breast cancer biomarkers [18–20], LC-SRM assessment of phosphorylation using phosphatase treatment [21], and tryptic peptide immunoprecipitation followed by LC-SRM quantification [22]. Additional approaches include discovering phosphorylation sites by liquid chromatography-tandem mass spectrometry (LC-MS/MS) prior to LC-SRM assay development [23–26]. All of these strategies focus on improving the measurement of the specific biomarkers or generating additional information about molecular switching events relevant to the function of the protein. Our goal is to generate in-depth molecular information about these clinical biomarkers and begin to relate them to other disparate aspects of the proteome.

1.1 Experimental Design

Many choices are available for the design of multiplexed biomarker panels, which incorporate different strategies for protein and peptide separation as well as different types of mass spectrometry (see Fig. 1). Here, we will apply different protein separations (SDS-PAGE and immunoprecipitation) to assess multiple markers across a wide range of protein expression, describing in detail the methods for the research previously reported in *Proteomics* [27] as well as some minor changes to improve the data quality particularly for phosphopeptides. SDS-PAGE and immunoprecipitation (IP)-SDS-PAGE are used to create two samples for LC-SRM analysis (see Fig. 2a). The first analysis evaluates high-abundance proteins (ALBU, HBA, HBB, ACTB, ENOA, G3P, LMNA, and VIME [28–30]) from in-gel digestion of whole cell lysate or tumor homogenate; the second analysis focuses on expression and

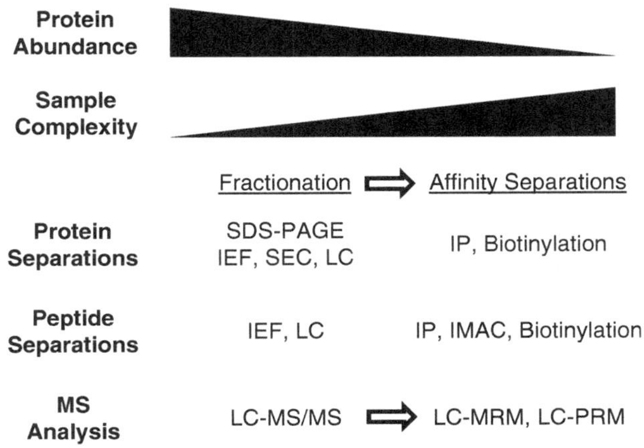

Fig. 1 Options for assay development in targeted proteomics. The analytical strategy can employ different types of protein and peptide separations as well as mass spectrometry methods, based largely on the abundance of the protein's expression and the complexity of the matrix, as shown above. Each assay will have specific constraints that require particular approaches. Here, we illustrate different approaches: using SDS-PAGE fractionation and protein immunoprecipitation (IP) prior to LC-MRM quantification

phosphorylation of ERα, PR, and HER2 after enrichment by immunoprecipitation and in-gel digestion. This second assay includes quantification of protein expression and levels of phosphorylation at nine sites (S154, S167, S294, Y537, S554, and S559 from ERα; S162 from PRβ; Y735 and Y877 from HER2). The development of the assays has two stages: (1) peptide selection and characterization and (2) examination of biological test cases using cell line models. Then, the assays are applied to frozen tumors from breast cancer patients. Our goal is to develop this platform to gain more molecular data about breast cancer biomarkers that is currently obtained in the clinic with additional data to assess tissue quality and content to inform about the robustness and reliability of the biomarker measurements.

2 Materials

All reagents and solutions should be prepared fresh each time using the highest quality chemicals and solvents (preferably LC-MS or similar grade). All safety precautions should be taken, and disposal of waste should be in accordance with local regulations. Personal protective equipment (e.g., safety glasses, lab coats, and gloves) should be properly utilized to eliminate risk of injury or chemical exposure.

2.1 Cell Culture and Antibody Reagents

1. Breast cancer cell lines: MCF7 (ER^+) [31] and BT474 ($HER2^+$) [32] from the American Type Culture Collection (ATCC, Manassas, VA).

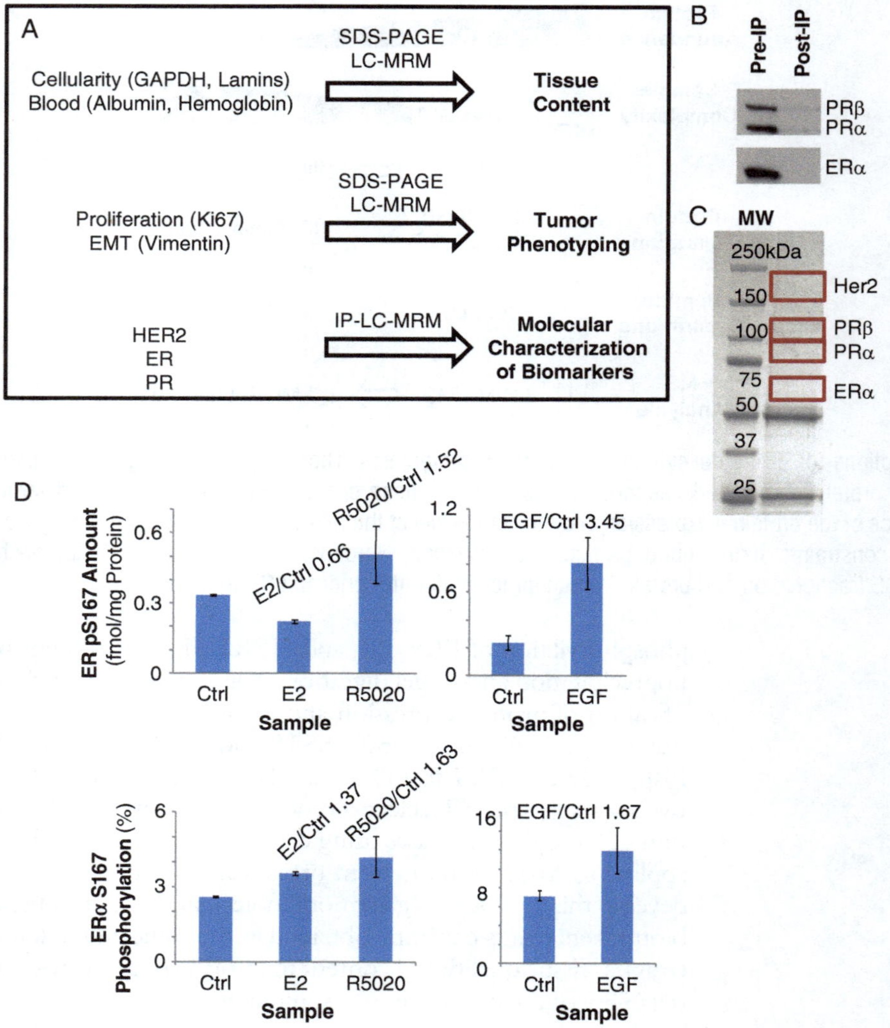

Fig. 2 Assay development for multiplexed biomarker quantification in breast cancer. Different strategies are employed for proteins expressed at different levels (**a**). To prepare each assay, protein capture during immunoprecipitation is assessed by immunoblots (**b**). The strategy for ER, PR, and HER2 analysis utilized IP followed by SDS-PAGE to isolate the proteins of interest from the antibody (**c**). Biological test cases using stimulation and drug treatment in MCF7 cells are used to test the assays (**d**) prior to clinical implementation

2. Dulbecco's modified Eagle's medium (D5796, Sigma, St. Louis, MO).
3. RPMI-1640 culture media (R8758, Sigma, St. Louis, MO).
4. DMEM, 1× (cat# 17-205-CV, Cellgro).
5. 15 cm tissue culture dishes.
6. Phosphate-buffered saline.

7. Phenol red-free Dulbecco's modified Eagle's medium (17-205-CV, Cellgro, Corning Life Sciences).
8. Charcoal stripped FCS (12676-029, Gibco, Thermo).
9. Fetal bovine serum (SH30910.03, Thermo).
10. Penicillin/streptomycin (P0781 Sigma, St. Louis, MO).
11. Cell lifter (cat# 3008, Corning).
12. PBS, pH7.4 (cat# 10010023, Gibco).
13. NaCl (cat# S271, Fisher Scientific).
14. PlusOne Tris (cat# 17-1321-01, GE Healthcare).
15. NP-40 (cat# 492016, Calbiochem).
16. NaF (cat# S7920, Sigma-Aldrich).
17. Na_3VO_4 (cat# 81104, Alfa Aesar).
18. Na_2MoO_4 (cat# 243655, Sigma-Aldrich).
19. Phosphatase inhibitor cocktail 3 (cat# P0044, Sigma-Aldrich).
20. Okadaic acid potassium salt (cat# O7885, Sigma-Aldrich).
21. Deoxyribonuclease I (cat# D2821, Sigma-Aldrich).
22. Roche mini-complete protease inhibitor cocktail (cat# 11836153001, Roche Applied Science, Indianapolis, IN).
23. β-Estradiol (E2758).
24. EGF (100-15, PeproTech, Rocky Hill, NJ).
25. Lapatinib ditosylate (S1028, Selleck, Houston, TX).
26. Promegestone or R5020 (NLP004005MG, PerkinElmer, Boston, MA).
27. ERα antibody (F-10, Santa Cruz Biotechnology).
28. PR antibody (6A1, 3172, Cell Signaling Technologies).
29. HER2 antibody (OP15, Calbiochem).
30. Pierce Protein A/G UltraLink Resin (53132, Thermo).
31. ERα (DYC5715-2), PR (DYC5415-2), and HER2 (DYC1129-2) DuoSet IC ELISA kits (R&D Systems, Minneapolis, MN).

2.2 Tissue Processing

1. Optimal cutting temperature (OCT) medium.
2. Microscope slides and cover slips.
3. Ethanol.
4. Reagents for hematoxylin and eosin staining.

2.3 Consumables, Solutions, and Buffers for SDS-PAGE and LC-MRM

1. Lysis buffer [100 mM NaCl, 20 mM Tris (pH 7.5), 0.5% NP40, 100 mM NaF, 10 mM Na_3VO_4, 50 mM Na_2MoO_4, 10 μL/mL phosphatase inhibitor cocktail 3 (P0044, Sigma-Aldrich, St. Louis, MO), 320 nM okadaic acid (Cat# O7885, Sigma-Aldrich), 100 U/mL deoxyribonuclease I (Cat# D2821, Sigma-Aldrich), and 1 tablet/

10 mL Roche mini-complete protease inhibitor cocktail (Cat# 11836153001, Roche Applied Science, Indianapolis, IN)].

2. Bradford Protein Assay Kit (Pierce #23236, Thermo).
3. Water (cat# AH365-4, Honeywell).
4. Methanol (cat# AH230-4, Honeywell).
5. Acetic acid (cat# A6283, Sigma-Aldrich).
6. Criterion XT 4–12% Bis-Tris precast gel (cat# 345-0123, 345-0124, Bio-Rad).
7. Precision plus protein dual color standards (cat#161-0374, Bio-Rad).
8. Gel-loading pipet tips (1–200 μL, cat# 102-707-138, Fisherbrand).
9. Sample loading buffer (Criterion XT, Bio-Rad).
10. 1.25 M aqueous dithiothreitol (#7900, Cell Signaling).
11. Running buffer (Criterion XT MOPS, Bio-Rad).
12. Criterion staining/blotting trays (cat# 345-9920, Bio-Rad).
13. Destain: Aqueous 10% methanol and 5% glacial acetic acid (A6283, Sigma-Aldrich). Prepare in a fume hood.
14. Gel staining solution: Dissolve 5 mg Coomassie Brilliant Blue G (B5133 Sigma-Aldrich) in 100 mL gel destain (above).
15. Stainless steel disposable scalpel (cat# 73-0121, Personna).
16. Stainless steel disposable scalpel (cat# 371611, Bard-Parker).
17. Ammonium bicarbonate for in-gel tryptic digestion. Weigh out 0.79 g ammonium bicarbonate (A6141, Sigma-Aldrich) and dissolve in 80 mL distilled, deionized water, confirm pH = 8, and bring volume to 100 mL. Dilute with distilled, deionized water to make lower concentrations.
18. Gel dehydration buffer: Aqueous 50 mM ammonium bicarbonate with 50% methanol. Mix equal volumes of 100 mM ammonium bicarbonate (above) and methanol.
19. Reducing agent: To make 10× stock (20 mM), dissolve 5.7 mg of tris(2-carboxyethyl)phosphine hydrochloride (C4706, Sigma-Aldrich) in 1 mL of distilled, deionized water. Store in the dark at 4 °C.
20. Alkylating reagent: To make 10× stock (200 mM), dissolve 37 mg Iodoacetamide (RPN6302V, GE Healthcare, Chicago, IL) in 1 mL distilled, deionized water. Store in the dark at 4 °C.
21. Trypsin for in-gel digestion: Sequencing grade modified trypsin (V5111, Promega, Madison, WI) is prepared at 20 ng/μL. Dissolve 20 μg in 1 mL of aqueous 50 mM acetic acid, aliquot, and store at −80 °C.
22. Peptide extraction buffer: Aqueous 50% acetonitrile with 0.01% trifluoroacetic acid (TFA).

23. Acetonitrile (cat# AH015-4, Honeywell).
24. Formic acid (cat# 94318, Fluka).
25. LC-MRM solvent system: Loading solvent, autosampler syringe wash solvent and solvent A: aqueous 2% acetonitrile, 0.1% formic acid in LC-MS grade water. Solvent B: aqueous 90% acetonitrile, 0.1% formic acid in LC-MS grade water. Degas before use. Prepare in a fume hood.
26. Gas cylinders (CO_2 for cell culture, nitrogen and argon for MS).

2.4 Equipment

1. General lab equipment: Source of distilled, deionized (18 MΩ) water, vortex mixer, orbital shaker, pH meter, UV-visible spectrophotometer, cell culture incubator, refrigerated microcentrifuge capable of 14,000 × *g*, sonicator for cell disruption, vacuum concentrator (Speedvac), water bath with temperature control, analytical balance, gel electrophoresis system including cell and power supply, and gel imaging system or digital camera and microplate reader.
2. Tissue processing and microscopy: Cryostat for tissue sectioning, automated tissue staining system, whole slide scanner with associated control and analysis software (Aperio), and Biopulverizer from pulverization of frozen tumor tissues (BioSpec, Bartlesville, OK).
3. LC-SRM MS: NanoUPLC system with refrigerated autosampler, column switching compartment with 10-port valve, loading pump, and nanoflow pump interfaced with a triple-quadrupole mass spectrometer (Quantiva, Thermo, or equivalent). The following columns were used for these experiments: trap column (C18 PepMap100, 100 μm ID × 2 cm, C18, 5 μm particle size, 100 Å pore size) and analytical column (C18 PepMap100, 75 μm ID × 25 cm, C18, 2 μm particle size, 100 Å pore size). The nanoelectrospray source uses emitters pulled from 20 μm ID fused silica capillary to form 10 μm ID tips.
4. Data analysis software: Skyline software for LC-SRM data analysis (or vendor-provided equivalent) (https://skyline.gs.washington.edu/labkey/project/home/software/Skyline/begin.view) [33], Cluster 3.0 (http://bonsai.hgc.jp/~mdehoon/software/cluster/software.htm) [33], and Java TreeView (http://jtreeview.sourceforge.net/) [34] for visualization using heat maps.

3 Methods

3.1 Assay Development

This section describes two different scenarios. In the first, the peptides are determined based on the biology of the specific biomarkers of interest, which require enrichment to enable measurement of expression and posttranslational modification. In the second, the peptides are selected for common "housekeeping proteins" that will report on the content and quality of the tumor tissue specimen, which are expected to be high abundance and readily measured in whole cell lysate or tumor homogenate.

1. Select peptides for each of the biomarker protein based on known biology as well as publically available MS data. Peptides for monitoring protein expression should not have mutation sites or the possibility of posttranslational modifications. Phosphopeptides will be specifically determined based on the amino acid residues of interest. See Table 1 for selected peptides for ER, PR, and HER2 expression and phosphorylation analysis.
2. For other proteins, use LC-MS/MS data or LC-SRM screening to select peptides to monitor expression [17]. For tissue QC housekeeping proteins, we chose the following peptides: VDPVNFK (HBA), LLVVYPWTQR (HBB), and LVNEVTEFAK (ALBU) report on blood infiltration or ascites, while AGFAGDDAPR (ACTB), SGAQASSTPLSPTR (LMNA), VGVNGFGR (G3P), and YISPDQLADLYK (ENOA) assess cellularity. In addition, SLYASSPGGVYATR and QDVDNASLAR were used to monitor vimentin, which is relevant to epithelial-to-mesenchymal transition in breast cancer.
3. Synthesize and characterize each peptide, as previously described [35]. Dissolve each peptide in LC-MS grade water or LC solvent A in a volume resulting in 1 mM peptide concentration. If needed, add acetonitrile or formic acid to help dissolve hydrophobic peptides. Store peptide stock solutions at −80 °C.
4. Infuse 1 μM peptide solutions in aqueous 20% acetonitrile to select and optimize transitions (pairs of parent and fragment ions as well as collision energy values maximizing fragment ion production).
5. Develop a method using Skyline [36]. See **Note 1**.
6. Prepare a dilution series to test peptide assay performance through calibration curves in buffer using blanks and spiked amounts of peptide from 20 amol to 1 pmol injected on column using three technical replicates at each concentration point.
7. Determine the appropriate amounts of each standard based on performance of calibration curves in buffer or matrix. See **Note 2**.
8. Additional assay characterization can be performed as needed; the National Cancer Institute's Clinical Proteomics Tumor Assessment Consortium has provided guidelines for systematic assay characterization [37]. See **Note 3**.

3.2 Cell Culture and Biological Test Cases

The rationale for this section is to define specific model systems that can serve as positive controls for the analytical process. These

Table 1
Peptides and LC-MRM transitions for monitoring breast cancer biomarkers

Analyte	Peptide sequence	Transitions
ERα (402-412)	LLFAPNLLLD**R**	$y_7 - y_9$
ERα (582-595)	YYITGEAEGFPAT**V**	b_2, b_6, b_{10}, y_4
ERα (143-158)	EAGPPAFYRPNSDNR**R**	$y_{12}^{3+} - y_{14}^{3+}$
ERα pSer154	EAGPPAFYRPN**pS**DNR**R**	$(y_{12}-98)^{2+}, y_{13}^{3+}, y_{14}^{3+}$
ERα (143-157)	EAGPPAFYRPNSDN**R**	$y_{10}^{2+}, y_{11}^{2+}, y_{13}^{2+}$
ERα pSer154	EAGPPAFYRPN**pS**DN**R**	$y_{11}^{2+}, (y_{11}-98)^{2+}, y_{12}^{3+}$
ERα (165-180)	LASTNDKGSMAMESA**K**	$y_9, y_{14}^{2+}, y_{15}^{2+}$
ERα pSer167	LA**pS**TNDKGSMAMESA**K**	$y_9, (y_{14}-98)^{2+}, (M + 2H\text{-}98)^{2+}$
ERα (288-299)	AANLWPSPLMI**K**	b_5, y_7, y_8
ERα pSer294	AANLWP**pS**PLMI**K**	$b_5, y_7, (y_7-98)^+$
ERα (549-555)	LHAPTS**R**	b_2, y_4, y_5
ERα pSer554	LHAPT**pSR**	$b_2, (y_5-98)^+, (M + 2H\text{-}98)^{2+}$
ERα (556-581)	GGASVEETDQSHLATAGSTSSHSLQ**K**	$y_{21}^{3+} - y_{23}^{3+}$
ERα pSer559	GGA**pS**VEETDQSHLATAGSTSSHSLQ**K**	$y_{19}^{2+}, y_{21}^{2+}, (M + 3H\text{-}98)^{3+}$
PRβ (74-89)	TQDQQSLSDVEGAYS**R**	y_3, y_5, y_6, y_9
Total PRalpha and beta (561-567; 725-731)	QLLSVV**K**	$y_3 - y_6$
PRβ (160-167)	VLSPLMS**R**	$y_5, y_6, y_5^{2+}, y_6^{2+}$
PRβ pSer 162	VL**pS**PLMS**R**	$y_3, y_5, (y_6-98)^+, (y_6-98)^{2+}$
HER2 (546-552)	VLQGLPR	$b_3, y_4 - y_6$
HER2 (861-868)	ITDFGLAR	$y_4 - y_7$
HER2 (971-978)	ELVSEFSR	$y_3 - y_6$
HER2 (725-736)	VLGSGAFGTVYK	y_5, y_6, y_9, y_{10}
HER2 pTyr 735	VLGSGAFGTV**pYK**	y_2, y_5, y_6, y_{10}
HER2 (869-883)	**L**LDIDETEYHADGGK	$y_{11}^{2+} - y_{14}^{2+}$
HER2 pTyr 877	LLDIDETE**pY**HADGG**K**	$y_{11}^{2+} - y_{14}^{2+}$

controls can be integrated into analysis of clinical samples to serve as quality controls to assess IP efficiency, proteolysis, and LC-MRM instrument performance.

1. MCF7 (ER^+) [31] and BT474 ($HER2^+$) [32] breast cancer cell lines are obtained from the American Type Culture Collection (Manassas, VA) and grown under the recommended conditions. Grow cells to 80% confluence and plate 9.5×10^6 cells in 15 cm culture dishes and allow them to attach overnight.
2. Remove media 24 h prior to treatment, and wash three times with room temperature PBS.
3. For ligand stimulation, replace the media with phenol red-free Dulbecco's modified Eagle's media supplemented with 10% charcoal stripped FCS to "serum-starve" the cells by removing lipids, hormones, and selected growth factors.
4. Treat MCF7 and BT474 cell lines with estradiol (10 nM for 30 min), EGF (50 ng/mL for 10 min), estradiol plus EGF (10 nM and 50 ng/mL for 20 min), and promegestone (100 nM for 1 h).
5. For drug treatment, replace with the usual media. Treat cell lines with lapatinib (1 μM for 1 h). Rescue signaling by treatment by EGF (50 ng/mL for 10 min) for comparison.
6. After treatment, wash the cells once with room temperature PBS. From each 15 cm plate, cells are harvested on ice using a cell scraper and collected into 1.0 mL of cold cell lysis buffer.
7. Sonicate twice using 10 s pulses; cool on ice.
8. Pellet debris by centrifugation at $14{,}000 \times g$ for 15 min at 4 °C, and collect the supernatant for each cell lysate. An aliquot (5 μL) is retained for protein quantification using Bradford assays. A second aliquot (20 μL or 2% of the total volume) is set aside for Western blotting to monitor the pre-IP protein expression levels and enable qualitative assessment of immunoprecipitation (IP) efficiency (see Fig. 2b). The remaining bulk cell lysate is aliquoted into 1.5 mL microcentrifuge tubes (Eppendorf). For IP of ER and PR, 2 mg of total protein is recommended, while only 200 μg is used for HER2 IP, and 50 μg is used for whole cell lysate SDS-PAGE protein fractionation and LC-MRM. Use immediately.

3.3 Tissue Selection and Processing

1. Search available samples for retrospective analysis or plan prospective analysis (see **Note 4**). Examine the available data for diagnosis, molecular profiling information, tumor tissue availability and total wet weight, pathology review criteria, patient demographics, and confounding variables (e.g., time of collection).
2. Select samples that meet the study criteria. In this case, patients had to be in age between 18 and 89 at diagnosis, and tumor wet weight

had to be greater than 25 mg. We selected five patients from four different categories: ER^+ and ER^- as well as $HER2^+$ and $HER2^-$.

3. Embed each bulk tumor in OCT and section on a cryostat in 5–10 μm thickness.
4. Thaw mount each tissue section on slides.
5. Stain with hematoxylin and eosin using an autostainer (see Fig. 3a) and seal with a cover slip.
6. Scan each slide.
7. Have a board certified pathologist review the slide to confirm the diagnosis and characterize the tissue content (e.g., cellularity, malignancy, stroma, necrosis, immune infiltration, ascites, blood, etc.).
8. Cut away the bulk OCT, and wash with cold ethanol to remove the residual OCT while keeping the samples in the vapor phase of dry ice.
9. Pulverize each frozen tumor and resuspend the sample in lysis buffer.
10. Clarify by centrifugation as described for cell lysates, and retain and aliquot the supernatants for immediate use.

3.4 Immunoprecipitation, SDS-PAGE, and In-Gel Digestion

1. Divide the sample for ER/PR (2 mg total protein) and HER2 immunoprecipitation (200 μg total protein). See **Note 5**. In the first sample, ERα antibody (2 μg) is added to the cell lysate or tissue homogenate on ice. After 5 min of incubation, the PR antibody (13 μg) is added to the same aliquot of cell lysate or tissue homogenate. In the second sample, HER2 antibody (antibody 1 μg) is added to the cell lysate or tissue homogenates on ice. These samples will be processed in parallel and combined prior to gel electrophoresis.
2. Incubate the cell lysates or tissue homogenates with the antibodies using slow rotation on a shaker at 4 °C for 1.5 h.
3. Prewash Protein A/G UltraLink Resin (53132, Pierce, 20 μL of 50% slurry is needed per IP) with 1 mL 0.5% NP-40 solution by inverting and spinning down at 500 × *g* for 30 s. Discard the supernatant and repeat twice. After the second spin down, add enough 0.5% NP-40 solution to return to the original volume (20 μL slurry × # of IPs).
4. Add 10 μL of prewashed Protein A/G UltraLink Resin to the lysate, and incubate with slow rotation at 4 °C overnight.
5. The beads are then pelleted at 1000 × *g* for 30 s. The supernatant is removed, and an aliquot (2% or 20 μL) is retained for Western blotting (to compare pre-IP and post-IP levels to qualitatively assess capture efficiency).

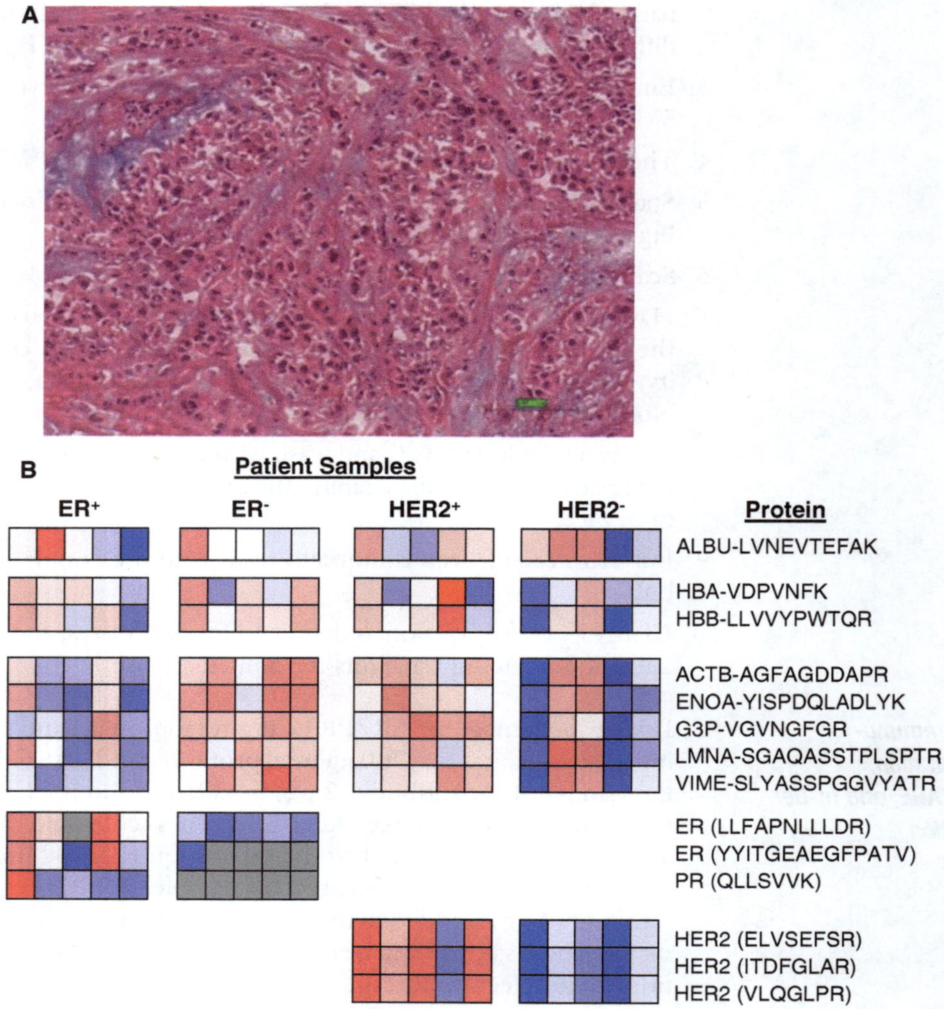

Fig. 3 Multiplexed biomarker quantification in frozen breast tumors. Hematoxylin and eosin (H&E)-stained slides (**a**) should be reviewed by a board certified pathologist to verify tumor content and assess other quality metrics (e.g., stromal, immune infiltration, necrosis, inter alia). Then, LC-MRM analysis is performed on each tissue specimen for biomarker quantification, represented here as heat maps (**b**). *Red* indicates highest expression and *blue* the lowest, and *gray* indicates that the protein was not detected. In both ER and HER2 pilot studies, one positive patient (by immunohistochemistry or fluorescence in situ hybridization performed in the clinic) was not detected as positive by the mass spectrometry assays, indicating the need to include additional metrics to assess tissue quality and content within each biomarker platform

6. Wash the immune-complex three times in 1 mL 100 mM Tris (pH 7.0) by gentle rotation for 10 min at 4 °C. Pellet beads at 1000 × *g* for 30 s after each wash, and decant the supernatant. After final wash step, mix the immune-complex with SDS-PAGE loading buffer and 1 μL of DTT, and heat the samples at 95 °C for 10 min to denature the proteins.

7. Cool the samples on ice for at least 2 min, and load the samples [the maximum volume per number of lanes depends on the selection of gel (e.g., 30 μL on an 18-lane gel)]. Load all samples by pipetting and add protein standards to specific lanes.
8. Proteins are separated using a running voltage of 150 V over an 80-min time period. Separation is visually inspected using the prestained molecular weight markers; the gel can be stopped when 50 and 37 kDa protein markers are sufficiently resolved to enable excision of the ERα and PRα bands, which are the most difficult to separate from the antibody.
9. Stain the gel with Coomassie Brilliant Blue G (cat no B5133, Sigma), in an aqueous solution of 10% methanol/5% acetic acid and destain.
10. Excise the gel bands for ERα, PRα, and PRβ, as shown in Fig. 2. ERα (MW, 55–70 kDa), PRα (MW, 75–100 kDa), and PRβ (MW, 100–135 kDa) are excised separately prior to in-gel digestion. See Fig. 2c.
11. Dice gel bands into 1 mm^3 pieces, and wash with 1 mL 18 MΩ water shaking gently at room temperature for 30 min (Vortex mixer, Fisher Scientific).
12. Destain and dehydrate gel slices twice in 1 mL 50 mM aqueous ammonium bicarbonate/50% methanol for 20 min.
13. Gel slices are rehydrated for disulfide bond reduction using 100 μL total volumes, which contain 10 μL of 20 mM Tris (2-carboxyethyl)phosphine hydrochloride (TCEP) and 90 μL of 50 mM ammonium bicarbonate. The final concentration of TCEP is 2 mM. Perform the reduction step twice at 37 °C for 15 min.
14. Alkylation of free cysteine in the samples is also achieved using 100 μL total volumes containing 10 μL 200 mM Iodoacetamide (IAA) and 90 μL of 50 mM ammonium bicarbonate. The final concentration of IAA is 20 mM. This step is also done twice, but in the dark at room temperature for 20 min.
15. Dehydrate the gel slices with 200 μL 50 mM ammonium bicarbonate/50% methanol three times, 15 min each.
16. Add 20 μL of 20 ng/μL trypsin and after 10 min sufficient volumes (~100 μL) of 30 mM ammonium bicarbonate to each sample and incubate overnight at 37 °C. Make sure the gel cubes are completely rehydrated and submerged.
17. Collect the digest solution into a new 0.5 mL microcentrifuge tube, and extract gel slices twice with 100 μL aqueous 50% acetonitrile/0.01%TFA while shaking at room temperature for 20 min.

18. Concentrate the extracted peptide solutions to dryness by vacuum centrifugation.
19. For analysis of the tissue QC markers panel, perform gel electrophoresis as described above loading only 10 μg of total protein, and run the gel only long enough to capture the proteins.
20. Excise the gel plug containing all of the proteins and digest as above. If the gel is developed by more than 2–3 mm, cut the gel into sections for parallel digestion, and recombine the digested peptides to maximize recovery.

3.5 Addition of Standards and LC-MRM Analysis

1. Redissolve the dry peptides from each sample in 30 μL internal standards mix (1× IS stock), and transferred to autosampler vials.
2. Inject 5 μL of each sample, which is loaded on the trapping column 4–6 μL/min and desalted by washing for 5 min with aqueous 2% acetonitrile/0.1% formic acid or aqueous 2% acetonitrile/0.05% heptafluorobutyric acid to improve loading/capture of hydrophilic peptides due to the more hydrophobic counterion (loading solvent).
3. Program the LC so that a gradient from 5 to 50% solvent B delivered at 300 nL/min over 35 min in order to achieve peptide separation.
4. Set the instrument parameters for mass spectrometry data acquisition as follows: 2400 V nanoelectrospray voltage with 250 °C ion transfer tube temperature. Set Q1 mass resolution to 0.4 resolution or 0.7 (if scheduling is used, see **Note 6**) and Q3 to 0.7 resolution. Fragmentation is obtained with 1.5 mTorr argon. Each transition is monitored for 20 ms.
5. Enter all Q1 (parent) and Q3 (product) *m/z* values into the MRM method, and add the optimized q2 collision voltages.
6. Acquire data in triplicate for each sample.
7. Analyze data with Skyline by importing the raw data and verifying peak assignments.
8. Compare the fragmentation pattern (transition ranking) and retention time of endogenous and synthetic peptides, which should be highly similar (e.g., <5% variation). Remove transitions with higher levels of variability or interference observed in the transition ratios. If signal is low, be aware that the detection of the lowest transitions may be lost; however, quantification is still possible using the remaining transitions with acceptable signal.
9. Export the peak area to a .csv file for calculations using Microsoft Excel. Calculate average, standard deviation, and relative standard deviation to evaluate the levels of each peptide and phosphopeptide in each sample as well as the overall assay performance.

10. Perform hierarchical clustering of the data using Cluster 3.0. Filter the data for missingness, if desired. This software refers to genes and arrays, which in this experiment correspond to peptide measurements and samples, respectively. In other words, each peptide quantification is in a row, and each sample is in a column. Log transform, center, and normalize the "genes," which are the peptide measurements. Then, perform clustering of "genes" using the average linkage. Clustering of "arrays" or samples can also be performed, but in this dataset we wanted to show the differences of biomarker positive and negative samples, rather than perform unsupervised clustering of the data.
11. Visualize the results as a heat map in Java Treeview. Load the CDT file created from Cluster 3.0. Adjust the pixel settings to maximize the range of color. Set positive values to red, negative values to blue, 0 to white, and use gray for missing data.
12. Export data as images or postscript files.
13. Compare tabular data and heat maps to assess cell line or tumor tissue experiments. Examples of data from cell line experiments are shown in Fig. 2d; clinical sample analysis is presented in Fig. 3b.
14. If desired, compare LC-MRM quantification with an ELISA kit, according to the manufacturer's instructions. Dilute cell lysates and tissue homogenates to the working range of each ELISA kit. Measure standards and cell line samples in duplicate and patient specimens in triplicate with a microplate reader, and quantify the protein expression levels in each sample by calculation using the standard curves.

In summary, this protocol illustrates molecular analysis of breast cancer biomarkers, but should be generalizable to the study of other human diseases.

4 Notes

1. The developers have prepared extensive documentation and tutorials for Skyline use; check their website for details.
2. Calibration curves varying the amount of light synthetic peptide with a fixed amount of stable isotope-labeled peptide can be performed in buffer; however, reverse calibration curves in biological matrix that use fixed amounts of light synthetic peptide and variable amounts of SIS peptide may be preferable.
3. For these studies, we used process replicates of the IP-LC-MRM process in a 3 × 3 design (three replicates prepared and analyzed from a bulk specimen on each of 3 days) to examine the reproducibility and robustness of the assays.

4. At our institution, we significantly benefit from the Total Cancer Care protocol that consents patients to donate specimens for research as well as to lifetime follow-up to maximize clinical data, medical history, and outcome information. Using that umbrella protocol, pretreatment frozen breast tumor specimens were obtained under protocols that have been approved by as nonhuman subjects research after internal review (MCC 17239 and MCC 17241).
5. Initially, we had reported that 50 μg of tissue homogenate was separated by SDS-PAGE to isolate HER2 for LC-MRM analysis. This method was replaced when the IP-LC-MRM was shown to be more effective in detecting phosphorylation sites. If HER2 expression is the endpoint of interest, SDS-PAGE of the smaller amount would be sufficient.
6. Scheduling refers to setting time windows for detection of each peptide to maximize the sampling and subsequently the data quality. This step is only necessary for large numbers of peptides (>30, when considering both light/proteolytic peptides from the endogenous protein and SIS with 3–5 transitions for each).

Acknowledgments

This project was supported as an academic-industrial partnership using a sponsored research agreement with Proteome Sciences plc. This work has been supported in part by the Tissue, Collaborative Data Services, Analytic Microscopy, and Proteomics Cores at the H. Lee Moffitt Cancer Center & Research Institute, an NCI-designated Comprehensive Cancer Center (P30-CA076292). Instrument purchases have been funded through the Moffitt Foundation and the Bankhead-Coley Cancer Research program of the Florida Department of Health (06BS-02-9614, 09BE-04). Amino acid analysis was performed by Virginia Johnson and Larry Dangott at the Texas A&M Protein Chemistry Laboratory. We would like to thank Jason Held and Chris Benz for discussions of model systems and test cases.

References

1. Le Romancer M, Poulard C, Cohen P, Sentis S, Renoir JM, Corbo L (2011) Cracking the estrogen receptor's posttranslational code in breast tumors. Endocr Rev 32:597–622
2. Anbalagan M, Huderson B, Murphy L, Rowan BG (2012) Post-translational modifications of nuclear receptors and human disease. Nucl Recept Signal 10:e001
3. Clemm DL, Sherman L, Boonyaratanakornkit V, Schrader WT, Weigel NL, Edwards DP (2000) Differential hormone-dependent phosphorylation of progesterone receptor A and B forms revealed by a phosphoserine site-specific monoclonal antibody. Mol Endocrinol 14:52–65
4. Hopp TA, Weiss HL, Hilsenbeck SG, Cui Y, Allred DC, Horwitz KB, Fuqua SA (2004) Breast cancer patients with progesterone

receptor PR-A-rich tumors have poorer disease-free survival rates. Clin Cancer Res 10:2751–2760
5. Colomer R, Montero S, Lluch A, Ojeda B, Barnadas A, Casado A, Massutí B, Cortés-Funes H, Lloveras B (2000) Circulating HER2 extracellular domain and resistance to chemotherapy in advanced breast cancer. Clin Cancer Res 6:2356–2362
6. Asgeirsson KS, Agrawal A, Allen C, Hitch A, Ellis IO, Chapman C, Cheung KL, Robertson JF (2007) Serum epidermal growth factor receptor and HER2 expression in primary and metastatic breast cancer patients. Breast Cancer Res 9:R75
7. Cho SH, Jeon J, Kim SI (2012) Personalized medicine in breast cancer: a systematic review. J Breast Cancer 15:265–272
8. Bartlett JM, Rea D, Rimm DL (2011) Quantification of hormone receptors to guide adjuvant therapy choice in early breast cancer: better methods required for improved utility. J Clin Oncol 29:3715–3716
9. Subik K, Lee JF, Baxter L, Strzepek T, Costello D, Crowley P, Xing L, Hung MC, Bonfiglio T, Hicks DG, Tang P (2010) The expression patterns of ER, PR, HER2, CK5/6, EGFR, Ki-67 and AR by Immunohistochemical analysis in breast cancer cell lines. Breast Cancer (Auckl) 4:35–41
10. Konecny G, Pauletti G, Pegram M, Untch M, Dandekar S, Aguilar Z, Wilson C, Rong HM, Bauerfeind I, Felber M, Wang HJ, Beryt M, Seshadri R, Hepp H, Slamon DJ (2003) Quantitative association between HER-2/neu and steroid hormone receptors in hormone receptor-positive primary breast cancer. J Natl Cancer Inst 95:142–153
11. Sun CF, Wu TL, Tsao KC, Wu JT (2001) Development of two ELISA for estrogen and progesterone receptor with sufficient sensitivity for fine needle aspirate and core biopsy. J Clin Lab Anal 15:138–143
12. Hoofnagle AN, Becker JO, Wener MH, Heinecke JW (2008) Quantification of thyroglobulin, a low-abundance serum protein, by immunoaffinity peptide enrichment and tandem mass spectrometry. Clin Chem 54:1796–1804
13. Hoofnagle AN, Becker JO, Oda MN, Cavigiolio G, Mayer P, Vaisar T (2012) Multiple-reaction monitoring-mass spectrometric assays can accurately measure the relative protein abundance in complex mixtures. Clin Chem 58:777–781
14. Kirkpatrick DS, Gerber SA, Gygi SP (2005) The absolute quantification strategy: a general procedure for the quantification of proteins and post-translational modifications. Methods 35:265–273
15. Gerber SA, Kettenbach AN, Rush J, Gygi SP (2007) The absolute quantification strategy: application to phosphorylation profiling of human separase serine 1126. Methods Mol Biol 359:71–86
16. Chen Y, Gruidl M, Remily-Wood E, Liu RZ, Eschrich S, Lloyd M, Nasir A, Bui MM, Huang E, Shibata D, Yeatman T, Koomen JM (2010) Quantification of beta-catenin signaling components in colon cancer cell lines, tissue sections, and microdissected tumor cells using reaction monitoring mass spectrometry. J Proteome Res 9:4215–4227
17. Remily-Wood ER, Liu RZ, Xiang Y, Chen Y, Thomas CE, Rajyaguru N, Kaufman LM, Ochoa JE, Hazlehurst L, Pinilla-Ibarz J, Lancet J, Zhang G, Haura E, Shibata D, Yeatman T, Smalley KS, Dalton WS, Huang E, Scott E, Bloom GC, Eschrich SA, Koomen JM (2011) A database of reaction monitoring mass spectrometry assays for elucidating therapeutic response in cancer. Proteomics Clin Appl 5:383–396
18. Sprung RW, Martinez MA, Carpenter KL, Ham AJ, Washington MK, Arteaga CL, Sanders ME, Liebler DC (2012) Precision of multiple reaction monitoring mass spectrometry analysis of formalin-fixed, paraffin-embedded tissue. J Proteome Res 11:3498–3505
19. Sellappan S, Blackler A, Liao WL, O'Day E, Xu P, Thyparambil S, Cecchi F, Hembrough T, Catenacci DV (2016) Therapeutically induced changes in HER2, HER3, and EGFR protein expression for treatment guidance. J Natl Compr Cancer Netw 14:503–507
20. Nuciforo P, Thyparambil S, Aura C, Garrido-Castro A, Vilaro M, Peg V, Jimenez J, Vicario R, Cecchi F, Hoos W, Burrows J, Hembrough T, Ferreres JC, Perez-Garcia J, Arribas J, Cortes J, Scaltriti M (2016) High HER2 protein levels correlate with increased survival in breast cancer patients treated with anti-HER2 therapy. Mol Oncol 10:138–147
21. Domanski D, Murphy LC, Borchers CH (2010) Assay development for the determination of phosphorylation stoichiometry using

multiple reaction monitoring methods with and without phosphatase treatment: application to breast cancer signaling pathways. Anal Chem 82:5610–5620

22. Schoenherr RM, Whiteaker JR, Zhao L, Ivey RG, Trute M, Kennedy J, Voytovich UJ, Yan P, Lin C, Paulovich AG (2012) Multiplexed quantification of estrogen receptor and HER2/Neu in tissue and cell lysates by peptide immunoaffinity enrichment mass spectrometry. Proteomics 12:1253–1260
23. Britton DJ, Scott GK, Schilling B, Atsriku C, Held JM, Gibson BW, Benz CC, Baldwin MA (2008) A novel serine phosphorylation site detected in the N-terminal domain of estrogen receptor isolated from human breast cancer cells. J Am Soc Mass Spectrom 19:729–740
24. Atsriku C, Britton DJ, Held JM, Schilling B, Scott GK, Gibson BW, Benz CC, Baldwin MA (2009) Systematic mapping of posttranslational modifications in human estrogen receptor-alpha with emphasis on novel phosphorylation sites. Mol Cell Proteomics 8:467–480
25. Held JM, Britton DJ, Scott GK, Lee EL, Schilling B, Baldwin MA, Gibson BW, Benz CC (2012) Ligand binding promotes CDK-dependent phosphorylation of ER-alpha on hinge serine 294 but inhibits ligand-independent phosphorylation of serine 305. Mol Cancer Res 10:1120–1132
26. Zhang Y, Beck CA, Poletti A, Edwards DP, Weigel NL (1994) Identification of phosphorylation sites unique to the B form of human progesterone receptor. In vitro phosphorylation by casein kinase II. J Biol Chem 269:31034–31040
27. Chen Y, Britton D, Wood ER, Brantley S, Magliocco A, Pike I, Koomen JM (2017) Quantitative proteomics of breast tumors: tissue quality assessment to clinical biomarkers. Proteomics 17(6). https://doi.org/10.1002/pmic.201600335
28. Hassanein M, Huiart L, Bourdon V, Rabayrol L, Geneix J, Nogues C, Peyrat JP, Gesta P, Meynard P, Dreyfus H, Petrot D, Lidereau R, Noguchi T, Eisinger F, Extra JM, Viens P, Jacquemier J, Sobol H (2013) Prediction of BRCA1 germ-line mutation status in patients with breast cancer using histoprognosis grade, MS110, Lys27H3, Vimentin, and KI67. Pathobiology 80:219–227
29. Karihtala P, Auvinen P, Kauppila S, Haapasaari KM, Jukkola-Vuorinen A, Soini Y (2013) Vimentin, zeb1 and Sip1 are up-regulated in triple-negative and basal-like breast cancers: association with an aggressive tumour phenotype. Breast Cancer Res Treat 138:81–90
30. Yamashita N, Tokunaga E, Kitao H, Hisamatsu Y, Taketani K, Akiyoshi S, Okada S, Aishima S, Morita M, Maehara Y (2013) Vimentin as a poor prognostic factor for triple-negative breast cancer. J Cancer Res Clin Oncol 139:739–746
31. Soule HD, Vazguez J, Long A, Albert S, Brennan M (1973) A human cell line from a pleural effusion derived from a breast carcinoma. J Natl Cancer Inst 51:1409–1416
32. Lasfargues EY, Coutinho WG, Redfield ES (1978) Isolation of two human tumor epithelial cell lines from solid breast carcinomas. J Natl Cancer Inst 61:967–978
33. de Hoon MJ, Imoto S, Nolan J, Miyano S (2004) Open source clustering software. Bioinformatics 20:1453–1454
34. Saldanha AJ (2004) Java Treeview—extensible visualization of microarray data. Bioinformatics 20:3246–3248
35. Chen Y, Fisher KJ, Lloyd M, Wood ER, Coppola D, Siegel E, Shibata D, Chen YA, Koomen JM (1647) Multiplexed liquid chromatography-multiple reaction monitoring mass spectrometry quantification of cancer signaling proteins. Methods Mol Biol 2017:19–45
36. MacLean B, Tomazela DM, Shulman N, Chambers M, Finney GL, Frewen B, Kern R, Tabb DL, Liebler DC, MacCoss MJ (2010) Skyline: an open source document editor for creating and analyzing targeted proteomics experiments. Bioinformatics 26:966–968
37. Abbatiello SE, Schilling B, Mani DR, Zimmerman LJ, Hall SC, MacLean B, Albertolle M, Allen S, Burgess M, Cusack MP, Gosh M, Hedrick V, Held JM, Inerowicz HD, Jackson A, Keshishian H, Kinsinger CR, Lyssand J, Makowski L, Mesri M, Rodriguez H, Rudnick P, Sadowski P, Sedransk N, Shaddox K, Skates SJ, Kuhn E, Smith D, Whiteaker JR, Whitwell C, Zhang S, Borchers CH, Fisher SJ, Gibson BW, Liebler DC, MacCoss MJ, Neubert TA, Paulovich AG, Regnier FE, Tempst P, Carr SA (2015) Large-scale interlaboratory study to develop, analytically validate and apply highly multiplexed, quantitative peptide assays to measure cancer-relevant proteins in plasma. Mol Cell Proteomics 14:2357–2374

Methods in Molecular Biology (2018) 1788: 269–277
DOI 10.1007/7651_2017_78

Published online: 05 October 2017

Quantitative Proteomic Analysis of Mass Limited Tissue Samples for Spatially Resolved Tissue Profiling

Paul D. Piehowski, Rui Zhao, Ronald J. Moore, Geremy Clair, and Charles Ansong

Abstract

Traditionally, proteomic studies have been carried out on whole tissues or organs enabling the profiling of thousands of proteins within a single LC-MS analysis. A disadvantage of this approach is that proteomes generated from whole tissues are an "average" that represents a blend of cell types and distinct anatomical regions which can obscure important biological phenomena. Laser capture microdissection (LCM) is an elegant method that allows tissue features of interest, as small as a single cell, to be identified and isolated for downstream analysis. Herein we describe an approach that utilizes an immobilized enzyme reactor (IMER) coupled directly to nanoLC-MS/MS for highly sensitive, automated, quantitative proteomic analysis of the microscopic tissue specimens generated by LCM.

Keywords Immobilized enzyme reactor, Laser capture microdissection, Mass spectrometry, NanoLC, Nanoproteomics, Proteomics

1 Introduction

Quantitative global proteomics analyses are now broadly applied across the field of biological research to study an organism (s) protein expression in response to perturbation or phenotype of interest [1]. In order to produce high quality quantitative data, efficient and reproducible sample preparation workflows are crucial, especially when employing label-free methods [2, 3]. This becomes increasingly challenging as sample size decreases due to the protein losses incurred in the multiple processing steps required for producing peptides amenable to MS analysis [4]. This challenge has limited the effective application of proteomics to many sample types of great interest, including LCM dissected tissues, FACS sorted cells, circulating tumor cells, and developing embryos.

The potential significance of these measurements has driven the development of a wide variety of approaches to address the inherent limitations of proteomic sample handling. These methodologies employ three main approaches, or a combination thereof, to reducing sample losses: substitute denaturants, proteomic reactors, and online sample handling. Sample extraction and protein denaturation, a critical first step in all proteomic workflows, is achieved using chaotropes, most often high concentrations of urea or detergents. Both of these reagents are incompatible with LC-MS analysis and need to be removed resulting in significant protein losses. By substituting these reagents with volatile solvents or acid-cleavable detergents, all sample manipulations can be made within the same vessel greatly reducing losses [5–9]. This approach has the clear advantage of being the most straightforward to implement and has been used robustly with LCM tissue samples [7, 8, 10]. Nevertheless, many sample isolations require the use of salts or detergents which make this approach incompatible. The proteomic reactor approach employs a solid phase support to immobilize proteins and/or peptides, allowing protein chemistry, digestion, and buffer exchange without sample transfers. This approach has been successfully applied in both manual and online configurations [11–13]. Proteomic reactor approaches offer greater reagent flexibility; however, they require many manual sample handling steps and long preparation times challenging reproducibility and requiring expert practitioners [14, 15]. Online sample handling borrows the technologies of the microscale separations community to carry out proteomics sample preparation within enclosed systems, i.e., fused silica capillaries, controlled by chromatography pumps and fluidic systems [16–18]. Most commonly, immobilized enzyme reactors (IMER) are used to achieve rapid protein digestions and eliminate the need for frequent additions of fresh protease [19–25]. These configurations allow for the miniaturization of reaction vessels and solid phase supports, reducing losses to nonspecific binding while facilitating reproducible handling of nanoliter sample volumes in an automated fashion. In theory, online handling represents the ideal solution for nanoproteomics, but in practice online handling systems are highly complex creating large numbers of potential fail points, resulting in significant downtimes and precluding their application to biological problems requiring large sample numbers to achieve sufficient statistical power.

To address this challenge, we have developed a simplified online platform capable of reproducibly handling small protein quantities and allowing for label-free proteomics analysis, termed the Simplified Nano-Proteomics Platform (SNaPP) [19, 26]. Prior to injection, proteins are extracted, denatured, and reduced by the addition of a small volume of extraction buffer. The SNaPP system utilizes immobilized trypsin beads packed into a fused silica capillary to create a high-efficiency enzyme reactor with minimal surface area [27, 28]. This reactor is directly interfaced with a capillary solid phase extractor (SPE) for online desalting of the resultant peptides. The SPE is then switched in line with a 50 μm I.D. analytical column for LC-MS/MS analysis. This system has been successfully applied for quantitative analysis of blastocysts, LCM dissected tissues, sorted cells, glomerular filtrate, and exosomes.

2 Materials

Prepare buffer solutions using Milli-Q purified water and analytical grade reagents. Mobile phase A and B should be premixed HPLC grade solvents. Dithiothreitol (DTT) can be purchased from Sigma in 7.7 mg preweighed individually sealed aliquots to ensure freshness and accuracy.

1. Tris buffer: 50 mM Tris, 1 mM $CaCl_2$, pH 8.0. Weigh out 767 mg Tris–HCl and 925 mg Tris-Base, and then transfer to 250 mL graduated cylinder. Next, weigh out 27.5 mg $CaCl_2$ and add to the cylinder. Add water to bring the total volume to 250 mL and check the pH to verify (*see* **Note 1**).
2. DTT stock: 500 mM dithiothreitol in Tris buffer (*see* **Note 2**). Empty the contents of a DTT preweigh tube into 0.6 mL microcentrifuge tube (*see* **Note 3**). Add 100 μL of Tris buffer and vortex for 1 min to solubilize DTT.
3. IMER column wash: Mix 25 mL mobile phase A with 25 mL of mobile phase B.

3 Methods

3.1 Sample Preparation

1. Prepare the homogenization buffer by weighing out 480 mg of urea into a 1.5 mL microcentrifuge tube. Add 650 μL of Tris buffer and 20 μL of DTT stock to dissolve urea (*see* **Note 4**). Once dissolved bring the final volume to 1 mL with Tris buffer.
2. Deposit 10 μL of homogenization buffer on top of the dissected tissue mounted on the LCM cap. Allow the buffer to absorb into the tissue for 10 min.
3. After absorption, the tissue will be become very soft. Pipette the entire volume of homogenization buffer volume up and down 5–10 times to remove the tissue from the cap and transfer vial to LC-MS vial (*see* **Note 5**).
4. Rinse the LCM cap with a 20 μL aliquot of Tris buffer and transfer wash to LC-MS vial.
5. Incubate sample in orbital shaker at 60 °C for 30 min to denature and reduce the proteins.
6. Spin samples at 15,000 rcf for 5 min to pellet any remaining particulates and remove condensed solvent from the walls of the vial (*see* **Note 6**).

3.2 SNaPP Analysis

1. Load samples into the autosampler set at 4 °C. (*See* **Note 7**).
2. LCMSnet software is used to build the sample queue and initiate SNaPP analysis, as explained in Fig. 1. (*See* **Note 8**).
3. Sample is processed by SNaPP system, a diagram of the process is shown if Fig. 2.

3.3 Data Analysis

1. Place all the Thermo ".raw" files and a fasta file containing the protein sequences for your organism(s) in a single folder.
2. Estimate the retention time shifts in your analyses by checking the retention times of multiple peptides using MS1 masses across the different datasets (e.g., using Xcalibur for viewing and exploring Thermo Raw files).
3. If necessary install Maxquant 1.5.2.8 or higher, open it and configure the fasta file(s) of interest in the Andromeda search engine.
4. Load your Raw files click on the "No fraction" button then set your experiment names.
5. In the "Group-specific parameters" and the "Global parameters" tabs, keep all the values to default except for the following: enable the LFQ quantification; use "Trypsin" as digestion enzyme in "Specific" mode use "Oxidation(M)" as "Variable modification" and no fixed modification; enable "re-quantify" and "Match between runs" and use your estimation of the time shifts (Step 2) as "match time window [min]"; for the identification set the "Min. peptide length" at 6, and the "Min. peptides," "Min razor + unique peptide" and "Min unique peptides" at 2. (*See* **Notes 13** and **14**).

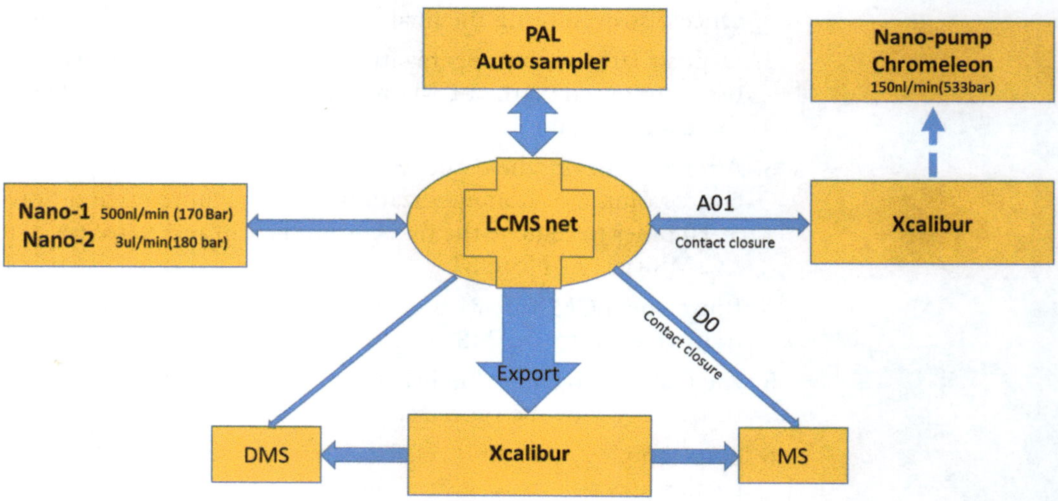

Fig. 1 Instrument control diagram. All sample retrieval, valve positioning, and method switching is controlled by LCMSnet software

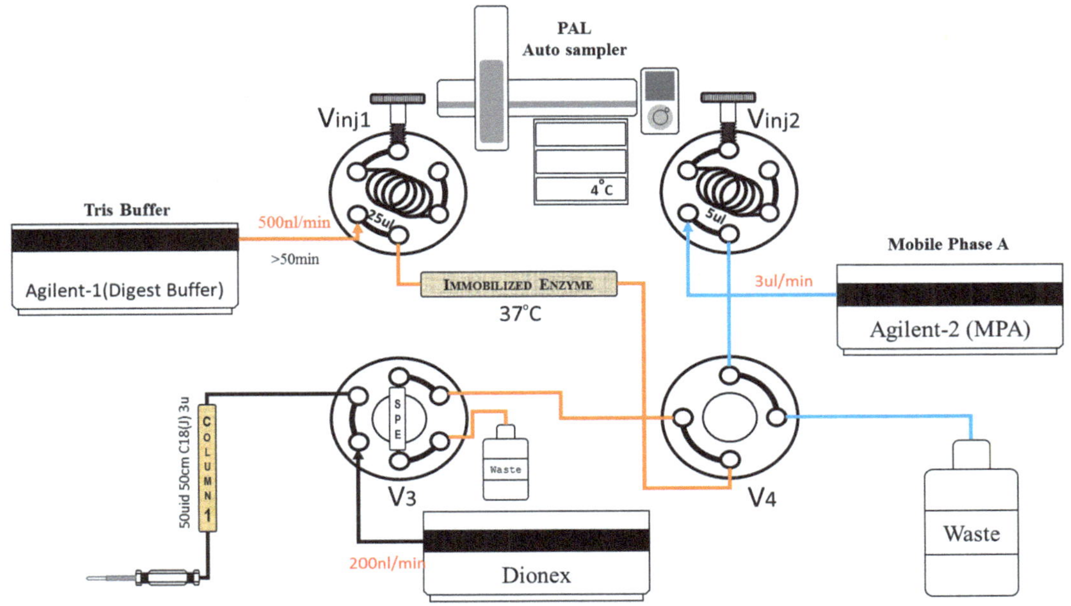

Fig. 2 The SNaPP system was constructed based on the previously reported configuration [19]. Two Agilent 1200 series nano pumps are used to deliver the digestion buffer (Agilent-1) and SPE wash buffer (Agilent-2). (*See* **Note 9**). A Dionex UltiMate 3000 RSLCnano pump is used to deliver the RPLC separation gradient. Two 6 ports injection valves were used for loading, SNaPP samples (V_{inj1}), and predigested peptide samples (V_{inj2}). The third 6 port valve (V_3) is used to set up the SPE (5 μm C18, 150 μm × 4 cm) and analytical column (1.7 μm AQ C_{18}, 50 μm × 25 cm) and one 4 port valve was used for switching between SNaPP or peptide methods (*See* **Note 10**). The system is coupled to a Thermo QExactive Mass spectrometer by a replaceable custom ESI emitter (150 μm O.D. × 20 μm I.D.); (*see* **Notes 11** and **12**)

6. Click on the start button.
7. When the search is done, go to the folder containing your raw files. MaxQuant should have created the subfolders "combined/txt/." In this folder, use the files "peptide.txt" and "proteinGroups.txt" for the peptide and protein information respectively. Use the columns starting with "Intensity" in the "peptide.txt" file and the column starting with "LFQ intensity" in the "proteinGroups.txt" file. (*See* **Note 15**).
8. For the peptide table, remove the rows containing a "Reverse" match or a "Potential contaminant" match if appropriate, for the protein table remove the rows containing a "Reverse" match a "Potential contaminant" match or a "Only identified by site" match.
9. Perform the normalization of the dataset and the statistics with appropriate statistical standards.

4 Notes

1. The pH of Tris buffer solutions is temperature dependent. This recipe is designed to give pH 8.0 at 37 °C. This is used because the digestion is carried out at this temperature. Tris buffer solutions are good for about 2 months if stored at 4 °C when not in use. Buffer should be kept no more than 2 weeks if stored at room temperature.
2. DTT is a reducing agent and its strength decreases over time when in solution. This solution should be made up fresh immediately prior to sample extraction.
3. DTT pellet can get stuck in the bottom of the preweigh tube and it is sometimes necessary to break up the pellet by pinching the bottom of tube to transfer. When this happens, it is essential that the Tris buffer used for solubilization is used to rinse preweigh tube to ensure quantitative transfer of the DTT.
4. Solvation of urea is an endothermic process and 8 M is a nearly saturated solution. Solubilization can be greatly sped up by shaking at 37 °C.
5. We have found that most tissues will lift off the cap easily; in some cases, it is necessary to use the pipette tip to dislodge the tissue slice by scraping the cap. Any transferred solids will be broken down further during incubation.
6. If there is a visible pellet, supernatant should be carefully transferred to a fresh LC-MS vial.
7. Samples should be spun upon thawing to check for protein precipitation. Avoid keeping undigested samples in autosampler at 4 °C for more than ~3 days as protein precipitation may occur.
8. LCMSnet software and tutorials are available at https://github.com/PNNL-Comp-Mass-Spec/LCMSNet.
9. We use Poroszyme™ Bulk for packing our columns because it is commercially available, highly consistent and low cost. IMER columns are slurry packed using a sixfold dilution in digestion buffer. For column prep we use a 10 cm long × 150 μm I.D. × 360 μm O.D. capillary section, prefitted with a 2 mm sol-gel frit. Slurry is loaded into a 500 μL syringe with 100 μL of slurry and connected to the capillary via metal union. Transfer approximately 2 cm of trypsin beads to the capillary. We have found that 2 cm length columns can handle protein loadings of up to 1.5 μg in the current cart configuration.
10. The second injection port is not critical to the operation of the SNaPP system but it is highly recommended for troubleshooting as it allows the digestion column to be easily isolated.

11. The SNaPP system is largely agnostic to the mass spectrometer it is coupled to and thus is compatible with all mass spectrometers capable of data-dependent MS/MS acquisition. Preliminary experiments suggest that the Thermo Lumos Orbitrap provides increased sensitivity compared with a QExactive.
12. Depending on the protein mass of the sample, proteomic data collection can be carried out with default settings. As the sample gets smaller, the MS2 maximum IT becomes important. Increasing this time to 100 ms can greatly increase the peptide identification rate. Longer ion times can decrease the instrument duty cycle, so when using 100 ms we recommend using a TOPN of no more than 12.
13. For the MaxQuant search, we have tested multiple options for the identification parameters. In our hands with the FDR set at 0.01 for the identification, with a "Min. peptide length" of 6 with 2 unique peptides and a maximum number of missed cleavages set at 2 provided the most identifications; however, depending on your specific setup and your trypsin digestion efficiency, you might want to test and adjust these parameters.
14. The method described in this paper resulted in a higher number of identifications, a better reproducibility and a lower number of contaminant peptides compared to the FASP method [12], both at the peptide and protein level [26]. However, if you have a specific interest in cysteine containing peptides, be aware that the method described here will provide a reduced cysteine containing peptide coverage compared to FASP as the method does not block the DTT reduced cysteines with a carbamidomethyl group.
15. Using MaxQuant we have compared label-free quantification using iBAQ, LFQ (ref: https://www.ncbi.nlm.nih.gov/pmc/articles/PMC4159666/), and spectral counting methods. For sample-limited analyses performed by the SNaPP approach, the LFQ quantification approach performed the best.

Acknowledgment

Portions of this research were supported by grants from the National Heart Lung Blood Institute of NIH (U01 HL122703), National Institute of General Medical Sciences of NIH (P41 GM103493) and the PNNL Laboratory-Directed Research and Development (LDRD) program. Work was performed in W. R. Wiley Environmental Molecular Sciences Laboratory (EMSL), a Department Of Energy (DOE) office of Biological and Environmental Research (BER) national user facility located at Pacific Northwest National Laboratory (PNNL).

References

1. Nilsson T, Mann M, Aebersold R, Yates JR III, Bairoch A, Bergeron JJ (2010) Mass spectrometry in high-throughput proteomics: ready for the big time. Nat Methods 7:681–685
2. Piehowski PD, Petyuk VA, Orton DJ, Xie F, Moore RJ, Ramirez-Restrepo M, Engel A, Lieberman AP, Albin RL, Camp DG (2013) Sources of technical variability in quantitative LC–MS proteomics: human brain tissue sample analysis. J Proteome Res 12:2128–2137
3. Domon B, Aebersold R (2010) Options and considerations when selecting a quantitative proteomics strategy. Nat Biotechnol 28:710–721
4. Altelaar AFM, Heck AJR (2012) Trends in ultrasensitive proteomics. Curr Opin Chem Biol 16:206–213
5. Wang N, MG X, Wang P, Li L (2010) Development of mass spectrometry-based shotgun method for proteome analysis of 500 to 5000 cancer cells. Anal Chem 82:2262–2271
6. Wang HX, Qian WJ, Mottaz HM, Clauss TRW, Anderson DJ, Moore RJ, Camp DG, Khan AH, Sforza DM, Pallavicini M, Smith DJ, Smith RD (2005) Development and evaluation of a micro- and nanoscale proteomic sample preparation method. J Proteome Res 4:2397–2403
7. Braakman RBH, Tilanus-Linthorst MMA, Liu NQ, Stingl C, Dekker LJM, Luider TM, Martens JWM, Foekens JA, Umar A (2012) Optimized nLC-MS workflow for laser capture microdissected breast cancer tissue. J Proteomics 75:2844–2854
8. Liu NQ, Braakman RBH, Stingl C, Luider TM, Martens JWM, Foekens JA, Umar A (2012) Proteomics pipeline for biomarker discovery of laser capture microdissected breast cancer tissue. J Mammary Gland Biol Neoplasia 17:155–164
9. Kulak NA, Pichler G, Paron I, Nagaraj N, Mann M (2014) Minimal, encapsulated proteomic-sample processing applied to copy-number estimation in eukaryotic cells. Nat Methods 11:319–324
10. De Marchi T, Braakman RBH, Stingl C, van Duijn MM, Smid M, Foekens JA, Luider TM, Martens JWM, Umar A (2016) The advantage of laser-capture microdissection over whole tissue analysis in proteomic profiling studies. Proteomics 16:1474–1485
11. Chen W, Wang S, Adhikari S, Deng Z, Wang L, Chen L, Ke M, Yang P, Tian R (2016) Simple and integrated spintip-based technology applied for deep proteome profiling. Anal Chem 88:4864–4871
12. Wisniewski JR, Ostasiewicz P, Mann M (2011) High recovery FASP applied to the proteomic analysis of microdissected formalin fixed paraffin embedded cancer tissues retrieves known colon cancer markers. J Proteome Res 10:3040–3049
13. Ethier M, Hou WM, Duewel HS, Figeys D (2006) The proteomic reactor: a microfluidic device for processing minute amounts of protein prior to mass spectrometry analysis. J Proteome Res 5:2754–2759
14. Liebler DC, Ham AJL (2009) Spin filter-based sample preparation for shotgun proteomics. Nat Methods 6:785–785
15. Nel AJ, Garnett S, Blackburn JM, Soares NC (2015) Comparative re-evaluation of FADP and enhanced FASP methods by LC-MS/MS. J Proteome Res 14(3):1637–1642
16. Tian RJ, Alvarez-Saavedra M, Cheng HYM, Figeys D (2011) Uncovering the proteome response of the master circadian clock to light using an AutoProteome system. Mol Cell Proteomics 10
17. Lopez-Ferrer D, Petritis K, Robinson EW, Hixson KK, Tian ZX, Lee JH, Lee SW, Tolic N, Weitz KK, Belov ME, Smith RD, Pasa-Tolic L (2011) Pressurized pepsin digestion in proteomics. Mol Cell Proteomics 10
18. Lopez-Ferrer D, Petritis K, Lourette NM, Clowers B, Hixson KK, Heibeck T, Prior DC, Pasa-Tolic L, Camp DG, Belov ME, Smith RD (2008) On-line digestion system for protein characterization and proteome analysis. Anal Chem 80:8930–8936
19. Huang EL, Piehowski PD, Orton DJ, Moore RJ, Qian W-J, Casey CP, Sun X, Dey SK, Burnum-Johnson KE, Smith RD (2016) SNaPP: simplified nanoproteomics platform for reproducible global proteomic analysis of nanogram protein quantities. Endocrinology 157:1307–1314
20. Hustoft HK, Brandtzaeg OK, Rogeberg M, Misaghian D, Torsetnes SB, Greibrokk T, Reubsaet L, Wilson SR, Lundanes E (2013) Integrated enzyme reactor and high resolving chromatography in "sub-chip" dimensions for sensitive protein mass spectrometry. Sci Rep 3:3511
21. Hustoft HK, Vehus T, Brandtzaeg OK, Krauss S, Greibrokk T, Wilson SR, Lundanes E (2014) Open tubular lab-on-column/mass spectrometry for targeted proteomics of nanogram sample amounts. PLoS One 9

22. Jiang S, Zhang Z, Li L (2015) A one-step preparation method of monolithic enzyme reactor for highly efficient sample preparation coupled to mass spectrometry-based proteomics studies. J Chromatogr A 1412:75–81
23. Yamaguchi H, Miyazaki M (2013) Enzyme-immobilized reactors for rapid and efficient sample preparation in MS-based proteomic studies. Proteomics 13:457–466
24. Safdar M, Sproß J, Jänis J (2014) Microscale immobilized enzyme reactors in proteomics: latest developments. J Chromatogr A 1324:1–10
25. Colquhoun DR, Feild BJ (2015) Automated, online sample preparation for LC-MS analyses: affinity capture, digestion, and clean-up. State-of-the-art and emerging technologies for therapeutic monoclonal antibody characterization volume 3. Defining the next generation of analytical and biophysical techniques. Am Chem Soc 1202:335–356
26. Clair G, Piehowski PD, Nicola T, Kitzmiller JA, Huang EL, Zink EM, Sontag RL, Orton DJ, Moore RJ, Carson JP (2016) Spatially-resolved proteomics: rapid quantitative analysis of laser capture microdissected alveolar tissue samples. Sci Rep 6:39223
27. Slysz GW, Lewis DF, Schriemer DC (2006) Detection and identification of sub-nanogram levels of protein in a nanoLC-trypsin-MS system. J Proteome Res 5:1959–1966
28. Slysz GW, Schriemer DC (2009) Integrating accelerated tryptic digestion into proteomics workflows. Methods Mol Biol 492:241–254

Methods in Molecular Biology (2018) 1788: 279–287
DOI 10.1007/7651_2017_87

Published online: 26 October 2017

High-Throughput Proteomic Analysis of Fresh-Frozen Biopsy Tissue Samples Using Pressure Cycling Technology Coupled with SWATH Mass Spectrometry

Yi Zhu and Tiannan Guo

Abstract

In the era of precision medicine, there is an increasing need to measure several thousand proteins expressed in minimal amount of fresh-frozen biopsy tissue samples from clinical cohorts. Here we present the detailed protocol for high-throughput proteomic analysis of less than 1 mg fresh-frozen tissue sample using pressure cycling technology (PCT), SWATH mass spectrometry, and OpenSWATH workflow. The PCT procedures enable simultaneous analysis of 16 biopsy tissue samples per batch, followed by SWATH analysis of about 15 samples per mass spectrometer per working day. The thus generated SWATH maps can be perpetually analyzed in silico, offering a practical solution for digital biobanking.

Keywords Biopsy tissue, Digital biobanking, Pressure cycling technology, SWATH mass spectrometer

1 Introduction

High-throughput genomic and transcriptomic analysis of fresh-frozen tissue samples have been the main drivers for precision medicine. Proteins are the gears of life activities; however, high-throughput proteomic analysis of biopsy-level fresh-frozen samples remains a challenge. Firstly, the conventional tissue lysis and protein extraction methods that are routinely employed for relatively large tissue (>10 mg) are not practically applicable for biopsy-level samples owing to lack of reproducibility, low extraction efficiency, and substantial sample loss. Secondly, two-dimensional chromatographic separation that is indispensable for most shotgun proteomic workflows is a major bottleneck for high-throughput analysis.

PCT recently emerged as a technology that can be used to facilitate various steps in proteomic sample preparation [1]. Tissue lysis and protein extraction are realized using programmed cycles oscillating between ultrahigh and ambient pressure in the presence of lysis buffer. In addition, protein digestion is enhanced and accelerated under pressurized condition. In Ruedi Aebersold's group, we have developed a PCT-based protocol integrating tissue lysis, protein extraction, and digestion (Fig. 1) in a seamless manner [2]. Briefly, a small piece of wet biopsy tissue (about 1 mg) was placed in a pressure-resistant PCT-MicroTube with protein lysis buffer, followed by pressure cycling to perform protein lysis and extraction. Proteolytic enzymes were then added to the solution, to enable protein digestion under cycling pressures. This permits analysis of minimal amount of sample as well as precise control (by barocycler) of the reaction solvent volume and incubation time for each step, effectively minimizing technical variation. The thus generated peptides were cleaned on reversed-phase columns, after which they were ready for sequential window acquisition of theoretical fragment ion (SWATH)-mass spectrometry (MS) analysis. We further evaluated the minimal amounts of samples required for the PCT-SWATH analysis and found that as few as 50,000 human cells and 0.2–0.5 mg of wet mouse and human tissues produced peptide sufficient for multiple optimal SWATH-MS analyses [3].

In order to further improve the performance of the PCT-SWATH method, we used a novel miniaturized device, namely, the PCT-MicroPestle, which works as a disposable

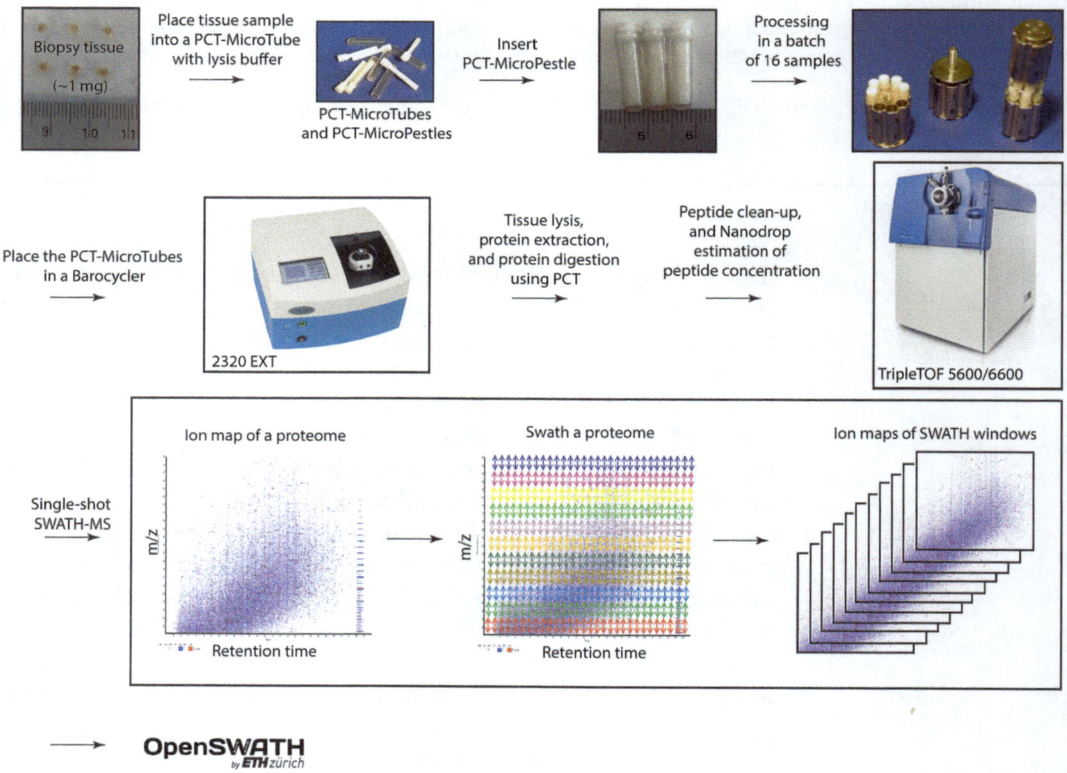

Fig. 1 Workflow for PCT-SWATH. Biopsy tissue samples are placed in PCT-MicroTubes, lysed with PCT-MicroPestle and lysis buffer under cycling pressure in a batch of 16 samples using a barocycler. Peptides are cleaned, quantified, and analyzed using single-shot SWATH-MS. Data are analyzed using OpenSWATH

mechanical tissue homogenizer inside the PCT-MicroTube sample container (Fig. 1) [4]. Under repeated pulses of high hydrostatic pressure, the PCT-MicroTube walls collapse around the essentially less compressible Teflon™ insert, functioning as a micro-homogenizer. Pressure cycling also acts directly on the sample, helping to disrupt membranes and solubilize proteins. This contributes to more efficient tissue lysis and extraction. Results showed that PCT-MicroPestle yielded 20–40% more MS-ready peptide mass from various tissues with better reproducibility compared to PCT processing without the PCT-MicroPestle [4].

2 Materials

All chemical reagents are of analytical grade. All solvents used for HPLC-MS are of HPLC grade. Prepare and store all reagents with ultrapure water at room temperature unless indicated otherwise.

2.1 OCT (Optimal Cutting Temperature) Compound Washing

1. Milli-Q water.
2. 70% ethanol in Milli-Q water: Mix 70 ml of ethanol with 30 ml of Milli-Q water.
3. 85% ethanol in Milli-Q water: Mix 85 ml of ethanol with 15 ml of Milli-Q water.

2.2 Tissue Cutting and Weighing

1. A hard plate of 20 cm × 20 cm size (glass/ceramic).
2. Small SDS PAGE cutting blades (Sigma).
3. 70% ethanol in Milli-Q water: Mix 70 ml of ethanol with 30 ml of Milli-Q water.
4. Forceps for clinical lab use.
5. Analytic balance (METTLER TOLEDO, XS205).

2.3 PCT-MicroPestle-Assisted Tissue Lysis

1. Barocycler 2320-EXT, PCT-MicroTubes, PCT-MicroPestle, and PCT-MicroCaps of 50, 100, and 150 μl of size, respectively. All from Pressure BioSciences, Inc., South Easton, MA.
2. Lysis buffer: 8 M urea in 100 mM ammonium bicarbonate (NH_4HCO_3) (pH 8).
3. Buffer stock: 100 mM NH_4HCO_3, pH 8, freshly prepare.
4. Bicinchoninic acid kit (BCA) (Sigma-Aldrich, St. Louis, MO).

2.4 Protein Reduction and Alkylation

1. Reducing reagent stock: 100 mM Tris-2-carboxyethyl-phosphine (TCEP) in 100 mM NH_4HCO_3, freshly prepare.
2. Alkylation reagent stock: 800 mM iodoacetamide (IAA) in 100 mM NH_4HCO_3, freshly prepare, and protect from light by wrapping with aluminum foil.

2.5 PCT-Assisted Digestion

1. Lys-C (mass spectrometry grade, Wako, Richmond, VA, USA).
2. Trypsin (sequencing grade modified, Promega, Madison, WI, USA).
3. 20% trifluoroacetic acid (TFA).

Lys-C and trypsin were stored in −80 °C before use.

2.6 Peptides Cleaning Up and Concentration Measurement

1. MicroSpin Column C18 Silica (5–200 μl loading, 5–60 μg capacity, The Nest Group Inc.).
2. Speedvac vacuum concentrator.
3. Buffers used for C18 cleaning up: 100% CH_3OH; 80% ACN/0.1% TFA in water; 2% ACN/0.1% TFA in water.
4. Peptide dissolving buffer: 0.1% FA/2% ACN in HPLC water.
5. iRT peptide cocktail (Biognosis, Zurich, Switzerland).

2.7 LC-MS/MS Analysis

1. Eksigent 1D+ nano-liquid chromatography (LC) systems (Eksigent, Dublin, CA, USA).
2. TripleTOF 5600 or 6600 mass spectrometer.
3. Buffer A, 2% ACN and 0.1% FA in water, and buffer B, 98% ACN and 0.1% FA in water. Both buffer A and B are of HPLC grade.

3 Methods

Carry out all procedures at room temperature unless otherwise specified.

3.1 OCT Compound Washing (See Note 1)

1. Take out the frozen tissue samples from the storage container. Place each tissue block in a 1.5-ml Eppendorf tube.
2. Add 1 ml of 70% ethanol into each tube, and wash the tissues by shaking at 800 rpm for 30 s. Discard the supernatant.
3. Add 1 ml of 100% water into each tube, and wash the tissues by shaking at 800 rpm for 30 s. Discard the supernatant.
4. Add 1 ml of 70% ethanol into each tube, and wash the tissues by shaking at 800 rpm for 5 min. Discard the supernatant. Repeat this step one more time.
5. Add 1 ml of 85% ethanol into each tube, and wash the tissues by shaking at 800 rpm for 5 min. Discard the supernatant. Repeat this step one more time.
6. Add 1 ml of 100% ethanol into each tube, and wash the tissues by shaking at 800 rpm for 5 min. Discard the supernatant. Repeat this step one more time.

3.2 Tissue Cutting and Weighing

1. Further cut each tissue block into small pieces (of about 1 mm^3 size) on a hard plate (glass/ceramic) in a biosafety cabinet (BSC) (see Note 2).
2. Take out enough fresh Eppendorf tubes, and label them according to the number and type of tissue samples. Weigh each empty tube and record the value.
3. Transfer each well-cut tissue piece into the corresponding tube by simply using a small pipette tip (200 μl size) or a pair of forceps. Close the cap of the tube. Weigh again and record the value. The net weight of each tissue piece can be calculated from the two readings (see Note 3).
4. Put the tissue samples at 4 °C after weighing.

3.3 PCT-MicroPestle-Assisted Tissue Lysis

1. Label the PCT-MicroTubes appropriately. Add 30 μl (V1) of lysis buffer into each PCT-MicroTube. Transfer the tissue pieces from the Eppendorf tubes into PCT-MicroTubes by using a small pipette tip (200 μl size) or a pair of forceps. Make sure the piece of tissue is soaked in the lysis buffer (see Note 4).
2. Close each tube with PCT-MicroPestle cap according to manufacturer's protocol. Sonicate the closed PCT-MicroTubes with tissue samples in a water bath for three times, each time 10 s, to fragment DNAs.
3. Go to PCT-assisted lysis step (see manufacturer's protocol) with 60 pressure cycles, each consisting of 25 s of 45,000 p.s.i. high pressure plus 10 s of atmospheric pressure, at 25 °C in barocycler (see Note 5).
4. Randomly check the protein concentration of several samples using BCA assay referring to the manufacturer's protocol (see Note 6).

3.4 Protein Reduction and Alkylation

1. Take out the tissue samples and set the barocycler to 33° C to preheat for subsequent digestion steps. Put each PCT-MicroTube into a 1.5 ml of Eppendorf tube and briefly spin down at 1000 rpm for 1 min.
2. Open the PCT-MicroPestle, and make sure there is no tissue attaching to the pestle. Discard the pestle.
3. Add 5 μl of 100 mM TCEP stock and 2.5 μl of 800 mM IAA stock into each PCT-MicroTube containing the tissue lysate. Top up with 100 mM NH_4HCO_3 to a final volume of 50 μl (V2). The final concentrations of TCEP and IAA are 10 and 40 mM, respectively.
4. Close the tubes with a new set of PCT-MicroCaps of 50 μl size. Incubate the solution in darkness with gentle shaking (600–800 rpm) at room temperature for 30 min (see Note 7).

3.5 PCT-Assisted Digestion

1. Simply spin down the PCT-MicroTubes at 1000 rpm for 1 min. Open the caps.
2. Add appropriate amount (about 2–3 μg) of lysyl endoproteinase Lys-C (enzyme to substrate ratio 1:40) to the protein solutions based on the BCA reading.
3. Further dilute the tissue lysates with 100 mM NH_4HCO_3 to a volume of 100 μl (V3). The urea concentration in this step is diluted to 2.4 M. Close the tubes with a new set of PCT-MicroCaps of 100 μl size.
4. Go to the PCT-assisted Lys-C digestion step. Perform the digestion in the barocycler at 33 °C using 45 cycles, each consisting of 50 s at 20,000–45,000 p.s.i. high pressure and 10 s at atmospheric pressure.
5. Take out the samples from the barocycler. Simply spin down the tubes at 1000 rpm for 1 min. Open the caps.
6. Add appropriate amount (about 2–3 μg) of trypsin (enzyme to substrate ratio 1:50) to the protein solutions based on the BCA reading.
7. Further dilute the solution with 100 mM NH_4HCO_3 to a final volume of 150 μl (V4). The urea concentration in this step is diluted to 1.6 M. Close the tubes with a new set of 150 μl PCT-MicroCaps (see Note 8).
8. Go to the PCT-assisted trypsin digestion step. Perform the digestion in the barocycler at 33 °C using 90 cycles, each consisting of 50 s at 20,000 p.s.i. high pressure and 10 s at atmospheric pressure (see Note 9).
9. Take out the samples from the barocycler. Simply spin down the tubes and open the caps. Open the caps and add 8 μl of 20% trifluoroacetic acid to each tube at a final concentration of around 1% to stop enzymatic digestion (final pH 2–3).

3.6 Peptides Cleaning Up and Concentration Measurement

1. Clean the peptides with the C18 microspin columns according to the manufacturer's protocol. Collect the peptide elution in a new set of 1.5 ml Eppendorf tubes and put them into a Speedvac concentrator. Dry the peptides under vacuum (see Note 10).
2. Redissolve the peptides in 50 μl of 0.1% formic acid (FA) and 2% acetonitrile (ACN) in HPLC water. Sonicate the peptide solution in an icy water bath for 5 min to facilitate peptide solubilization.
3. Measure the peptide concentration with NanoDrop 1000 spectrophotometer at 280 nm (1 Ab = 1 mg ml^{-1}) according to the manufacturer's protocol.
4. Further adjust peptide concentration to 0.3 μg $μl^{-1}$ using 0.1% FA and 2% ACN in HPLC water. Spike 1μl of iRT peptide cocktail (Biognosis, Zurich, Switzerland) into every 9 μl of peptide sample before LC-MS analysis.

3.7 LC-MS/MS Analysis

The LC-MS/MS analysis is done in TripleTOF 5600 or 6600 mass spectrometry in SWATH mode (see Note 11).
All buffers for LC-MS/MS analysis are of HPLC grade.

1. Inject 2–4 μl (0.6–1.2 μg) of the above prepared peptide (spiked with iRT peptide) into an Eksigent 1D+ nano-liquid chromatography (LC) systems (Eksigent, Dublin, CA, USA) for SWATH-MS in technical duplicate.
2. Separate peptide mixtures online by a HPLC using a nano-bored C18 column (75 μm i.d. × 20 cm, 3 μm 200 Å Magic C18 AQ resin).
3. Use buffer A 2% ACN and 0.1% FA in water and buffer B 98% ACN and 0.1% FA in water for HPLC gradient. Separate the peptides by using a linear gradient of 2–35% buffer B over 60 min at a flow rate of 300 nl min^{-1}.
4. Acquire SWATH data in a 5600 TripleTOF mass spectrometer (Sciex, Concord, Canada). Isolation window is 25 Da (containing 1 Da for the window overlap) with an ion accumulation time of 100 ms per SWATH scan. There are a set of 32 overlapping SWATH windows to cover the precursor mass range of 400–1200 Da. The collision energy for each window is set on the basis of the collision energy for a 2+ ion centered in the respective window with a collision energy spread of 15 eV. Accumulation time for each fragment ion scan and for the survey scans acquired at the beginning of each cycle is 100 ms, and the total cycle time is 3.3 s. All ionized peptide precursors in this mass range are fragmented, generating comprehensive fragment ion maps (*See* **Note** 12).

3.8 SWATH Data Analysis (See Note 13)

1. Build the appropriate libraries for the study [2, 3].
2. Convert the SWATH wiff files into mzXML files using ProteoWizard msconvert [5].
3. Convert the mzML file using OpenMS tool FileConverter [6].
4. Perform OpenSWATH with input files including the mzXML file, the TraML library file, and the TraML file for iRT peptides [7].
5. Identify the assays with a false discovery rate below 0.1%. Using the algorithm TRansition of Identification Confidence (TRIC) to align high confidence peptide features from different samples [8]. The parameters for the feature_alignment.py are as follows: max_rt_diff = 30, method = global_best_overall, nr_high_conf_exp = 2, target_fdr = 0.001, use_score_filter = 1.
6. Perform label-free protein-level quantification using the function ProteinInference from the R package aLFQ [9] an output from OpenSWATH. The parameters that differ from the default for ProteinInference are peptide_method = "top,"

peptide_topx = 3, transition_topx = 5, consensus_ proteins = FALSE, consensus_peptides = FALSE, and consensus_transitions = FALSE.

4 Notes

1. Tissues embedded in OCT compounds must be washed to eliminate OCT before PCT-SWATH analysis.
2. To maintain a cold environment and to prevent protein degradation, the plate should be pre-chilled in a freezer and then placed firmly on ice during cutting.
3. This step is essential when one establishes the method for an unknown tissue type to estimate the amount of the proteolytic enzyme such as trypsin and to calculate peptide yield (the peptide yield for prostate cancer tissue is around 60 μg peptide per mg tissue on average based on our experience). After the workflow is set up, one can roughly estimate the tissue weight based on the size of the tissue.
4. The small tissue piece that may be sticky to the wall of the PCT-MicroTube depends on the tissue type, which can be solved by pipetting up and down with the lysis buffer for one or two times. Don't pipette too much because bubbles will be generated.
5. Make sure the barocycler is working properly all the way from lysis to digestion according to manufacturer's protocol.
6. Proteolytic enzyme amounts will be estimated based on the result of protein concentrations.
7. 8 M urea lysis buffer could be prepared in a big stock and aliquoted, stored in −20 °C for usage; TCEP, IAA, and ammonium bicarbonate should be prepared freshly.
8. The maximum volume of the PCT-MicroTube is 150 μl. There are four volume values mentioned in this protocol, namely, V1, V2, V3, and V4, for each relevant step. This is decided by the size of the PCT-Microtube/MicroPestle themselves and the concentration of urea. The PCT-MicroPestle allows for a capacity of 50 μl, which included the sample, the lysis buffer, and the volume of air necessary for optimal tube compression and sample homogenization; thus V1 is 30 μl at this step, and the working concentration of urea during the lysis step is 8 M. Increasing lysis buffer volume will result in less tube compression and will reduce the effectiveness of the PCT-MicroPestle homogenization. Later the reaction volume increases at each step by adding reagents as well as topping up with 100 mM NH_4HCO_3 buffer to reduce urea concentration for enzymatic digestion, as demonstrated in the protocol, is 2.4 M for Lys-C and 1.6 M for trypsin.

9. Trypsin/Lys-C Mix (Mass Spec Grade from Promega) could be used in this step to replace the sequential usage of Lys-C and trypsin, to reduce the enzymatic reaction time to 60 min according to unpublished results.
10. Peptides can be stored in −80 °C for several months before SWATH-MS analysis.
11. The mass spectrometry used here is TripleTOF 5600. One can change the instrument settings based on the type of mass spectrometry.
12. SWATH windows can also be customized to construct various windows scheme for specific tissue type to achieve higher degree of sensitivity. In that case, ion accumulation time for each SWATH cycle can be adjusted down to 30 ms so that the entire cycling time is about 2–3 s.
13. The MS data analysis method demonstrated here is specific for SWATH data analysis. Other methods can be chosen accordingly to different datasets generated by different types of mass spectrometers. For specific commands of the latest version of the above mentioned software tools, please refer to the website http://openswath.org/.

References

1. Powell BS et al (2012) Pressure cycling technology in systems biology. Methods Mol Biol 881:27–62
2. Guo TN et al (2015) Rapid mass spectrometric conversion of tissue biopsy samples into permanent quantitative digital proteome maps. Nat Med 21(4):407–413
3. Shao S et al (2015) Minimal sample requirement for highly multiplexed protein quantification in cell lines and tissues by PCT-SWATH mass spectrometry. Proteomics 15(21):3711–3721
4. Shao S et al (2016) Reproducible tissue homogenization and protein extraction for quantitative proteomics using MicroPestle-assisted pressure-cycling technology. J Proteome Res 15 (6):1821–1829
5. Kessner D et al (2008) ProteoWizard: open source software for rapid proteomics tools development. Bioinformatics 24(21):2534–2536
6. Sturm M et al (2008) OpenMS—an open-source software framework for mass spectrometry. BMC Bioinformatics 9:163
7. Rost HL et al (2014) OpenSWATH enables automated, targeted analysis of data-independent acquisition MS data. Nat Biotechnol 32(3):219–223
8. Rost HL et al (2016) TRIC: an automated alignment strategy for reproducible protein quantification in targeted proteomics. Nat Methods 13 (9):777–783
9. Rosenberger G et al (2014) aLFQ: an R-package for estimating absolute protein quantities from label-free LC-MS/MS proteomics data. Bioinformatics 30(17):2511–2513

Methods in Molecular Biology (2018) 1788: 289–296
DOI 10.1007/7651_2017_82

Published online: 05 October 2017

Straightforward Protocol for Gel-Free Proteomic Analysis of Adipose Tissue

Yvonne Pasing, Armin Schniers, and Terkel Hansen

Abstract

After conducting systematic and quantitative comparisons of different sample preparation techniques regarding their capability to efficiently and reproducibly recover proteins from biopsies, we present here our superior protocol for extracting proteins from low amounts of adipose tissue. Adipose tissue as a matrix in bottom-up proteomics is challenging due to the extremely high lipid content.

The lysis buffer utilized contains the detergent sodium deoxycholate, which does not impair the activity of trypsin and therefore enables direct digestion without detergent removal steps. The resulting workflow is time saving, cost efficient, easy to perform, and it can also be applied to other hydrophobic samples.

Keywords Adipose tissue proteome, Detergents, Hydrophobic samples, Sample preparation, Sodium deoxycholate, Sodium laurate

1 Introduction

Adipose tissue (AT) is not just a fat depot but also a metabolically highly active endocrine organ with a central role in energy balance and glucose homeostasis [1]. An excess of AT has been connected with many pathophysiological processes as for example insulin resistance, Type II diabetes, cardiovascular and fatty liver disease, and pulmonary and muscular pathologies [2]. Analyzing the proteome of adipose tissue is challenging, as the high lipid content impedes the protein extraction, solubilization, and chromatographic separation of peptides [3]. AT consists of a variety of different cell types and has therefore a high dynamic range of protein abundances. Buffers and workflows for proteomic analysis of highly hydrophobic samples like AT have to be chosen and tested carefully. The majority of the published studies apply gel-based approaches for proteomic analyses of adipose tissue. For gel-free proteomics, trypsin friendly surfactants are interesting alternatives to classical approaches with detergents like sodium dodecyl sulfate and removal methods like protein precipitation or filter-aided sample preparation, which need to be done before enzymatic digestion. Our results indicate advantages of using workflows without detergent removal steps prior to digestion

[4, 5]. These protease compatible surfactants offer the opportunity of direct removal after digestion due to cleavage under acidic conditions, as the resulting cleaving products are either precipitating or being LC-MS compatible. Avoiding multistep detergent removal protocols is time saving and reduces variances and inconsistencies due to sample handling. Some protease compatible surfactants are marketed for proteomics applications, with the benefit of detailed protocols for their use but the drawback of high costs for the application [6]. Sodium deoxycholate (SDC) [7–9] and sodium laurate [6, 10] offer low-cost alternatives. Both have similar properties and are available as low-cost bulk chemicals. In a comparative study testing nine different sample preparation strategies for the gel-free proteome analysis of AT, we identified the most efficient and reproducible detergent-based workflow for protein quantification [4]. Thereby, we established a straightforward, easy to use, and cheap workflow, which we later applied for the analysis of subcutaneous adipose tissue biopsies from an intervention study. We found that the low-cost alternatives SDC and SL are outperforming their expensive counterparts used specifically in proteomics. The presented protocol is based on the use of SDC, which was superior in our study [4]. This protocol is not only suitable for adipose tissue; we also encourage the use for all hydrophobic samples like membrane rich organelles (e.g., mitochondria, endoplasmic reticulum, and Golgi apparatus) and other hydrophobic tissues (e.g., brain biopsies). The method has, with some modifications, also been shown to be highly efficient and reproductive with regard to other troublesome biopsy samples [5].

2 Material

Prepare all solutions using ultrapure water and analytical grade reagents.

2.1 Biopsy Handling and Storage

Phosphate buffered saline (PBS), cryogenic sample vials, liquid nitrogen bucket or tank, tweezers, and scalpel.

2.2 Cell Lysis

Lysis buffer: 1% Sodium deoxycholate (SDC) in 50 mM ammonium hydrogen carbonate (ABC) buffer (**Notes 1–4**).

MagNA Lyser Instrument and MagNA Lyser Green Beads (Roche Diagnostics AG, Rotkreuz, Switzerland), tweezers and scalpel.

2.3 Tryptic Digest

Pierce™ BCA Protein Assay Kit (Thermo Fisher Scientific) or comparable.

0.5 M solution dithiothreitol (DTT), 1 M solution iodoacetamide (IAA) and 50 mM solution $CaCl_2$ (**Note 5**).

50 mM ammonium hydrogen carbonate (ABC) buffer (**Note 6**).

Trypsin.

Trifluoroacetic acid (TFA).

C18 Bond Elut OMIX pipette tips (Agilent, Santa Clara, USA) or comparable.

2.4 LC-MS/MS

High-resolution mass spectrometer with fragmentation capabilities coupled to a nano-liquid chromatography system, e.g., Q Exactive coupled to an EASY-nLC 1000 (Thermo Fisher Scientific, Bremen, Germany).

3 Methods

3.1 Biopsy Handling and Storage

1. Wash the obtained tissue directly after biopsy taking three times with phosphate buffered saline to remove contaminations like peripheral blood (**Note 7**).
2. Freeze biopsies immediately after washing in liquid nitrogen and store them at −70 °C until further sample preparation.

3.2 Cell Lysis

1. Prepare all necessary buffers and solutions. Keep the lysis buffer in the fridge until use.
2. Fill 50 mg of frozen biopsy and 500 μL of cooled lysis buffer in a MagNA Lyser Green Bead tube (Roche Diagnostics AG, Rotkreuz, Switzerland) and keep the tube on ice until cell lysis is completed (**Note 8**).
3. Perform cell lysis in the MagNA Lyser Instrument (Roche Diagnostics AG, Rotkreuz, Switzerland) in two cycles of 60 s and 6500 rpm (**Note 9**).
4. Let the tubes rest on ice for 30 min and centrifuge 5 min at 16,000 × *g*, 4 °C. (**Note 10**).
5. Harvest approximately 250 μL of cell lysate with a syringe from the aqueous layer underneath the lipid layer formed. Keep the tip of the needle above the beads.
6. Freeze the cell lysates at −70 °C until further sample preparation or continue immediately.

3.3 Tryptic Digest

1. Dilute 25 μL of cell lysates with 25 μL ABC buffer to a final concentration of 0.5% SDC.
2. Perform a BCA assay (Pierce™ BCA Protein Assay Kit, Thermo Fisher Scientific) according to the manufacturer's protocol in order to estimate the protein concentration in the cell lysate. Use the same buffer (50 mM ABC buffer with 0.5% SDC) for the standard curve (**Notes 11** and **12**).

3. Perform reduction of disulfide bridges: Add 1 μL DTT solution (final concentration 10 mM DTT) and incubate at 56 °C for 30 min.
4. Perform alkylation: Add 1 μL IAA solution (final concentration of 20 mM IAA) and incubate for 30 min in the dark.
5. Add 1 μL DTT solution to remove excess IAA.
6. Add 1 μL $CaCl_2$ solution to a final concentration of 1 mM calcium chloride.
7. Perform protein digest under gentle agitation for 14 h at 37 °C with trypsin in an enzyme-to-protein ratio of 1:20 (w/w) (**Note 13**).

3.4 Removal of SDC

1. Acidify protein digests with 10% TFA to a final concentration of 1% TFA and pH $\leq$ 2. Check the pH and add more TFA if necessary to obtain pH $\leq$ 2.
2. Incubate for 5 min at RT and centrifuge at 16,000 × *g* for 15 min.
3. Carefully collect the supernatants for desalting (**Notes 14** and **15**).
4. Desalt the samples, e.g., with C18 Bond Elut OMIX pipette tips (Agilent, Santa Clara, USA) according to the manufacturer's protocol.
5. Evaporate the samples until dryness in a vacuum concentrator and redissolve in 20 μL 0.1% TFA in water for the following LC-MS/MS analysis.

3.5 LC-MS/MS

Conduct nano-LC-MS/MS analysis, for example on a Q Exactive mass spectrometer coupled to an EASY-nLC 1000 system (both from Thermo Fisher Scientific, Bremen, Germany).

Inject 1.5 μg peptides per sample (**Note 16**) and preconcentrate them on an RP trapping column (e.g., Acclaim PepMap® 100, ID 75 μm, length 2 cm, nanoViper, C18 particle size 3 μm, pore size 100 Å, Thermo Fisher Scientific) with 0.1% formic acid (FA) at a flow rate of 20 μL/min. Separate the peptides on an RP main column (for example EASY-Spray, PepMAP® RSLC, ID 75 μm, length 50 cm, C18 particle size 2 μm, pore size 100 Å, Thermo Fisher Scientific) using a binary gradient (solvent A: 0.1% FA, solvent B: acetonitrile, 0.1% FA) at a flow rate of 200 nL/min with a column temperature of 60 °C. Elute the peptides with a nonlinear gradient of 0% ACN to 5% ACN over 19 min, further to 30% at 180 min and to 100% at 200 min. Subsequently, regenerate the column with 100% ACN for additional 10 min. (**Note 17**).

For MS analysis the Q Exactive mass spectrometer should be run in positive mode, the global settings are as follows:

Chromatographic peak width 15 s, default charge state 2, full MS survey scans from 400 to 2000 m/z, resolution of 70,000, AGC target value 3e6, maximum injection time 100 ms for MS scans. The 15 most intense peaks should be subjected to MS/MS with a resolution of 17,500 and the following setting: dynamic exclusion 10 s, underfill ratio 1%, charge states +2, +3, and +4, exclude isotopes, normalized collision energy 28, isolation width 2 m/z, AGC target value 5e4, and maximum injection time 50 ms (**Note 18**).

3.6 Data Evaluation

Process the raw data for example using the MaxQuant software with label-free quantification (LFQ). Download the newest version of the UniProt human database and search against it with the built-in search engine (Andromeda) using the following MaxQuant parameters:

Enzyme: Trypsin/P (specific), fixed modifications: Carbamidomethyl (C), variable modifications: Oxidation (M) and Acetyl (Protein N-term), max. 2 missed cleavages, decoy mode: revert, special AAs: KR, include contaminants: TRUE, MS/MS tol. (FTMS): 20 ppm, Top MS/MS peaks per 100 Da. (FTMS): 12, MS/MS deisotoping (FTMS): TRUE, PSM FDR: 0.01, Protein FDR: 0.01, Site FDR: 0.01, Use Normalized Ratios For Occupancy: TRUE, Min. peptide length: 7, Min. score for unmodified peptides: 0, Min. score for modified peptides: 40, Min. delta score for unmodified peptides: 0, Min. delta score for modified peptides: 6, Min. unique peptides: 0, Min. razor peptides: 1, Min. peptides: 1, Use only unmodified peptides and: TRUE, Modifications included in protein quantification: Oxidation (M); Acetyl (Protein N-term), Peptides used for protein quantification: Razor, Discard unmodified counterpart peptides: TRUE, Min. ratio count: 2, Use delta score: FALSE, iBAQ: FALSE, iBAQ log fit: FALSE, Match between runs: TRUE, Matching time window [min]: 0.7, Alignment time window [min]: 20, Find dependent peptides: FALSE, Site tables: Oxidation (M)Sites.txt, Decoy mode: revert, Special AAs: KR, Include contaminants: TRUE, RT shift: FALSE, Advanced ratios: TRUE, AIF correlation: 0, First pass AIF correlation: 0, AIF topx: 0, AIF min mass: 0, AIF SIL weight: 0, AIF ISO weight: 0, AIF iterative: FALSE, AIF threshold FDR: 0.

The results can be visualized with Perseus. Delete hits only identified by site, reverse hits, potential contaminants and hemoglobin. Perform log2 transformations and treat missing values as NaN.

4 Notes

1. Application of sodium laurate (SL) instead of SDC in the lysis buffer achieved nearly the same good results in our study, with a high average protein and peptide identification rate.

However, it gave higher standard deviations, higher percentages of missed cleavages, lower sequence coverages, and a lower reproducibility when compared to SDC, but it extracted higher proteins amounts than SDC [4]. Therefore, it is possible to use SL instead of SDC for the presented protocol and we recommend testing both surfactants when other samples than AT should be analyzed.

2. SDC has previously been tested for membrane proteomics and gave the best results when applied in a concentration of 5% in the lysis buffer, and diluting it to 1% SDC for digestion [9]. We could not confirm this for AT after testing 1%, 2.5%, and 5% SDC in the lysis buffer, but recommend testing different concentration in case the protocol is applied to another sample type.
3. If isobaric labeling (e.g., TMT) is desired, 100 mM triethylammonium bicarbonate (TEAB) should be used instead of ABC buffer [5]. Fractionation of labeled peptides with strong cation exchange or high-pH reverse phase is highly recommended. The settings for LC-MS/MS and data evaluation need to be adapted to the applied labeling technique.
4. If necessary, also high concentrations of urea can be used combined with higher concentrations of SDC. Five percent SDC can be dissolved in 8 M urea in 100 mM TEAB. The samples need to be diluted to ≤1 M urea before tryptic digestion and should contain 1 mM calcium chloride during digestion (*See* **Note 13**).
5. For 0.5 M DTT solution, dissolve 77.13 mg dithiothreitol in 1 mL ABC buffer. The solution can be stored in aliquots at −20 °C until use. For 1 M IAA solution, dissolve 184.96 mg iodoacetamide in 1 mL ABC buffer. Prepare the solution fresh and keep it in the dark. For 50 mM $CaCl_2$ solution, dissolve 110.98 mg calcium chloride in 20 mL water. The solution can be stored at 4 °C until use.
6. For 50 mM ABC buffer dissolve 198 mg ammonium hydrogen carbonate in 50 mL water. The buffer should not be older than four hours before use. Addition of protease inhibitors is optional
7. We recommend to place the biopsy on a tissue sieve or comparable for washing with PBS buffer.
8. Process the biopsies under cooled conditions and as fast as possible. It is recommended to cut the still frozen biopsies with a cooled scalpel for partitioning.
9. If a MagNA Lyser instrument is not available, the use of a bead mill is an alternative, which needs to be tested.

10. During the resting time, a lipid layer will form on top of the cell lysate. Wait until this process is finished before harvesting the cell lysate underneath with a syringe without destroying the lipid layer. Avoid contaminating the cell lysate with the lipids.
11. Make sure the buffer composition is exactly the same in all the standards as in the samples, especially if some of the buffer components inflict with the BCA reagents, e.g., TEAB. We recommend to dilute the TEAB concentration to 5 mM for all standards as well as samples.
12. The protein concentration for the tryptic digest solution should be around 1 mg/mL and might need to be adjusted before further sample preparation is performed. In this case, the volumes and amounts needed in the following steps need to be adjusted.
13. If necessary, a predigest with Lys-C can be conducted. If so, dilute the sample to maximum 6 M urea and 3.75% SDC. Lys-C digestion is then performed for 8 h at 37 °C with 1:100 enzyme to protein ratio. Afterward, dilute the sample to 1 M urea for tryptic digestion (14 h, 1:20 enzyme to protein ratio). Maintain a calcium chloride concentration of 1 mM during all digestions.
14. SDC will precipitate under acidic condition and can be removed by subsequent centrifugation. The pellet might be invisible, and hence a very careful removal of the supernatant is necessary.
15. Beside acid precipitation, a phase-transfer removal of SDC or SL with ethyl acetate is possible. Even though phase-transfer removal was evaluated as superior in previous studies [8, 11], we observed a high loss of protein and a low reproducibility. We recommend precipitation of SDC or SL, as we could not detect any disadvantages related to this procedure.
16. Measure the peptide concentration for example with a NanoDrop 2000 (Thermo Fisher Scientific, Bremen, Germany) at a wavelength of 205 nm and adjust the concentration for injection into the LC-MS/MS system according to the measurements.
17. Wash gradients should be run between the samples for minimizing memory effects. Standard samples with a known composition should be injected between approximately every 10th sample in order to control peak shape and retention time.
18. Optimization of the parameters for LC-MS/MS analysis according to the available system is recommended in order to increase the proteome coverage. Also fractionation approaches like strong cation exchange chromatography can be applied, especially when TMT labeling with multiplexing is chosen as quantification approach.

Acknowledgment

We would like to thank Prof. Rolf Jorde, Sayda Colnoe, Allan Didriksen, and Jack-Ansgar Bruun. Our work was supported by a grant from the North Norway Regional Health Authorities.

References

1. Galic S, Oakhill JS, Steinberg GR (2010) Adipose tissue as an endocrine organ. Mol Cell Endocrinol 316(2):129–139
2. Matsuzawa Y (2006) Therapy insight: adipocytokines in metabolic syndrome and related cardiovascular disease. Nat Clin Pract Cardiovasc Med 3(1):35–42
3. Peral B, Camafeita E, Fernandez-Real JM et al (2009) Tackling the human adipose tissue proteome to gain insight into obesity and related pathologies. Expert Rev Proteomics 6 (4):353–361
4. Pasing Y, Colnoe S, Hansen T (2017) Proteomics of hydrophobic samples: fast, robust and low-cost workflows for clinical approaches. Proteomics 17(6)
5. Schniers A, Anderssen E, Fenton CG et al. The proteome of ulcerative colitis in colon biopsies from adults—optimized sample preparation and comparison with healthy controls. Proteomics Clin Appl. Epub ahead of print. Doi: https://doi.org/10.1002/prca.201700053
6. Lin Y, Huo L, Liu Z et al (2013) Sodium laurate, a novel protease- and mass spectrometry-compatible detergent for mass spectrometry-based membrane proteomics. PLoS One 8(3):e59779. https://doi.org/10.1371/journal.pone.0059779
7. Zhou J, Zhou T, Cao R et al (2006) Evaluation of the application of sodium deoxycholate to proteomic analysis of rat hippocampal plasma membrane. J Proteome Res 5(10):2547–2553
8. Leon IR, Schwammle V, Jensen ON et al (2013) Quantitative assessment of in-solution digestion efficiency identifies optimal protocols for unbiased protein analysis. Mol Cell Proteomics 12(10):2992–3005
9. Lin Y, Wang K, Liu Z et al (2015) Enhanced SDC-assisted digestion coupled with lipid chromatography-tandem mass spectrometry for shotgun analysis of membrane proteome. J Chromatogr B 1002:144–151
10. Fu SL, Li JL, Chen J et al (2015) Extraction and identification of membrane proteins from black widow spider eggs. Zool Res 36 (4):248–254
11. Masuda T, Tomita M, Ishihama Y (2008) Phase transfer surfactant-aided trypsin digestion for membrane proteome analysis. J Proteome Res 7(2):731–740

Methods in Molecular Biology (2018) 1788: 297–312
DOI 10.1007/7651_2017_114

Published online: 10 December 2017

MALDI Imaging Combined with Laser Microdissection-Based Microproteomics for Protein Identification: Application to Intratumor Heterogeneity Studies

Rémi Longuespée, Deborah Alberts, Dominique Baiwir, Gabriel Mazzucchelli, Nicolas Smargiasso, and Edwin De Pauw

Abstract

Matrix-assisted laser desorption ionization (MALDI) imaging is widely used for in situ proteomic mapping and finds multiple applications in pathology. However, low fragmentation yields in MALDI avoid an optimal identification of peptides from tissues. On the other hand, LMD-based microproteomic analyses allow for the identification of hundreds to thousands of proteins from small tissue regions. Herein, we present the combination of MALDI imaging and LMD-based microproteomic approaches for parallel identification. We illustrate the workflow with an application to intratumor heterogeneity studies.

Keywords Formalin-fixed and paraffin-embedded tissues, Histopathology, MALDI imaging, Intratumor heterogeneity, Protein identification, Laser microdissection, Microproteomics

1 Introduction

MALDI imaging is now recognized as a method of choice for applications in fundamental and clinical pathology [1]. It is now on the process to be used for routine diagnosis in pathological institutes. However, MALDI imaging suffers from inefficient fragmentation yields that limit its potential for peptide/protein identification. In parallel, other methods have been designed for the identification of thousands of proteins from small tissue regions. In particular, LMD-based microproteomics have been created for proteomic biomarker discovery from FFPE tissue regions containing less than 3000 cells [2]. Microproteomic dataset could be correlated to MALDI imaging ones for parallel protein identification [3, 4]. The combination of approaches is particularly adapted when the identification of compounds detected from small tissue regions by MALDI imaging is necessary. Laser microdissection also provides the unique advantage to allow one to collect tissue regions of any shape for further chemical processing. Identification of compounds with LMD-based microdissection method is then possible in any histological context. The combination of methods presented here is dedicated to formalin-fixed and paraffin-embedded (FFPE) tissues.

As MALDI imaging allows for molecular analyses in respect with microanatomical context, it is particularly adapted to study intratumor heterogeneity [5]. Segmentation algorithm allows for in situ classification of tissue subclusters on the basis of their molecular differences [5]. Hereafter, we illustrate the combination of MALDI imaging and LMD-based microproteomics for parallel identification from a single section, in the context of intratumor heterogeneity (ITH) studies [4].

This combination of methods also holds good promise for future biomarker discovery assays related to ITH studies.

2 Material

2.1 Material and Solutions

1. Indium tin oxide conductive glass slide (Bruker Daltonics, Bremen, Germany)
2. Citric acid (CA) 50 mM, CA 10 mM, pH 6.0
3. Tipp-Ex (Tipp-Ex, Frankfurt am Main, Germany)
4. Tissue scanner
5. α-Cyano-4-hydroxycinnamic acid (CHCA) solution 5 mg/mL
6. Spraying system. In this protocol, SunCollect sprayer (Sun-Chrom, Friedrichsdorf, Germany)
7. MALDI mass spectrometer. In this protocol: UltrafleXtreme TOF/TOF (Bruker Daltonics, Bremen, Germany)
8. Peptide calibration standard I (Bruker Daltonics, Bremen, Germany)
9. Methanol
10. Polyethylene naphthalate (PEN) membrane glass slides for laser microdissection (Leica Microsystems, Wetzlar, Germany)
11. Laser microdissection device: Leica LMD 7000 (Leica Microsystems, Wetzlar, Germany)
12. Milli-Q H_2O
13. NaOH 1 M
14. RapiGest surfactant (Waters, Milford, MA) 0.1%
15. NH_4HCO_3 (ammonium bicarbonate) 500, 100, 50, and 25 mM
16. Dithiothreitol (DTT) 500, 131, and 171.6 mM
17. Iodoacetamide (IAM) 500 and 194.67 mM
18. Aluminum foils
19. Hydrochloric acid (HCl) 10 mM
20. Trypsin Gold, Mass Spectrometry Grade (Promega, Fitchburg, WI) 1 and 0.5 μg/μL
21. Acetonitrile (ACN) 100%
22. Trifluoroacetic acid (TFA) 10%

23. Microtubes 0.6 mL
24. Kimwipes disposable (Kimberly-Clark, Dallas, TX)
25. Centrifuge
26. Thermoshaker
27. Antistatic gun Zerostat 3 (Sigma-Aldrich, St. Louis, MO)
28. Binocular microscope
29. Sonicator
30. Cleaning paper
31. Speed vacuum centrifuge
32. ZipTip (ZT) cartridge C18 2 μg, 5 μg (Millipore, Billerica, MA)
33. NH_4HCO_2 (ammonium hydroxide) 20 mM, pH 10
34. Formic acid 0.1% in water, 0.1% in acetonitrile
35. MassPREP digestion standard (MPDS) mixtures, mix 1 and 2 (Waters, Milford, MA)
36. Speed vacuum (Thermo Scientific, Waltham, MA)
37. Ultraperformance liquid chromatography (UPLC) nanoACQUITY 2D (Waters, Milford, MA)
38. Reverse phase (RP) XBridge BEH C18 5 μm column (300 μm × 50 mm) (Waters, Milford, MA)
39. Trap column Symmetry C_{18} 5 μm (180 μm × 20 mm) (Waters, Milford, MA)
40. Analytical column BEH C_{18} 1.7 μm (75 μm × 250 mm) (Waters, Milford, MA)
41. Orbitrap mass spectrometer Q Exactive or Q Exactive Plus (Thermo Fisher Scientific, Waltham, MA)

2.2 α-Cyano-4-hydroxycinnamic Acid Solution 5 mg/mL

1. Weight 5 mg of CHCA in a tube. Add 700 mL of acetonitrile and 300 mL of mQ water with 0.1% TFA.

2.3 Citric Acid Solutions

1. Citric acid stock solution (50 mM): weight 0.96 g of citric acid and dissolve in 99.94 mL Milli-Q water.
2. NaOH 1 M: weight 40 g of NaOH and dissolve in 1 L Milli-Q water.
3. Citric acid 10 mM, pH 6.0: transfer 20 mL of the stock solution (50 mM) to a 100 mL bottle and adjust to 50 mL with Milli-Q water. Adjust the pH to 6.0 by adding 1 M NaOH. Adjust the final volume to 100 mL with Milli-Q water.

2.4 NH_4HCO_3 (Ammonium Bicarbonate) Solutions

1. 500 mM. Weight 39.53 g of NH_4HCO_3 and dilute in 1 L Milli-Q water. Store at 4 °C in a glass bottle.
2. 100 mM. Dilute 5× the NH_4HCO_3 500 mM solution, NH_4HCO_3 500 mM/Milli-Q water 1:4. Store at 4 °C in a glass bottle.
3. 50 mM. Dilute 10× the NH_4HCO_3 500 mM solution, NH_4HCO_3 500 mM/Milli-Q water 1:9. Store at 4 °C in a glass bottle.
4. 25 mM. Dilute 20× the NH_4HCO_3 500 mM solution, NH_4HCO_3 500 mM/Milli-Q water 1:19. Store at 4 °C in a glass bottle.

2.5 DTT Solutions

1. 500 mM.

   Weight 0.7713 g of DTT. Solubilize with 10 mL of Milli-Q water. Aliquot in tubes with 100–500 μL and cover with aluminum foil. Freeze at −20 °C until use.
2. 131 mM in 50 mM NH_4HCO_3.

   Mix 10 μL of DTT 500 mM with 9.1 μL of Milli-Q water and 19.1 μL of 100 mM NH_4HCO_3 in a 0.6 mL microtube. Cover with an aluminum foil.
3. 171.6 mM in 50 mM NH_4HCO_3.
   Mix 10 μL of DTT 500 mM with 4.57 μL Milli-Q water and 14.57 μL of 100 mM NH_4HCO_3 in a 0.6 mL microtube. Cover with an aluminum foil.

2.6 IAM Solutions

1. 500 mM in water.

   Weight 1.85 g of IAM. Solubilize with 20 mL Milli-Q water. Aliquot tubes with 100–500 μL. Freeze at −20 °C until use.
2. 194.67 mM in 50 mM NH_4HCO_3.
   Mix 20 μL of IAM 500 mM with 5.68 μL Milli-Q water and 25.68 μL of 100 mM NH_4HCO_3 in a 0.6 mL microtube. Cover with an aluminum foil.

2.7 HCl Solutions

1. 1 M

   Commercial solutions of HCl are usually 12 M. Dilute 12 × the 12 M HCl solution with Milli-Q water to obtain a 1 M stock solution: HCl 12 M/Milli-Q water 1:11.
2. 10 mM
   Dilute 100× the 1 M HCl solution with Milli-Q water: HCl 1 M/Milli-Q water 1:99.

2.8 Trypsin Solutions

1. 1 μg/μL

   Reconstitute 100 μg of lyophilized trypsin in 100 μL of 10 mM HCl. Aliquot in tubes with 1, 2, or 3 μL depending on further use, and store at −20 °C until use.

2. 0.5 μg/μL
   Dilute 2× a trypsin stock solution (1 μg/μL) in 25 mM NH_4HCO_3. Depending on the number of samples to process, the amount of stock trypsin solution should be adapted.

2.9 RapiGest SF Surfactant 0.1%

1. Reconstitute one vial (1 mg) in 1 mL NH_4HCO_3 50 mM. Aliquot in tubes with 100 μL and store at −20 °C before use.

2.10 TFA 10%

1. Mix 10 mL of TFA with 90 mL of Milli-Q water. Store at 4 °C in a glass bottle.

2.11 ZipTip (ZT) Solutions

1. ZT Solution 1 (ACN + 0.1% TFA): mix 999 μL ACN with 1 μL TFA in a tube.
2. ZT Solution 2 (ACN/H_2O 1:1 + 0.1% TFA): mix 499.5 μL ACN with 499.5 μL Milli-Q H_2O and 1 μL TFA in a tube.
3. ZT Solution 3 (H_2O + 0.1% TFA): mix 999 μL H_2O with 1 μL TFA in a tube.
4. Larger volumes can also be prepared and stored in glass bottles at 4 °C.

2.12 20 mM NH_4HCO_2 (Ammonium Formate) Solution pH 10

1. 200 mM ammonium formate pH 10 stock solution: NH_4OH (ammonium hydroxide) commercial solutions are usually 25%. For 1 L of solution, use a 1 L graduated cylinder and place 850–900 mL of Milli-Q H_2O. Add 15.4 mL of NH_4OH 25%. Mix well. Add 1.62 mL HCO_2H (formic acid). Mix well. Adjust the pH to 10 with NH_4OH 25% or formic acid. Adjust the volume to 1 L with Milli-Q H_2O. Store in a glass bottle at 4 °C.
2. 20 mM ammonium formate pH 10: for 1 L of solution, mix 100 mL of 200 mM ammonium formate pH 10 stock solution with 900 mL of Milli-Q H_2O. Check the pH. Store in glass bottles at 4 °C.

2.13 0.1% Formic Acid in Water

Mix 1 mL of formic acid with 999 mL of Milli-Q H_2O.

2.14 0.1% Formic Acid in ACN

Mix 1 mL of formic acid with 999 mL of ACN.

3 Methods

3.1 MALDI Imaging

3.1.1 Tissue Sectioning

1. After surgery, immerse the tissue overnight in 10% formalin, dehydrate and embed in paraffin using well-known procedures widely used among institutes of pathology [6]. Cut 5 μm-thick sections from the FFPE block and deposit on one ITO glass slide on the conductive side.
2. Perform three serial 5 μm sections and deposit on LMD membrane slides.

3.1.2 Antigen Retrieval

1. Place the ITO glass slide with the section to analyze by MALDI imaging in a histological holder.
2. Fill a recipient that can contain glass histological holder with CA 10 mM pH 6.
3. Preheat at 98–99 °C (near boiling buffer) and immerse the histological holder with the glass slides for 60 min.
4. Remove the histological holder and place it in a recipient full of mQ water. Replace the water three times.
5. Remove the sections from the histological holder and let the sections dry.

3.1.3 Tissue Scanning

1. Apply teach marks using Tipp-Ex on corners around the tissue.
2. Perform high-resolution scan (1200–2400 dpi).

3.1.4 Trypsin Spraying and Incubation

1. Wash the system with the mixture ACN/mQ water 7:3.
2. Spray 200 μL of 25 mM NH_4HCO_3.
3. Use an aliquot of 10 μL of trypsin 1 μg/μL and dilute in 200 μL of 25 mM NH_4HCO_3.
4. Spray the volume of trypsin solution over the tissue section on a surface equivalent to the half of an ITO section at a flow rate of 10 μL/min.
5. Place the section in a humidity chamber preheated at 37°C and place the humidity chamber in an incubator at 37 °C for 4 h (see **Note 1**).

3.1.5 Matrix Spraying

1. Fill the spaying system with CHCA 5 mg/mL. Spray the surface of the tissue at a flow rate of 20 μL/min for 15 passages. Remove the section and deposit 0.5 μL of peptide calibration standard on an area covered with matrix.
2. Wash the system with the mixture ACN/mQ water 7:3.

3.1.6 Data Acquisition

1. Hold the glass slide to the tissue target and insert it in the MALDI instrument.
2. In FlexControl, select the method recommended by the constructor for peptide mass range in the positive mode (see **Note 2**).
3. Calibrate the method with the peptide calibration standard and save it.
4. In FlexImaging, create a new imaging run, select the calibrated method, and select the scan of the tissue section and perform teaching. Adjust the raster step to the tissue of interest as well as the number of laser shots per position.
5. Define an area covering the tissue section and perform the run.

3.1.7 Matrix Removal and Histological Staining

1. Remove CHCA from the tissue section analyzed by MALDI imaging by immersion in two successive baths of methanol. Let the section dry for 2 min.
2. Perform hematoxylin and eosin (HE) staining from the section as described before [2] and perform a high-resolution scan of the section.

3.1.8 Image Segmentation and Determination of Heterogeneous Compounds (See **Note 3**)

1. Load the images in the last version of the software SCiLS Lab (SCiLS, Bremen, Germany) and perform baseline removal using the convolution algorithm.
2. Normalize the spectra using total ion count (TIC) algorithm.
3. Perform peak peaking from the area of interest.
4. Perform segmentation. Analyze all spectra using bisecting k-means algorithm. Select the peaks picked as a mass range, 0.2 Da as interval width value and TIC as a normalization method. The process will result in a segmented image as shown in Fig. 1 where the molecular subclusters are highlighted in different colors.
5. In SCiLS Lab, create regions from the heterogeneous clusters and increase the transparency. Load the HE scan after MALDI and superimpose with the MALDI segmented dataset. This will result in the correlation of histological morphologies with molecular clusters as shown in Fig. 2a.
6. Create from the MALDI imaging dataset a small region in the heterogeneous clusters. Perform a principal component analysis (PCA) to compare the molecular content between the two regions. Use the peak picked as m/z interval and the following parameters: TIC normalization, interval width value 0.2 Da, five PC for PCA, and analysis from individual spectra.
7. Select m/z of interest at the edges on component 1. These m/z values are the most discriminant between the regions the spectra used for PCA (Fig. 2c).
8. Verify the intensity trends of the peaks of interest between the selected clusters by MALDI imaging and from the average spectra of the clusters (Fig. 2d and 2e).

At this step, enough information is collected from MALDI imaging to proceed to LMD-based microproteomic procedure.

3.2 MALDI Imaging-Guided LMD-Based Microproteomics (See Note 4)

3.2.1 Laser Microdissection

1. From the HE-stained section after MALDI imaging, perform cell counting using digital pathology software such as QuPath (https://qupath.github.io/) or Cytomine (http://www.cytomine.be/) [7]. Cytomine allows sharing online annotations and comments of the tissues between different institutions.
2. Localize the regions of interest (heterogeneous regions) by comparing the MALDI segmented images superimposed with

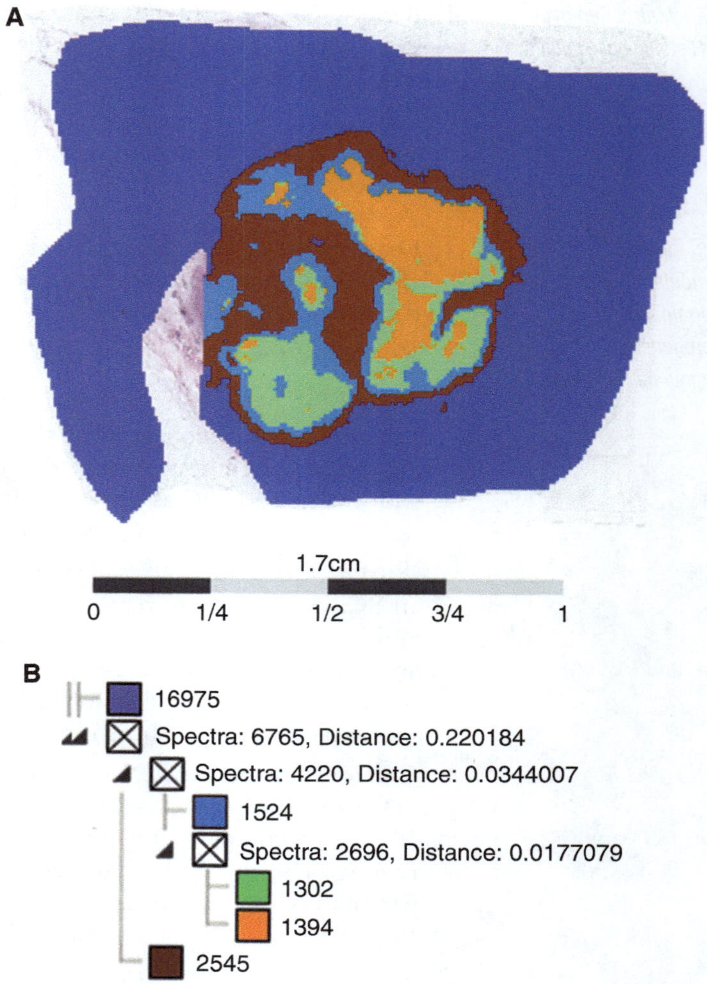

Fig. 1 MALDI imaging segmented dataset. (**a**) Molecular cluster from the breast cancer tissue section. (**b**) Hierarchy of the clusters. The *dark blue* cluster corresponds to the areas around the tumor. The *brown* cluster corresponds to the tumor. A subcluster in *light blue* is composed of two subclusters in *orange* and *green*, selected figure 3, inset **a**. Adapted with permission from [4]

the HE scan in SCiLS Lab, and the morphology of the tissue in bright field observation through the laser microdissector (see **Note 5**). Select areas from the clusters of interest containing the same number of cells on the basis of the counts from QuPath. A number of 3500 cells for cancer samples are recommended.

3. Perform microdissection: we used the Leica LMD 7000 (Leica Microsystems, Wetzlar, Germany) with the following laser settings: laser wavelength, 349 nm; pulse energy, 52J; numerical aperture, 15; specimen balance, 14; laser head current, 100%; pulse frequency, 174 Hz; and focus offset, 50. The settings

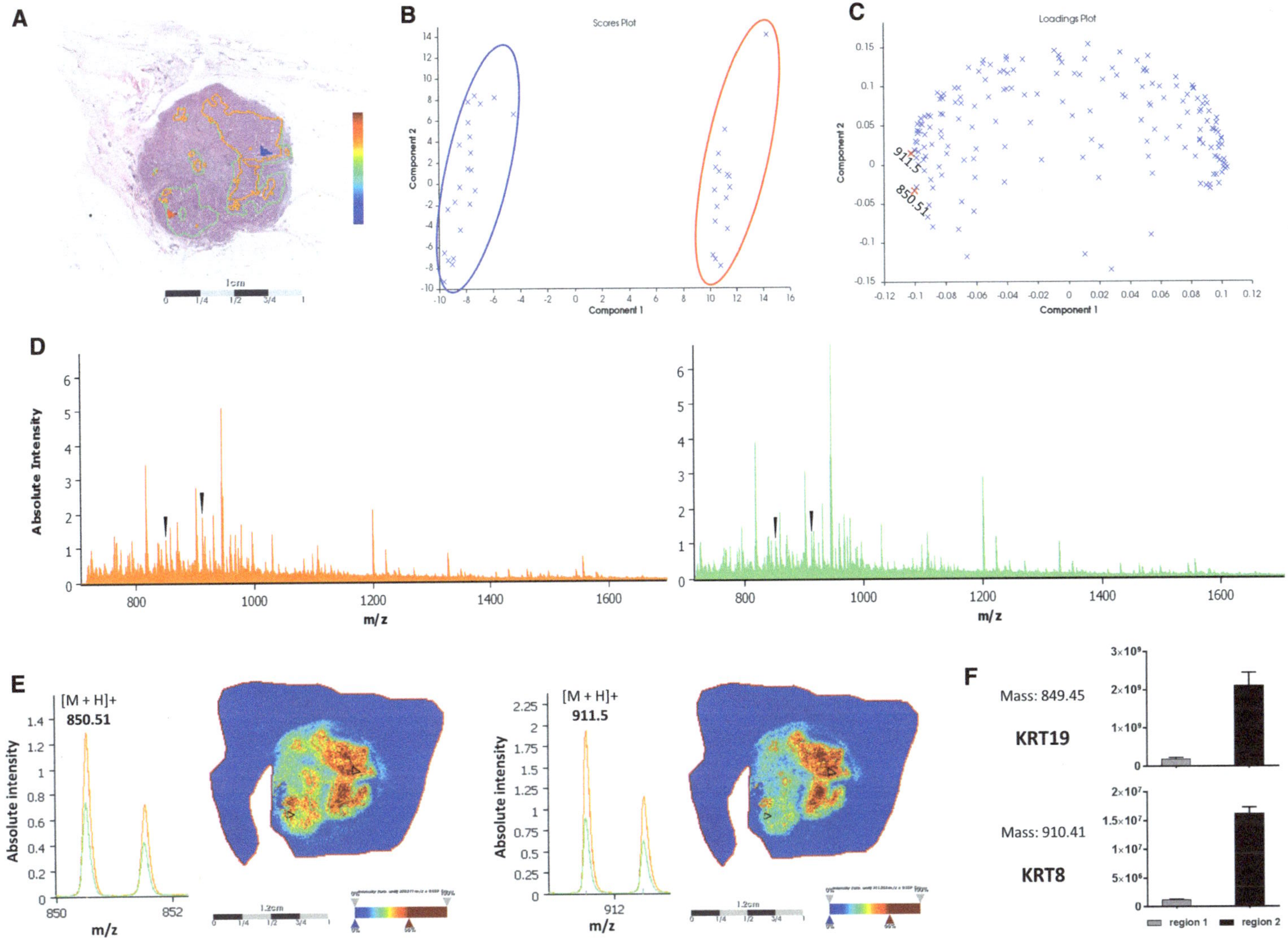

Fig. 2 Identification of heterogeneous peptide is a breast tumor. (**a**) Superimposition of the HE section after MALDI with the segmented dataset. The *green* and *orange* clusters are outlined. (**b**) Scores plot of the PCA analysis of the subset of spectra in the heterogeneous clusters. The selected area corresponds to the one selected for parallel

should be adapted to the tissue type. Three collection replicates should be performed, from the serial sections collected on LMD membrane slides (see **Note 6**).

3.2.2 Preprocessing

This step consists in collecting the tissue pieces from the cap to the bottom of the tubes to improve monitoring during processing.

- During laser microdissection, the tissue pieces are collected in tube caps. Store the tubes at 4 °C before use, in a clean environment.
- Use the antistatic gun to get rid of static electricity during sample preprocessing.
- Centrifuge the tube at 15,000 × *g* during 5 min at room temperature (RT) and dispose the tubes in a clean holder.
- Check for the presence of the tissue piece(s) at the bottom of the tubes using a binocular microscope.

3.2.3 Antigen Retrieval

This step consists in getting rid of the methylene bridges linking proteins induced by formalin fixation.

- Gently add 20 μL of 10 mM pH 6.0 citrate buffer on the tissue piece at the bottom of the tube without touching it with the tip.
- Centrifuge at 15,000 × *g* for 5 min at RT. If it is still visible, ensure yourself that the tissue is properly immersed in citric acid.
- Sonicate the immersed tissue pieces for 10 min in a holder that avoids overheating of the tubes. This step allows removing air bubbles that may stay at the surface of the tissue.
- Incubate in a thermoshaker at 99 °C for 1 h at 800 RPM. Quickly spin the tubes every 10 min in order to collect the condensed water from the cap to the bottom of the tubes.
- Collect the tubes in a clean holder and let these cool at RT.
- Add 2.2 μL of RapiGest 0.1% in the samples to obtain a final 0.01% concentration and spin the samples at 21,000 × *g* for 2 min. Quickly spin.
- Shake at 800 RPM for 10 min at RT. Quickly spin.
- Add 2 μL of 500 mM NH_4HCO_3 to all the samples. Agitate for 10 min at 800 RPM and centrifuge the tubes (21,000 × *g*, 1 min). Check that the pH of the samples is 7.

Fig. 2 (continued) laser microdissection. (**c**) Loadings plot of the PCA analysis. *m*/*z* 911.5 and *m*/*z* 850.5 are highlighted. (**d**) Average mass spectrum from the *orange* and *green* clusters. (**e**) MALDI images of the *m*/*z* 911.5 and *m*/*z* 850.5. (**f**) Identification of the *m*/*z* 911.5 and *m*/*z* 850.1 after correlation with the LMD-based microproteomic dataset. Reprinted with permission from [4]

3.2.4 Reduction-Alkylation-Reduction (See Note 7)

This step consists in getting rid of disulfide bridges still existing within proteins after antigen retrieval.

1. Reduction (reduction of the disulfide bridges and liberation of the thiol groups): the samples have now a volume of 24.2 μL. Add 2 μL of 131 mM DTT in 50 mM NH_4HCO_3 to obtain a final concentration of 10 mM DTT. Incubate the samples for 40 min at 56 °C while shaking at 800 RPM. Quickly spin.
2. Alkylation (blocking thiol groups with IAM): the samples have now a volume of 26.2 μL. Let the samples cool at RT and add 3 μL of 194.67 mM IAM in 50 mM NH_4HCO_3 to obtain a final concentration of 20 mM IAM. Incubate at RT for 30 min while shaking at 800 RPM. Quickly spin.
3. Reduction (eliminate IAM in excess): the samples have a volume of 29.2 μL. Add 2 μL of 171.6 mM DTT in 50 mM NH_4HCO_3 to add a concentration of 11 mM DTT. Incubate for 10 min at RT (24 °C) while shaking at 800 rpm. Quickly spin.

3.2.5 Digestion

This step consists in digesting the proteins from the tissue piece.

1. The samples have a volume of 31.2 μL. Add 4.26 μL of 0.5 μg/μL trypsin to obtain a final concentration of about 60 μg/mL. Quickly spin.
2. Incubate at 37 °C overnight while shaking at 800 rpm. Quickly spin.
3. The samples have a volume of 35.46 μL. Add 2.26 μL of 0.5 μg/μL trypsin to increase the concentration by about 30 μg/mL. Quickly spin.
4. The samples have now a volume of 37.72 μL. Add 150.88 μL of ACN to obtain a solution of 80% ACN. Quickly spin.
5. Incubate at 37 °C for 3 h while shaking at 800 RPM. Quickly spin.
6. The samples have a volume of 188.6 μL. Stop digestion by adding 10 μL of TFA 10% to obtain a final concentration of about 0.5% TFA in the tube. Agitate for 10 min at 800 RPM and centrifuge the tubes (21,000 × *g*, 1 min). Check the pH of the solution (pH < 3).
7. Incubate for 45 min at 37 °C while shaking at 800 rpm in order to cleave RapiGest.
8. Centrifuge the samples at 21,000 × *g* for 10 min, at 4 °C.
9. Collect the supernatant and transfer in new identified 0.6 mL tubes.
10. Evaporate the samples using the speed vacuum. Store the samples at 4 °C if necessary for 1 week maximum. If the storage time exceeds 1 week, store the samples at −20 °C.

3.2.6 Purification/Desalting/Concentration of the Samples

This step consists in purifying, desalting, and concentrating the proteolytic digests obtained from the tissues by solid phase extraction (SPE).

1. Prepare the three solutions described in Sect. 2.11.
2. If the samples were dried, resuspend in 20 μL Milli-Q H_2O with 0.1% TFA (ZT Solution 3). If the SPE is performed just after the digestion, transfer the necessary volume of sample in another 0.6 mL tube and adjust to 0.1% TFA, with maximum 5% of organic solvent, in an ideal volume of 20 μL. Depending on the expected amount of peptides resulting from the digestion process, use ZT C_{18} 5 μg or ZT C_{18} 2 μg (μZT).
3. Set the micropipette to 10 μL. Wash the ZT with Solution 1: take and withdraw in the waste. Repeat three times.
4. Activate the ZT by taking Solution 2. Withdraw in the waste. Repeat three times.
5. Equilibrate the ZT: take Solution 3 and withdraw in the waste. Repeat three times.
6. Load the sample. Adjust the sample to the used ZT. If the expected concentration of peptides in the sample is too high (>5 μg when ZT −5 μg are used, >2 μg when ZT −2 μg are used), dilute with a 0.1% TFA solution. Pipet ten times to load the peptides to the ZT. If the sample is diluted or if the volume of sample is higher than 20 μL, pipet 25 times.
7. Wash the samples: take Solution 3 and withdraw in the waste. Repeat five times.
8. Elute the sample: take Solution 2 and collect in a new microtube. The peptides are present in the solution in the new microtube. Repeat once. The proteolytic digest is then concentrated in 40 μL of ACN/H_2O 1:1 and 0.1% TFA.
9. Evaporate the sample completely with the speed vacuum.
10. Proceed to the subsequent LC-MS/MS analysis or store the sample at 4 °C for 1 week maximum, at −20 °C if the analysis cannot be performed within 1 week.

*3.2.7 Two-Dimensional NanoLC-Orbitrap Mass Spectrometry Analysis (See **Note 8**)*

1. Resuspend the samples in 10–11 μL of appropriate buffer for 2D LC-MS and MPDS mix in order to get a final concentration of 50 fmol alcohol dehydrogenase (ADH) per volume of injection (9 μL). Use different MPDS mix solutions (1 or 2) in samples of different types. For example, if two conditions are compared (A and B), use MPDS mix 1 for samples from condition A and MPDS mix 2 for samples from condition B. This will allow to further control that the ratio of proteins from MPDS mixes 1 and 2 is the one expected.

2. 9 μL of the samples is injected in the 2D LC-MS system of choice (UPLC nanoACQUITY 2D (Waters, Milford, MA) controlled by MassLynx in our case) coupled to a Q Exactive Plus, controlled by XCalibur.
3. The samples are then injected in a two-dimensional RP/RP system, with a first dimension in a high pH (pH 10) and a second dimension in a low pH (pH 3). The peptides are loaded on the high-pH column [XBridge BEH C18 5 μm (300 μm × 50 mm)] at 2 μL/min (20 mM ammonium formate solution adjusted to pH 10), and three elution steps (15 min each) with the following percentages of ACN are realized: 13.3% (fraction 1), 19% (fraction 2), and 65% (fraction 3).
4. The eluate from the "high-pH" column is then diluted ten times with acidified water before being loaded on the trap column [Symmetry C18 5 μm (180 μm × 20 mm)] and separated on the "low-pH" analytical column [BEH C18 1.7 μm (75 μm × 250 mm)]. The gradient on the low-pH column is 140 min long with the following settings: flow rate of 250 nL/min, solvent A (0.1% formic acid in water), and solvent B (0.1% formic acid in acetonitrile) with a linear gradient as follows: 0 min, 99% A; 5 min, 93% A; and 140 min, 65% A. Cleaning and re-equilibration steps then take place during the following 40 min (total run of 180 min).
5. The LC eluent is then directly electrosprayed from the analytical column at 2.1 kV through the liquid junction of the nanospray source. The chromatography system was coupled to a Thermo Scientific Q Exactive Plus hybrid quadrupole-Orbitrap mass spectrometer (Thermo Fisher Scientific, USA), programmed for data-dependent acquisition mode, with the following settings:
   - Top ten (data-dependent acquisition)
   - Parameters for MS: mass range, *m/z* 400 to 1750; resolution, 70,000; AGC target, 1×10^6; maximum injection time, 200 ms
   - Parameters for MS/MS: isolation window, *m/z* 2.0; stepped normalized collision energy (NCE), 21.2, 25, 28.8; resolution, 17,500; AGC target, 1×10^5; maximum injection time, 200 ms; underfill ratio, 1.0%; dynamic exclusion, 10 s

3.3 Data Processing for Protein Identification

The data processing we propose is adapted for the identification of compound from a single section (see **Note 9**).

3.3.1 Label-Free Quantification (LFQ)

1. Use the last version of MaxQuant software. Load the samples from the two heterogeneous regions to compare. Three technical replicates per region of interest should be processed. For identification, use Andromeda search engine with the last release of Uniprot human database for interrogation. Use the following

parameters, N-ter acetylation and oxidation of methionines, as variable modifications and carbamidomethylation of the cysteines (see **Note 10**) as fixed modification. Set the maximum number of miscleavages at two and the minimal length for identification at seven amino acids and at least two peptides required for identification, including one unique peptide. Check "LFQ" for data normalization. Set the maximum ratio count for LFQ at 2. Set the main search tolerance at 4.5 ppm. Set peptide spectrum match (PSM) and false discovery rate (FDR) at 0.01.

2. For detailed explanations on the different parameters for MaxQuant and an informed choice of these for data processing setup, please refer to [8].

3.3.2 Correlation of Peptide Masses with Identifications

1. After MaxQuant analysis, a "combined" folder is created and located in the folder of the first raw file loaded in the software.
2. In the subfolder "txt," open the file "peptides" in Excel.
3. In the column "Mass," search for the mass of interest without the decimals (e.g., 910 for 910.5) (see **Note 11**).
4. Report the corresponding values from the column "Intensity."
5. Calculate the mean intensities and the standard deviation between the triplicates.
6. Compare the intensity trends between the heterogeneous regions with the one obtained by MALDI imaging, as shown in Fig. 2f (see **Note 12**).

4 Notes

1. Incubation chambers can have different forms such as a tip box or a lunch box. The chambers should be hermetically sealed. We recommend using tissue papers moistened with a fixed volume of water to create the humidity.
2. The optimal method parameters greatly depend on the instrument used. FT-ICR instruments require a specific training to select the appropriate ionization settings.
3. The segmentation process is described to study intratumor heterogeneity from a single tissue section. In the future, studies of intratumor heterogeneity would require finding common clusters between tissues and correlate those with prognostic values of patients, as described in [5].
4. As mentioned in **Note 3**, LMD-based microproteomic method can be used in the context of identification from single sections as presented here, but also to compare regions originating from large patient cohorts. In this case, a biomarker discovery assay should be performed, as described in [2, 9].

5. Usually, it is easy to recognize the limits of regions of interest in tissues such as breast tumors, based on the morphology. In the future, more histologically homogenous tissue would require more sophisticated tools to correlate the localization from MALDI imaging data to bright field observation through the laser microdissector.
6. The LMD-based microproteomic method may lead to some technical variability, as for any proteomic method and especially when less abundant peptides are quantified. For the identification of heterogeneous peptides from single sections, technical triplicates should be used.
7. In the context of a biomarker discovery assay, this step is recommended. In the present context, it can be skipped in order to mimic the chemical preparation for MALDI imaging. Indeed, no reduction/alkylation/reduction step is performed for MALDI imaging.
8. For a biomarker discovery assay, it is recommended to run pre-fractionated samples or to use 2D LC-MS/MS to increase the number of protein identifications. For identification from single sections, 1D LC-MS/MS is sufficient for correlation.
9. Data processing for a biomarker discovery assay is described elsewhere [2, 9].
10. This modification should not be included when no reduction/alkylation/reduction is performed. When this step is performed, the user has to make sure that the mass of the identified peptide for the correlation with MALDI imaging dataset does not bear this modification.
11. A slight mass shift is expected between values from MALDI imaging values and LC-MS/MS.
12. Many identification candidates are expected for a given peptide, but in most cases, only one will show the same intensity trends between tissues.

Acknowledgments

The authors would like to acknowledge Lisette Trzpiot and Nancy Rosiere for their efficient technical assistance.

Conflict of Interest

The authors have declared no conflict of interest.

References

1. Longuespee R, Casadonte R, Kriegsmann M, Pottier C et al (2016) MALDI mass spectrometry imaging: a cutting-edge tool for fundamental and clinical histopathology. *Proteomics Clin Appl 10*:701–719
2. Longuespee R, Alberts D, Pottier C, Smargiasso N et al (2016) A laser microdissection-based workflow for FFPE tissue microproteomics: important considerations for small sample processing. *Methods 104*:154–162
3. Longuespee R, Casadonte R, Kriegsmann M, Wandernoth P et al (2017) Proteomic investigation of human cystic echinococcosis in the liver. *Mol Biochem Parasitol 211*:9–14
4. Alberts D, Pottier C, Smargiasso N, Baiwir D et al (2017) MALDI imaging-guided microproteomic analyses of heterogeneous breast tumors—a pilot study. *Proteomics Clin Appl* (in press)
5. Balluff B, Frese CK, Maier SK, Schone C et al (2015) De novo discovery of phenotypic intratumour heterogeneity using imaging mass spectrometry. *J Pathol 235*:3–13
6. Fox CH, Johnson FB, Whiting J, Roller PP (1985) Formaldehyde fixation. *J Histochem Cytochem 33*:845–853
7. Maree R, Rollus L, Stevens B, Hoyoux R et al (2016) Collaborative analysis of multi-gigapixel imaging data using cytomine. *Bioinformatics 32*:1395–1401
8. Tyanova S, Temu T, Cox J (2016) The MaxQuant computational platform for mass spectrometry-based shotgun proteomics. *Nat Protoc 11*:2301–2319
9. Herfs M, Longuespee R, Quick CM, Roncarati P et al (2017) Proteomic signatures reveal a dualistic and clinically relevant classification of anal canal carcinoma. *J Pathol 241*:522–533

INDEX

Minnie M. Sarwal and Tara K. Sigdel (eds.), *Tissue Proteomics: Methods and Protocols*, Springer Protocols Handbooks, vol. 1788, https://doi.org/10.1007/978-1-4939-7854-0,

K

L

M

N

O

P

Q

R

W

FSC
www.fsc.org
MIX
Papier | Fördert gute Waldnutzung
FSC® C083411

Kolibri 360 GmbH
Ferdinand-Jühlke-Straße 7
99095 Erfurt, Deutschland
produktsicherheit@kolibri360.de